SCIENCE AND FUTURE CHOICE

SCIENCE AND FUTURE CHOICE

Volume II
Technological challenges for social change

Edited by

PHILIP W. HEMILY

Deputy Assistant Secretary General for Scientific Affairs of NATO

and

M. N. ÖZDAŞ

Assistant Secretary General for Scientific and Environmental Affairs of NATO

CLARENDON PRESS · OXFORD
1979

Oxford University Press, Walton Street, Oxford OX2 6DP

Oxford London Glasgow
New York Toronto Melbourne Wellington
Kuala Lumpur Singapore Jakarta Hong Kong Tokyo
Delhi Bombay Calcutta Madras Karachi
Ibadan Nairobi Dar Es Salaam Cape Town

© *North Atlantic Treaty Organization 1979*

British Library Cataloguing in Publication Data

Science and future choice.
    Vol.2: Technological challenges for social change
    1. Science
    I. Hemily, Philip W   II. Özdaş, M N
    III. North Atlantic Treaty Organisation.
    Science Committee
    500   Q158.5   79-40618
    ISBN 0-19-858169-6

*Set by Hope Services, Abingdon
and printed in Great Britain by
Lowe and Brydone Ltd, Thetford, Norfolk.*

Preface to Volume II

The remarkable achievements of twenty years of non-military scientific co-operation between the member states of the Alliance were celebrated at a Commemoration Conference in April 1978 at the Palais d'Egmont in Brussels. Rather than holding a self-congratulatory review of good works in basic science, the NATO Science Committee welcomed a provocative look at the interactions of science and society.

In order to provide guidance on future directions and needs, priorities and emphases, the Conference identified and addressed a number of important issues that will influence future patterns of political interaction, as well as social and economic progress in the years to come. Of fundamental concern to us all are the ways that can be found by our countries to overcome the problems that confront them in the fields of energy and natural resources. We must make the most effective use that we can of our scientific and social knowledge in order to narrow the wide differences in expert opinions as to the nature of the energy problem that I noted during the Conference discussions. We have no alternative but to develop solutions that are satisfactory from the social and environmental, as well as the economic standpoints.

During the last century, the results of scientific research and subsequent technological development have had determining effects on changes in social values, and the trend will undoubtedly continue. Emerging attitudes to work, education, leisure, consumption, and economic growth are being profoundly influenced by the increasingly information-oriented society in which we now live, and these issues are discussed in the papers presented in this second volume.

In these difficult yet challenging times, the conference discussions elicited a wide range of opinions on the nature and importance of the changes in progress, and particularly on the types of societies towards which we may be aiming. This is important, because the ways in which we envisage the future determine our views about the appropriate contributions and directions of scientific and technological effort. Equally, it affects the vital contribution of the scientific community both in improving our understanding of our problems and in providing us with the means to tackle them.

The Conference underscored the vital need for international cooperation in science and technology in an increasingly interdependent world. Our concepts of mutual security have been expanded to include a broad range of concerns of a global nature, including protection of the environment, management of natural resources, and the welfare of our peoples. Research, particularly including applications in the social sciences in joint effort with the natural sciences, can provide better guidance in the long term to our political leaders in tackling the complex interdisciplinary problems that seem so intractable within the constraints of our national and international institutions.

I firmly believe that in facing the challenges presented by a changing world our countries have the capacity, the expertise, and the will to adjust future expectations and objectives to new circumstances. This leads to critical questions where science and technology are of fundamental importance. If, for example, past patterns of growth are not simply going to be repeated, what kind of progress and economic transformation can we expect and what new patterns of science and technology will be appropriate? Perhaps we must work towards less energy-intensive patterns of activity having smaller environmental impacts. These patterns would have to be more oriented to individual participation and fulfilment and attuned to the problems of the wider world community. This also raises the vital question about the type of society that will adequately reflect the aspirations of our peoples, and towards which our energies should be directed. It also makes clear that we are in a position to *choose* our future, not simply to experience it.

Joseph M.A.H. Luns

Secretary General
North Atlantic Treaty Organization

Brussels, November 1978

Contents

List of contributors

Willem Albeda

Professor in the Faculty of Business Administration in Delft; currently Netherlands Minister for Social Affairs

Lewis M. Branscomb

Vice-president and Chief Scientist of IBM Corporation

Harvey Brooks

Benjamin Pierce Professor of Technology and Public Policy at Harvard University

Harlan Cleveland

Formerly United States Ambassador to NATO; currently Director of the Program in International Affairs of the Aspen Institute for Humanistic Studies, Princeton, NJ

Umberto Colombo

General Manager of the Montedison Research and Development Division in Milan; currently President of the European Industrial Research Management Association

André Danzin

Directeur, Institut de Recherche d'Informatique et d'Automatique, Rocquencourt, France; Président Comité Européen de Reserche et de Développement (Comité Consultatif auprès de la Commission des Communautés Européennes Bruxelles)

Duncan Davies

Chief Scientist of the Department of Industry, London

François de Rose

French Ambassador to NATO from 1970 to 1975

Richard S. Eckaus

Ford International Professor at the Massachusetts Institute of Technology

Denos Gazis

Research-Member of IBM Thomas J. Watson Research Laboratory

Wolf Häfele

Deputy Director of the International Institute for Applied Systems Analysis, Laxenburg, Austria

Johan Jorgen Holst

Under-secretary of State for Defence for Norway

Marcus C.B. Hotz	Director, Integrated Programs Branch of the Canadian Department of the Environment; currently on leave with the Scientific Affairs Division of NATO
Karl Kaiser	Director of the Research Institute of the German Society for Foreign Affairs
Alexander King	Chairman of the International Federation of Institutes for Advanced Study
Edmond Malinvaud	Directeur Général, Institut National de la Statistique et des Études Economiques in Paris
Mihaljo Mesarovic	Director of the Systems Research Center at Case Institute of Technology, Cleveland, Ohio
Sylvia Ostry	Chairman of the Economic Council of Canada
Aurelio Peccei	Chairman of the Club of Rome
Eduard Pestel	Professor and Director of the Institute of Mechanical Engineering at the Technical University of Hannover; currently Minister of Arts and Science for Lower Saxony
Wolfgang Sassin	Member of the Energy Systems Programme at the International Institute for Applied Systems Analysis; on leave from the Nuclear Research Centre at Karlsruhe
Eugene B. Skolnikoff	Director of the Center for International Studies at the Massachusetts Institute of Technology
Robert J. Uffen	Dean of the Faculty of Applied Science at Queen's University, Kingston, Ontario
Lord Vaizey of Greenwich	Professor of Economics and Head of the School of Social Sciences at Brunel University, Uxbridge

11

Introduction: an emerging context for science

Marcus C. B. Hotz

11.1 The Twentieth Anniversary Conference

The vagaries of shifting national priorities have always affected alliances between nations, tending to make them short-lived and of decreasing value, almost from the moment that they were signed. When the North Atlantic Treaty was written, this had come to be well understood, and in extending the concept of mutual security to include co-operation in matters of social, economic, and political concern, it sought to widen the common interest of the alliance nations by strengthening and monitoring the stability of their institutions. The political, economic, and social conditions of the mid-1950s were such that the North Atlantic Council felt it necessary to invoke this provision of the Treaty, and invited the then Foreign Ministers of Canada, Italy, and Norway to advise them on the advantages and opportunities for such co-operation.

In their report[1], Lester Pearson, Gaetano Martino, and Halvard Lange noted their belief that progress in the field of science and technology was so crucial to the future of the Atlantic Community that the member states should develop the means of collaboration to attempt to solve the problems that confronted them through constructive international programmes. The Council responded by agreeing to establish the NATO Science Committee, and appointing a Science Adviser, later to become the Assistant Secretary General for Scientific and Environmental Affairs.

The Science Committee consists of representatives of all the member states, and over the years, it has developed programmes that have been remarkably successful in improving the scientific bases of the Alliance countries. The Committee does not undertake scientific endeavours of its own, but provides support for unclassified non-military research and training activities. It uses its small budget for grants, meetings, teaching institutes, and exchange fellows, stressing the initiation of international collaboration. Several hundreds of books and many thousands of publications in scientific journals have resulted from activities that it has sponsored.

Periodically, the Science Committee has organized meetings on topics of special concern, the so-called Science Committee Conferences, and invited the participation of well-known scientists as speakers. With the approach of its twentieth anniversary, it decided to have a Commemoration Conference that would bring together an array of natural and social scientists, scientific administrators, and political leaders to review the contribution of the NATO science programme in promoting and facilitating the interactions between countries and their scientific institutions for meeting longer-term objectives.

The Committee felt that an assessment of major scientific achievements together with prospectives on promising future trends would form a good base on which to discuss the current interactions between science, technology, and key societal problems, emphasizing areas where scientific research may make a substantial contribution. Invited to the Conference were ministers of science, senior parliamentarians, ambassadors, senior officials responsible for science policy, heads of national research councils, representatives of the scientific press and other news media, and prominent industrialists and scientists, who had a rare opportunity for discussion and conversation with natural and social scientists and the politicians and officials whose decisions increasingly depend on accumulated technical knowledge.

The first volume contains a number of papers that review developments and future prospects in a number of areas of science that have advanced particularly rapidly in the twenty years since the NATO Science Committee was created, all of which are, or have the potential of becoming, determining factors with respect to the changing social systems of the industrial democracies. The fields chosen were biology, systems science, mathematics, electronics, materials science, environmental sciences, and astrophysics; and two chapters (one by John Maddox and the other by Sir Sam Edwards) are devoted to overall questions, trends, and implications.

In this second volume are the papers that were prepared to help Conference participants develop a better understanding of the scientific, political, social, economic, and technological relationships, whose changing roles and interactions become increasingly significant as the industrial democracies move from responding to the challenges and constraints of man's natural environment to meeting the need for more effective control and management of the technology that he has created.

11.2 Science and decisions

In bringing together a group of such diverse interests and backgrounds as that invited to the Conference, there is always a risk that the improved mutual understanding sought by the organizers will not take place. Indeed, this never occurs on its own, either in the meeting or the working environment, but must be carefully nurtured, focused, and developed. It is not simply a matter of finding a common language; reflecting on communication between politicians and scientists shows the problem to be much more deeply rooted.

Politicians do not, as a rule, view the role of science in society in the same way as scientists. By following a scientific method that is essentially reductionist and analytical, scientists have been remarkably successful in learning a great deal about narrowly defined and sharply focused problem areas. At the same time, they have not shown themselves to be effective in developing the integrative synthetic type of enquiry system that is required for the approaches to the complex problems of modern societies that are demanded by policy-makers. The fact that many of these problems owe their origins to the very societal institutions that are seeking their answers serves to complicate matters, since these problems are perceived to threaten their continued usefulness. However, this in no way diminishes the need to translate the information that is available into forms that are understandable to, and usable by those who need to use it. Unless this is done, and done well, the limits to governance described by Cleveland (Chapter 20) will come upon us only at an ever-increasing rate, with decisions postponed until situations reach crisis proportions, posing further dangers with respect to rational outcome. Thus he gives emphasis to the need to seek out and train scientists and technologists as integrators, whose function would be to help interpret complex systems in ways compatible with the needs of planners and decision-makers.

This is well illustrated by the handling of a crisis situation which requires scientific knowledge or information for its resolution. The scientists who are to advise on dealing with the problem know that a research programme can never be the answer to a crisis; such a crisis, after all, requires an immediate technological response in the sense that technology is the application of scientific knowledge that is already available. Nevertheless, by their unwillingness to extend what they believe might be uncertain knowledge bases into information that will be used for decisions, they sometimes give the impression that they are unable to provide a rapid answer to a question that seems simple to non-technical people. The failure to recognize the need to describe the situation in an easily intelligible form is frustrating to those who desperately need to know, and has done much to prejudice the view of the scientific community in the eyes of those who are ultimately responsible for decisions on its support.

Perhaps because it is not possible to predict discovery and its application, in the light of public ambivalence on the role of science (Chapter 1), policy-makers have tended to place 'basic' research in a cultural slot in setting priorities. They have tended to assign it very low priority — as something that is socially desirable, but not really all that essential, ignoring its true role as the base of the knowledge industry on which the emerging 'information society' rests, and the only way of providing the understanding for smoothing the rough road of discontinuous change down to a level of 'acceptable incrementalism'.

The shift towards an information society (Chapter 14) that is strongly service-oriented has been commented on by many, some decrying the loss of personal initiative through increasing social welfare, others criticizing the loss of individual freedom that information banks potentially entail, while yet others see inherent the collapse of the type of society that we have known, an aspect which some feel is clearly reflected in contemporary music and art forms. The threat to society touches the lives of most people, whereas only a minority seem to be aware of the threats to individual freedom and initiative. Nevertheless, the resulting alienation from the centres of decision-making and the social institutions that they represent is probably one of the most serious challenges that now confront us; this situation is not helped by the unrealistic expectations of science as the universal problem-solver that so many people have developed over the last thirty years.

It is relatively easy to identify the general areas in which we need to stimulate significant development of new knowledge, and the directions suggested by Hare, Danzin, Brooks and Skolnikoff, and Peccei and Mesarovic amongst others (Chapters 3, 13, 14, and 18, respectively), all have a common thread – that of an interdisciplinarity, involving more or less discrete systems or subsystems that can be understood only by integrated studies that involve examination and analysis by several traditional disciplines. This is one of the things that we tend to do very badly; for example, the study of environmental systems calls for the use of background knowledge on how systems behave, while decision-making requires this to be linked with trend information of an observational type that we are only painstakingly learning how to collect, let alone use.

In many areas we have started to reach the limits of knowledge and understanding for further technological development, just at a time when research, as distinct from work on quality control and process reproducibility, is declining. One such area is that of metals in materials science, in which Ashby shows clearly that materials problems represent limiting factors in areas ranging from energy generation to building materials. Molecular biology (Chapter 4), which surely holds the key to many problems of health, agriculture, nutrition, and biological processing, seems now to be beset by an anti-intellectualism that may well be more destructive than the potential evils from which it seeks to deliver us. Similarly, failure to ensure adequate support of research in some scientific areas related to medicine and man's wellbeing can have severe social effects; for example, more satisfactory knowledge of the physiology and sociology of human reproduction are surely significant factors in managing world population, the demographic features of which Vaizey (Chapter 15) relates to fundamental human values that determine the nature of society.

So we have to look carefully at the way in which science can and is being used in the decision-making systems of government, both national and local, of industry, and even at personal levels. What is evident is that the scientific information currently being obtained is not easily being transformed into the synthesis required for decisions that affect the integrated whole, systems that reflect the human condition and the use of the biophysical environment and the resources of this planet. This is in sharp contrast to the emphasis upon immediate satisfaction, which Ostry in her Commentary feels must give way to a more effective incorporation of changing values and attitudes within persistent social structures.

11.3 Coping with complex systems: the problematique

If emphasis is to be given to the use of scientific and technical information in decision-making processes that are of social significance, a concept that goes far beyond that of government action, several authors in these volumes imply that this must, in the ultimate, be subject to review dictated by demands of public policy.

In pointing to the Canadian Conserver Society Project as exemplifying the kind of thinking that is necessary, Uffen in his Commentary quite correctly asserts that this approach does not imply a halt to economic growth, but that conservation, recycling, and reuse require new concepts and directions of growth. Of course, the concept of a conserver society is not an end in itself, but must extend to foster development of an ethic which will continue to be internalized as we move into whatever kinds of industrial or post-industrial societies that are consistent with the individual roles and cultural and social aspirations that we may create in the future. Any attempt to promote such a concept as a static, achievable goal within a given timescale will probably be self-defeating; the essential feature is an approach to sustainability in terms of demand and consumption.

Uffen is sharply critical of Häfele and Sassin's (Chapter 12) view that 'we can safely conclude that the energy problem is the only resource problem that demands a major solution in the foreseeable future'. Linking Ashby's concept of resource half-life (Chapter 2) with geopolitical shifts, environmental and renewable and non-renewable resource policy issues, he pleads for a totally integrated view that spans all aspects of human activity.

In spite of this, the fundamental thermodynamics of mankind's activities and biophysical environment demand prime consideration for all aspects of energy production and use, which challenges the validity of 'more of the same' solutions. The sobering thought that comes from a combination of the views of Häfele and Sassin with those of Hare must surely be that concentration on coal is not an acceptable answer for the continuing energy demands of the industrial societies, let alone those of the developing countries in the future.

Thus the conflicting approaches of Peccei and Mesarovic (Chapter 13) on the one hand, and Häfele and Sassin on the other, have an important meeting ground when considering the long-term economic problems now being faced by the industrial countries. The questions that the conflict poses are not questions of demands and numbers; rather they are concerned with whether we understand our problems correctly and whether the solutions proposed are unsuccessful because of forces beyond our control or lack of consistent real effort to resolve them, or, more seriously, whether we are at all capable of understanding our problems in a rapidly changing world. The frightening consequence of the last-mentioned case is, of course, a total lack of ability in dealing with the situations that confront us. Both Danzin (Chapter 14) and Churchman (Chapter 8) allude to this, and with an optimism that is derived from differing philosophical bases, believe that a technological solution lies in systems research; Peccei and Mesarovic see enormous problems in the education of people to be able to understand the new systems with which they must learn to cope. This view is implicitly shared by Edwards (Chapter 10), who, from yet another standpoint, underscores 'a distressing lack of numeracy in the population'.

While these authors stress ability to cope with the demands of modern society, other dimensions need to be brought into the educational discussion. Thus Vaizey (Chapter 15) describes the changes in the role of education in society and relates the migration of highly trained manpower, both internationally and in the rural to urban shift, to the structure of the economy and growing social pressures. This is the starting-point for a description of the employment effects of technology on the labour market, in which Albeda (Chapter 16) and Davis (in his Commentary on Albeda's chapter) stress the complexity of the structural unemployment that has arisen in the industrial countries and the need for new directions in public policy to combat it.

In the uncertainties imposed on us by the complex world in which we now live, perhaps the only certainty on which we can rely is that of change – and change to which we are forced to respond. From a position of hindsight, one can always see that there were clear indications of the way in which things were going – the directions of change – and there were usually a few 'true prophets' around, to whom nobody listened. Why? Do we fail to see the indicators of change because we don't want to? Are we unconsciously burying our heads in the sand and trying to frustrate the operation of a new restatement of Murphy's famous law – if you don't want something to happen, it will? In other words, by cutting off or significantly reducing the fundamental research that is the only way through which we have shown ourselves to be capable of understanding and developing the technologies of responding to the major real problems that confront our world, we are in danger of treating just the symptoms in our ignorance. These problems are all long-term ones, for which Pestel (Chapter 22) asserts that we must define more comprehensive goals than those of material growth. He believes that more effective indicators are needed to monitor social wellbeing and guide policy development.

Perhaps no area shows the inherent conflict between long-term 'needs' and short-term 'realities' as does that of economics (see Malinvaud's commentary on Eckaus's Chapter). Responses to immediate political realities often foreclose options that may, in the long run, have proved to be more productive than the selected alternatives; what needs to be better understood is the cumulative impact of policy options that have different institutional origins. Thus, for example, education for employment in the future must be considered together with population trends (Chapter 15) and the labour-intensitivity of technology (Chapter 16). Equally, development of broader public policy in response to specific events must also be done in a holistic context (again, see Malinvaud). A typical example has been the need to restructure the economies of non-oil-producing countries in the wake of the price escalation of the past few years, for which Eckaus (Chapter 17) shows that the relationship between size of operations and high economic risk cannot be isolated from undesirable aspects of the centralization of decisions, impacts on demographic shifts, and changes in lifestyles.

Science has long been accepted as a major determinant of economic development, but its impact is largely seen in the replacement of labour by capital and in the imposition of environmental and health-related constraints on industry. New science policies must relate to the structural changes that are taking place in our societies, and, while recognizing wider social needs and desires, scientists and policy-makers must be cognisant of the fundamental need to protect the life-support systems of the biosphere through a long-term conservation ethic and the promotion of suitable low-stress industrial technologies (see Colombo's commentary). It is possible that our current overriding concern with man-made economic and social 'imperatives' in the short term might displace one of the few 'absolutes' that must be inherent in the policy process.

11.4 Some international issues

The future health of the economies of the industrial countries will be increasingly influenced by the demands of the Third World (Chapter 18), and Eckaus (Chapter 17) believes that the call for a new international economic order reflects a general view that the system of international commerce may not provide adequate encouragement for development. Many of these countries have come to view rapid industrialization as the panacea for their ills, but the scientific and managerial infrastructures of all countries are not uniformly well organized and developed, and even among advanced countries maximum benefit rarely flows quickly from advances already made. Furthermore, as Brooks and Skolnikoff point out, while the multinational corporation is one of the most successful instruments of technology transfer to developing nations, it is also the most deeply resented, and the recipient nations argue that technology sold has had its innovation cost amortized in the developed markets, so that the marginal cost of supplying it to them is very small.

So the impact of science and technology on developing countries continues to be seen essentially in terms of technology transfer from industrial nations. Recipient countries have often found to their regret that attempting to graft sophisticated technology on to a society that for cultural, education, or institutional reasons has difficulty in using it effectively does not improve the situation of large segments of the population (see the commentary by King). Rapid industrialization has been widely viewed as the only quick road to realizing potential national wealth, as seen through the evidence of donors, and well-understood domestic technologies have generally been ignored.

These indigenous technologies may often be amenable to innovation and development, using carefully selected imported technicians in ways that can vastly benefit the agricultural and industrial bases of the country at low cost without straining scarce foreign currency reserves. Even more importantly, local people can be encouraged to give rein to their technologies with respect to their needs in ways that allow them to understand its functioning, thus allowing them to repair and maintain equipment themselves.

Analysing the security of the countries of the NATO Alliance in terms of ten 'potential contingencies', Holst (Chapter 19) suggests that many of the longer-term economic problems facing the world today have dimensions of a social and political nature that are of greater concern in the context of mutual security than the threat of direct aggression. Looming large among these issues are the conflicts over ocean resources and their jurisdiction, and the tensions between states and people over societal aspirations regarding development and the distribution of goods and services. Kaiser in his Commentary reinforces this perspective by supporting a concept that provides an extremely broad interpretation of Article 2 of the North Atlantic Treaty,[1] wherein long-term defence planning is inseparable

---

[1] Article 2 reads: 'The Parties will contribute toward the further development of peaceful and friendly international relations by strengthening their free institutions, by bringing about a better understanding of the principles upon which these institutions are founded, and by promoting conditions of stability and well-being. They will seek to eliminate conflict in their international economic policies and will encourage economic collaboration between any or all of them.'

from economic and social security. Research in areas that relate to sociopolitical institutions and policy development is an essential part of this concept, and the classical role of defence is now seen to be a response to specific types of threat only.

Issues that transcend political boundaries proved to be a major theme of the Conference. Several contributions touched on different aspects of resources, ranging across agriculture, energy production and use, management of renewable and non-renewable resources, and problems inherent in exploitation of the oceans. Challenges that require international action for effective management of man's use of the global environment included the carbon-dioxide–climate problem, transfrontier pollution, monitoring long-term trends, and the need for wider understanding of ecosystem interactions. In a sense subsuming all of these were the discussions of the implications of the 'information society', which were examined from technological, political, and sociological points of view.

11.5 New directions

The work of biological and physical scientists has identified significant, interdisciplinary areas of knowledge that will influence long-term institutional planning and development. However, the very structures of existing institutions make the pursuit of such knowledge and the innovative development of institutional change difficult to support nationally. Many of the areas that science warns us that we neglect at our peril were discussed at the Twentieth Anniversary Conference and are repeated here in these volumes. They range widely across the usually identified preserves of the social as well as the natural sciences.

Realizing that many of our goals are never reachable, and that the progress of scientific knowledge must itself modify and change any long-term objectives that may be set, the low profile of the role played by scientists in the formation of policy based on their work is surprising. Although there are many reasons for this, it highlights the need for research on the effective use of scientific knowledge and information by decision-makers who are accustomed to different enquiry systems.

Unfortunately, much of what needs to be done falls into research areas 'peripheral' to the interests of funding organizations and agencies with specifically defined functions, and is therefore difficult to support, yielding place to less innovative and more conventional projects. Indeed, failure to implement substantive innovations has often been blamed on institutional inability to see the need for cross-programming with others and develop joint approaches to long-term policy formulation (Chapter 17, Malinvaud's Commentary).

For example, central to many of the problems of human health and welfare must certainly be continued biological research, more especially in relating the discoveries of molecular biology to physiological processes. More soundly based information will have a profound influence on the desperately needed shift of emphasis towards preventive health care. But also, better understanding of metabolic processes and immunological responses (Chapter 5), translated into demographic issues and nutrition, will surely be central to mankind's ability to choose acceptable lifestyles over the next few generations.

Concurrently, much work will have to be done in the adaptation to people of what has loosely been called an 'information society' (Chapter 14). The concepts involved will undoubtedly modify the human condition, and their technological realization is, in many instances, already with us. Although Branscomb in his Commentary optimistically believes many of the tensions and anxieties of today's world to be evidence of a readjustment process in which aspirations for a better life will be shared on a truly global basis, much research remains to be done in areas that deeply concern our ability to cope with such changes (Chapter 13). Alienation of people from seemingly meaningless social institutions is becoming a serious threat to industrial societies. Research that will enable existing or new institutions to better reflect and respond to societal needs will surely provide a vital contribution to this crisis in governance (Chapter 20).

Although the intuitive links between science, innovation, and the economy are obvious, there has been a notable failure to develop predictive modes that would permit what has come to be called science policy to be used as an effective instrument in support of the national interests of the industrial countries. The consequent contraction in resources to support research and promote innovation has become a matter of concern, even at the corporate level, where the health of industrial research has become a problem with respect to long-term planning (Chapter 10).

The gravity of this situation is compounded in the materials (Chapter 2), resources (Chapter 12 and Uffen's Commentary on it), and environmental fields (Chapter 3), where important developments are in many cases knowledge-limited. Substitution of common for scarce materials and the introduction of novel materials for both engineering and information-related hardware present enormous scientific challenges; new, environmentally appropriate, low-waste and energy-conserving process technologies present another set. Major scientific and technological responses will be required to develop the world's ocean resources of both minerals and food, not to mention the political and economic problems of resource ownership and exploitation.

Environmental decision-making – the knowledge base required and how to use it institutionally, the interface between environmental and social systems, monitoring and data handling, and understanding the interactions of ecological systems demand continued scientific effort (see Chapter 3). Problems related to climate, agriculture and the green revolution, such as vegetative cover, land use and energy consequences, the impact of the cutting of tropical forests, and the destruction of ocean plankton on the oxygen cycle cannot be ignored; work initiated now is not likely to produce results for several years to come, given the long-term behaviour of global systems.

Maintaining and improving the relationships between developing and the industrialized countries (see, for example, Chapters 17, 19, and 22 and Kaiser's Commentary) are fundamental to ensuring an evolutionary sense of management of our mutually interdependent resources for the good of all nations. Problems yet to be satisfactorily resolved include the role of technology transfer in the assistance process, training programmes and the long-term economics of both donors and recipients. The special problems of the moderately industrialized countries also demand attention.

In the past, nations used conflict as a way out of inextricable situations, but de Rose (Chapter 21) reminds us of the emergence of superpowers with comparable capability for devastating warfare on a global scale, so that the fruits of victory are no longer related to the risks of hostilities. Thus the world has to find other ways of dealing with its unresolved problems. Not only has science sharpened our vision of the issues that confront us, but it also has the capacity, indeed the responsibility, to enable us to respond to them with some measure of certainty not available otherwise. Our lifestyles may have to be modified, populations brought within supportable bounds, and resources conserved, but we should be able to achieve these goals in ways that are compatible with the democratic social systems that we have struggled so hard to achieve and maintain.

Loss of our scientific base by failing to support it adequately, and failing to use it effectively, will surely lead to disruptive changes in our society that might well destroy us, or at best lead to the rigid and authoritarian social systems that the scientist Skaife [2] so clearly, and the poet Marias [3] so eloquently described for the termite.

References

[1] Lange, H., Martino, G., and Pearson, L.B. *Non-military cooperation in NATO*, North Atlantic Treaty Organization, Paris (1957).
[2] Skaife, H.S., *Dwellers in Darkness*, Longmans Green, London (1955).
[3] Marais, E.N., *Die Siel van die Mier*, Van Schalk, Pretoria (1934).

12

Resources and endowments: an outline of future energy systems

Wolf Häfele and Wolfgang Sassin

12.1 Introduction

Science and technology have in part lost their credibility. In recent years, too many unexpected problems have come up with the application of scientific knowledge. More and more concerned observers are beginning to fear that the negative side-effects of technological solutions designed in the spirit of ever bigger size, ever faster pace, and ever more efficient use will eventually dominate. Technological progress is deeply questioned. A clear perception of what really determines man's evolution has been lost, which is expressed in the notion of a scientific crisis [1].

Thus one is led to reflect on the original position of science in society and how it has evolved. One might still maintain that the task of science is cognition, nothing else; then all other questions were left to the rest of society, and what is now called a scientific crisis would disappear by way of definition; prudence and science then were clearly separate. Prudence, indeed, and not science is needed to identify and master the consequences of cognition. In reality, however, sciences have somewhat unconsciously assumed the role of prudence by trying to manage their own consequences. The results of this effort are in question. How prudence will institutionalize its relation to cognition and solve that Gordian knot remains to be seen.

In order to perceive the revolution in the position within society which science has achieved through its offspring technology, it is helpful to recall a few key events of the past ten years.

The space programmes at the end of the 1960s characterized an era of euphoric, yet unspecified hopes. The progress of mankind clearly manifested itself as the progress of technology. Innovation and technological gaps influenced political thinking. Neil Armstrong's first step on the moon led mankind into a new age indeed, but not into the space age expected in the sixties. The most important result of the Apollo programme can today be ascribed to man's new and critical look back on to the earth. The technological adventure of a space mission televised to hundreds of million homes introduced global perspectives as a modern reality.

The wider horizon suddenly made one realize that humanity, traditionally in dissent of its basic goals, with all its problems is tied to a relatively small planet. The customary rationale behind man's billion daily microoperations was no longer convincing in a more global and longer-term perspective.

From 1970 onward environmental concern spread much like a rumour and, supported by quickly gathered evidence, gained significant political momentum. Only two years after the first moon landing, the Club of Rome foresaw limits to growth on 'spaceship earth'. Technology from then on rapidly inspired fewer hopes and more fears. Again these reactions were hardly specified.

Within another year the oil crisis provided an unprecedented, concrete example. It highlighted the global dimension of local political conflicts, a dimension that was introduced through the world-wide deployment of oil technologies. Scarce natural resources obtained an immediate strategic and political importance; their fundamental nature was unveiled gradually by the subsequent efforts to manage the oil crisis.

The North–South dialogue heated up on the dispute of sovereign control in the exploitation of natural resources, a principle introduced with the aim of changing the international economic order. After all, the differences of opinion among some developed nations about the proliferation of sensitive nuclear technologies must be seen in the same context.

The world is, gradually and painfully, becoming aware of a new and fundamental interdependence that stems from the interaction between a fast technological development and the limited natural resources for an ever growing world population; this fast technological development is both induced and constrained by the limited natural resources. Scientific efforts to analyse this phenomenon started only once the problems had sufficiently piled up, which may partly explain the roots of the present crisis of science.

These efforts, called applied systems analysis, constitute a young and still exploratory step. Applied systems analysis aims at a better understanding of the complex interactions between technological decisions and their consequences upon the future institutional, political, and social patterns. This is an ambitious, long-term task.

The energy issue, for a great many reasons, plays a central role among the key problems of mankind and is gradually becoming the battleground for material and intellectual conflicts. Because many studies and investigations have been launched throughout the world, a first outline of the long-term consequences of technological decisions of the coming years is emerging in this particular area.

It is important to stress that none of the various models and projections developed in this context have the quality of prediction; they mainly concentrate on technical and economic aspects. A synthesis of the possible features of a future global energy system which we will attempt thus must be viewed as a kind of intellectual map. Such a map is indispensable to prepare for the right outfit and to help foresee certain reactions of fellow travellers. It shall invite judgement and shall not provide simplistic answers. Prudence asks for co-operation rather than for independence between science and decision-making.

12.2 The imminent energy problem

Let us look sufficiently back into the past in order to become aware of the fundamental dynamics of the energy system. Figure 12.1 summarizes the primary energy

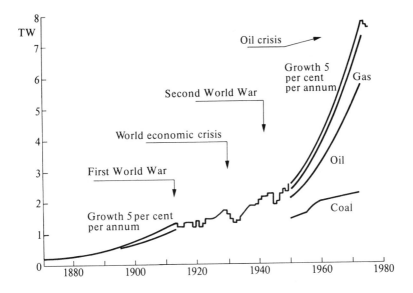

Fig. 12.1. Primary energy consumption of the past 100 years; data from [2].

consumption of the past 100 years. The growth period of 1870 to 1914 was essentially based on coal, *the* energy source of the first industrial revolution. During the extended period of world-wide crisis between the First World War and the end of the Second World War, the overall average growth rate of energy consumption dropped from originally 5 per cent per year to 1.7 per cent.

Between 1950 and 1970 the former 5 per cent growth was assumed on the basis of oil and, to a lesser extent, on gas. They altogether now account for slightly more than 70 per cent of global primary energy consumption. The present (1975) level is 7.5 TW, or 7.5×10^{12} W. One terawatt-year per year, or simply 1 TW, is roughly equivalent to 1000 million tons of hard-coal equivalent (we will use this unit throughout as a convenient global measure). 1.5 TW of the total of 3.5 TW of crude production oil are transported from the Persian Gulf over more than 10 000 km to the main consumer regions in Europe, Japan, and the east coast of the United States. The concentration of 20 per cent of the world's energy production in a small area, practically a point source, has contributed to the formation of OPEC, whose cartel was able to quadruple the world oil prices in 1973–4. Since then the worldwide economic recession, amplified by the high oil prices, has reduced oil consumption roughly by 10 per cent compared to 1973, yet the cartel has been able to maintain its high price level.

The really serious prospects for the coming two decades originate elsewhere, however; global oil reserves point to a physical supply bottleneck. The Workshop on Alternative Energy Strategies (WAES) in a two-year study investigated energy demand and various energy supply possibilities for the non-communist world until the year 2000 [3]. Figure 12.2, which is taken from this study, relates alternative

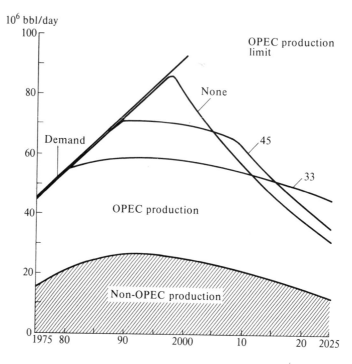

Fig. 12.2 Workshop on Alternative Energy Strategies (WAES): new and medium-term prospects for WOCA ('World Outside Communist Areas') [3].

projections of oil supply and demand; they are based on a fairly consistent set of assumptions, including economic growth rates, energy prices, government actions to stimulate the development of alternative energy sources as well as energy conservation, and investments in oil exploration to increase oil reserves. The general features of Fig. 12.2 also appear in various other WAES scenarios in conjunction with alternative assumptions. There is a high probability that the oil supply from non-OPEC sources will pass through a maximum well before the year 2000. An OPEC production limit, designed for a nearly constant supply of energy over an extended time period, would open a gap between demand and supply some time in the next ten years, perhaps quite soon. The consequences of a higher OPEC production limit are quite obvious as it could delay the occurrence of an oil-supply gap at the cost of having this gap widen more quickly and OPEC oil revenues decrease at a correspondingly faster rate. A responsible decision about the energy strategy to be adopted for a gradual reduction of the dependence on oil, and in particular of OPEC oil, thus clearly must consider a much longer time-horizon. It must also consider similar problems that might arise with other energy forms when taking over the present central role of oil in the global energy system. It is important to note – and Fig. 12.2 already points to it – that the difficulties now experienced through a politically supported economic cartel will quickly change into a physical bottleneck of oil supply, caused by the limited natural oil resources.

H.E. Goeller and A.M. Weinberg, in an early reaction to the material resource limitations postulated in the first study of the Club of Rome [4, 5], have tried to put the general resource problem into perspective. They defined 'Demandite' as the average mineral mined by man from non-renewable resource deposits. The average composition of elements in 'Demandite' produced and consumed for the reference year 1968 is shown in Fig. 12.3. Apart from the consumption of air, water, and bioproducts, which are recycled by nature, 'Demandite' accounts for the total material throughput of the world economy. Sixty-six mole per cent, or two-thirds of all atoms processed in this category come from hydrocarbons (and thus quantify the exceptional role of energy in quite an unusual way). Reduced carbon

$$(CH_{1 \cdot 71})_{0 \cdot 66} \ (SiO_2)_{0 \cdot 21} \ (CaCO_3)_{0 \cdot 08} \ Fe_{0 \cdot 01} \ X_{0 \cdot 04}$$

X = composite of all other elements each
contributing less than 1 mole per cent

Fig. 12.3 'Demandite': elements produced from non-renewable resources; world, 1968 [4].

in the form of coal, oil, and natural gas in the developed economies holds an even larger share, for example, in 1968 80 per cent in the USA. Comparing the use of all important elements to their abundance within the top 1000 m in the Earth's crust the authors concluded that only extractable hydrocarbons and phosphorus are scarce resources. Whereas phosphorus can, at least in principle, be recycled into the biosystem, reduced carbon is oxidized and forever lost as an energy source. Realizing that all other non-substitutable elements within deposits down to this depth could support man's present consumption rates for millions of years, we may safely conclude that fossil energy resources are the only resources which pose a problem that demands a principal solution in the future.

12.3 Long-term energy-demand prospects

Before assessing the various alternative energy sources which could and, sooner or later must, solve our energy-supply problems, it is necessary to outline the development of energy demand. We will first attempt to establish the orders of magnitude and later return to various feedback mechanisms which will modify the likely development of future energy consumption.

There are essentially three independent causes which stimulate the continuation of economic, and also energy, growth. They are world population growth, the development of less developed countries, and continued industrialization in developed countries, each of which have different weight and influence upon future energy demand over time.

By far the most important long-term stimulus for energy growth will result from the efforts to reduce the differences between developed and developing countries, for example, in terms of the uneven distribution of energy consumption per capita of countries. Figure 12.4 shows that more than 70 per cent of the world population

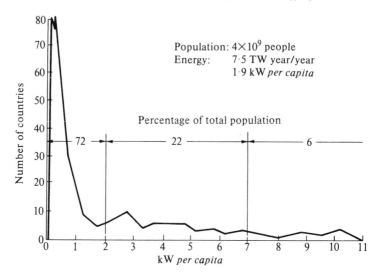

Fig. 12.4 Per capita energy consumption: an uneven distribution.

live with less than 2 kW *per capita,* and more than 80 countries have a consumption as low as 0.2 kW *per capita*, with only 6 per cent of the world population enjoying consumption levels greater than 7 kW *per capita*. Energy consumption is highly correlated to economic output [6], and substantive improvement of the economic conditions in the developing world will significantly increase the present global energy consumption level of 1.9 kW *per capita*. The present short- to medium-term planning in developed nations on the other hand points to a reduced but still continued growth of their consumption *per capita*. A set of simple scenarios should help us to anticipate the consequences of such developments upon the global, aggregate energy demand.

Figure 12.5 refers to a population projection published at the 1974 United Nations Population Conference in Bucharest [7]. Both higher and lower projections exist, and the recent general trend has been to project a somewhat lower increase [8]. Such revisions mainly reduce the asymptotic level to be reached but do not significantly change the demographics of the next 50 years. Interested in a broad overview, we have not chosen the lowest existing population projection for Fig. 12.5 that would, *ad hoc*, screen off potential problems.

This population projection as a basis leads to three alternative scenarios for global energy demand that differ in the energy consumption level for the average global citizen. The lowest — extreme — curve assumes an immediate freeze at the present level of 2 kW *per capita*. The solid and broken lines assume the process of economic development of the less developed countries (LDCs) to take 70 or 100 years, respectively, resulting in an average consumption of 5 kW *per capita*. This level corresponds to the present average in central Europe and the USSR; that of

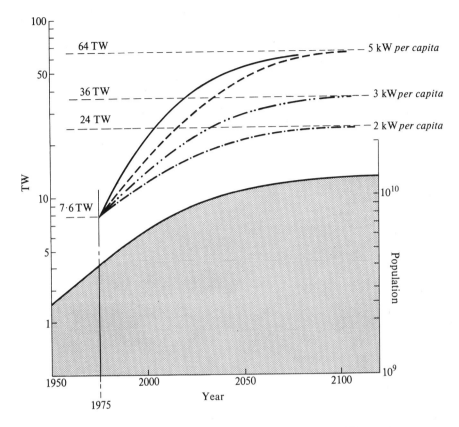

Fig.12.5. Global energy demand scenarios. (Population data from [7].)

the USA is 11 kW *per capita*. The scenario implies that the LDCs undergo the same transitional process from an essentially rural and agricultural society to an industrialized society as set in a hundred years ago for the developed countries; it also implies a successful energy conservation strategy in developed countries. The middle curve essentially extrapolates the growth rate of the global *per capita* consumption of 1914 to 1950, which was a period of growth under crisis conditions with an annual growth of 1.3 per cent. If this rate were resumed from now on it would lead to approximately 3 kW *per capita* around 2020; it was then deliberately set to zero.

This simple exercise shows that the present energy needs will multiply if the most basic goal of providing a still growing world population with acceptable material conditions is pursued. Even in the hypothetical case of an average zero energy growth *per capita*, today's 7.5 TW would triple over the next 100 years. Responsible planning must envisage still higher factors.

Against such demand projections we can now better evaluate the critical energy resource situation. Figure 12.6 shows the cumulative total consumption resulting from the scenarios of Fig. 12.5. At its last meeting in August 1977, the world energy conference attempted a reassessment of the economically recoverable energy reserves potentially available at higher prices [9]. Since higher energy prices affect the viability of more expensive production technologies for thinner deposits, the reserve figures stated now are significantly larger than previous data would indicate. To allow for the combined effects of further intense exploration and continuing price increase we have doubled the revised estimates of oil and gas reserves to a total of 400 TW years in Fig. 12.6.

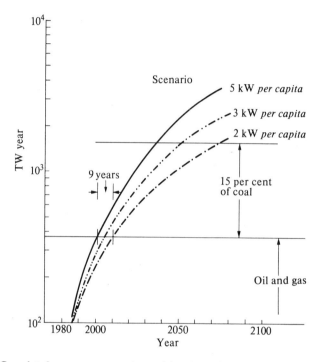

Fig. 12.6 Cumulated energy consumption and fossil energy reserves.

Similarly we have doubled the economically recoverable coal reserves to 1200 TW years, arriving at a fairly optimistic total of 1600 TW years of fossil energy reserves. According to Fig. 12.6 and irrespective of the scenario chosen, oil and gas would practically be exhausted within 30 to 40 years from now if their present total share in the supply balance were maintained. Including coal reserves, one finds that our present energy technologies based on fossil sources can support the 5 kW scenario of Fig. 12.5 for only another 60 to 70 years.

What does this imply? Without going into more details of the complex resource/ reserve question one may say that mankind is compelled to the fundamental transition of gradually changing its energy system to one based on non-fossil energy sources. And according to Fig. 12.6, even the lowest hypothetical scenario, which assumes an immediate switch to zero energy growth for the average individual, will not avoid exhaustion of fossil reserves. It would certainly extend the time horizon for a transition away from traditional fossil energy, but at the price of a drastically limiting economic progress. The *nine*-year difference in the exhaustion of oil and gas reserves between the highest and lowest scenarios of Fig. 12.5 quantifies the small degree of flexibility of long-term energy conservation. We stress long term because conservation measures in a short-term context ask for a different evaluation.

The deeper reason for our general resource dilemma thus is definitely not future economic growth. Instead, it is the quantum jump mankind has taken during the past one hundred years. After all, between 1880 and today, energy consumption has grown by a factor of 30 (Fig. 12.1) and population has only tripled. The idea of low or zero growth dangerously detracts from this fact. It is our present achievements and not an affluent future which are at stake in the long run.

12.4 Alternative supply options: a first-order inventory

With this warning let us briefly review the various alternative supply options. Two points must be noted. The transition period of some 50 to 70 years is not very sensitive to various conservation efforts. Therefore only those sources can substantially contribute to a substitution of traditional fossil energy which are somehow within reach of the present technology. Further, alternative sources have to match, at least approximately, in terms of their capacity a global energy demand of dozens of TW. The hypothetically low demand scenario of Fig. 12.5 leads to already 24 TW, and more realistic scenarios are in the 50–TW range.

Renewable energy sources are advocated these days as an alternative and, sometimes, also as a particularly attractive way of meeting energy demand with a view to the environment. An extensive and careful evaluation of localized renewable sources, made under the auspices of the Ministry for Research and Technology of the Federal Republic of Germany [10], is summarized in Table 12.1. Apart from solar energy, which requires separate consideration, all the sources listed in Table 12.1 are geographically localized.

Table 12.1 Localized renewable energy sources

Source	Technical potential TW(e) or equivalent	Technological maturity*
Hydropower	2.9	Mature
(Glacier Greenland)	0.1	Economic Potential 1.1 TW Presently utilized 13%
Wet geothermal	0.1	Installed 0.0013 TW
Wind	3	Feasible 1 TW (?)
Tidal power	0.04	Installed 0.0002 TW
Wave power	1 TW/35,000 km	To be developed
Ocean thermal gradient	0.35+	To be developed

*Note 1 TW = 1 TW year/year
+Within 10 km distance from coastline

Hydropower has a technical potential of almost 3 TW. This corresponds to an all-out effort where virtually every small river would have to be exploited through a cascade of dams and power stations. Accumulation of the melting waters from a significant portion of the Greenland ice (more than half of Southern Greenland), in spite of the environmental problems of covering the surface with a black absorptive substance and hundreds of dams and large power stations along the southern coastline, would give a potential energy supply of 0.1 TW[11]. The technical potential of hydropower by far exceeds its economic potential which is on the order of 1.1 TW, out of which 13 per cent are harvested today. Wind power appears to have a comparable technical potential. The environmental side-effects of wind power at the TW level are not at all negligible. To produce 3 TW of mechanical energy would require to set up arrays of windmills, up to 100 m high, over roughly 5 million km² in coastal areas not too distant from large consuming centres. The additional extremely difficult problem of energy storage and power levelling will, in contrast to hydropower, reduce its economic potential quite dramatically. Wet geothermal and tidal power are minor sources in this context. Wave power appears more attractive, but linear installations of almost the perimeter of the earth would be needed to obtain 1 TW. The ocean thermal gradient if harvested at the TW scale would raise severe environmental questions because of the necessary displacement of very large bodies of water. Its technical potential within 10 km distance from the coastline (a distance at which connection of smaller OTEC plants to the electric grid should not be too difficult) was estimated at 0.35 TW.

One may safely conclude from the multitude of studies that renewable energy sources, apart from solar energy, can technically contribute a few TW at best. Economic and local environmental constraints as well as competing land use requirements will further reduce their contribution to the global supply balance. There is a fundamental gap on the order of dozens of TW between the capacity of these sources and an anticipated demand. This gap justifies the development of all alternatives. Wherever there is a potential to use such renewable sources on the basis of favourable local conditions it should be exploited. The gap also illustrates the delusion that renewable natural sources would provide a real alternative technological choice for large nations.

Let us take a new look at coal. In search of sufficiently abundant energy sources in our global context we will deliberately put aside the well known economic, environmental, and social problems of coal, much in the way in which we have considered localized renewable sources. Needless to say, there are more hydrocarbons, of which coal is a specific type, than were introduced in Fig. 12.6 as economically accessible reserves. The proven and inferred 'geological resources' of coal, i.e. the deposits already known that can technically be mined by methods that have proven to be economic elsewhere are, at least in principle, very large. Fig. 12.7 relates global coal resources and fractions considered economically recoverable at three successive meetings of the World Energy Conference[12]. Both the amount of resources and the percentage of the reserves in terms of resources have increased since 1974. Technological progress and generally higher future oil prices will increase the amount of economically recoverable coal. To be on the safe side, the 1200 TW years of long-term coal reserves we assumed in Fig. 12.6 correspond to twice the amount shown by the small black portion of the 1977 bar in Fig. 12.7.

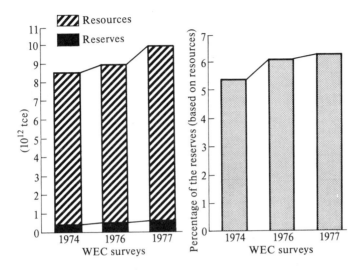

Fig. 12.7 Recent additions to coal resources/reserves [12].

To assess the ultimate potential of coal, the largest fossil resource known, we shall now hypothetically assume that by vigorous R and D efforts, together with the necessary economic adaptations to account for much higher coal prices, a still larger fraction of coal resources could be harvested.

In the past coal held the main share in the global energy balance, and we will try to sketch the possibilities and limitations of promoting coal into the role of the main primary energy source again. Figure 12.8 is a specific logarithmic plot of the contribution of the various energy sources to the global energy balance for the past hundred years [13]. It is striking to see that the substitution of the various sources follows a quite regular logistic pattern, which is well known in market theory and is also observed in many biological systems. The distorted behaviour during the periods of war and economic recession obviously influenced the overall growth rates (compare Fig. 12.1) but could not significantly alter the secular substitution phenomenon towards energy forms that are ever more versatile and easier to handle. With such a market penetration concept at hand it is quite easy to extrapolate past trends and to discuss the introduction of new alternative energy sources and their likely position in the market as a function of time.

Figure 12.9 is the result of such an exercise for 'new coal'. Whereas traditional uses of coal, according to Fig. 12.8, lead to a fast phasing out of coal, we here have assumed that a revival of coal with new mining and conversion technologies could be initiated in the next few years. Moving at the same speed at which oil penetrated

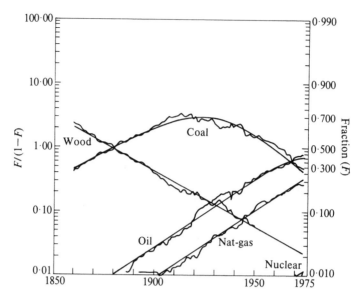

Fig. 12.8 Primary energy substitution: a logistic phenomenon.

the global market of the past, coal would obtain 80 per cent of the total market, its old maximum, in approximately one hundred years from now. The volume of the market was assumed to correspond to the 5-kW scenario (solid line) in Fig. 12.5.

In order to limit the burden on the coal resources, two additional assumptions

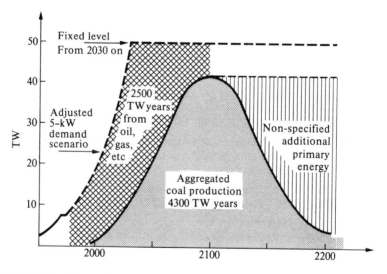

Fig. 12.9 Coal–a future primary energy option?

were introduced: the growth rate of global energy consumption was set at zero beginning in 2030, and a non-specified additional primary energy source would be commercially available in 2080. This non-specified source would then have to phase out coal at the same speed at which 'new coal' was forced into the global energy system. Coal production integrated over time then amounts to 4300 TW years, approximately 45 per cent of the geological coal resources. This is a staggering result. For simple technological reasons not all the coal present in its geological environment can be recovered, of course, an average recovery factor of 50 per cent of actually mined deposits is a reasonable upper estimate today. In our example the extremely large resources of coal would, rigorously exploited, at best support a new coal age for another century. But there is more to this example. The additional energy to make up for the gap between the expected total demand and the gradually increasing (hypothetical!) coal supply in Fig.12.9, aggregated between now and the year 2100, amounts to 2500 TW years. Even if we correct the demand scenario for a more gradual transition towards zero growth, this deficiency is equivalent to ten times the presently stated oil and gas reserves of 210 TW years.

A quick look at the carbon dioxide question should suffice to establish the dimension of problems that would pile up if mankind had to continue mainly on fossil energy for another 100 years. The climatic consequences of a further increase of the carbon dioxide content of the atmosphere are still debated by meteorologists [14]. There is even some uncertainty about the quantitative exchange of carbon dioxide between the various reservoirs. Various simulation models have shown that a doubling of the atmospheric carbon dioxide could occur around 2020, and, according to our insufficient knowledge today, this would mark a point where serious climatic changes can no longer be excluded [15]. A global primary energy option based on coal as in Fig.12.9, or one that includes unconventional fossil hydrocarbons, such as heavy crudes, tar sands, oil shales, or gas from geopressure zones [16] instead, will certainly depend on whether there is an effective way of carbon dioxide disposal other than the release to the atmosphere. Compare for example, the carbon contents of the various natural carbon dioxide reservoirs and their exchange rates (Fig.12.10) with the coal consumption in Fig.12.9. The combustion of 4300 TW years of coal (the aggregated coal production in Fig.12.9) would mean to transfer 4500×10^9 t of carbon in the form of carbon dioxide through the mixed surface layers of the ocean into the deep sea. This definitely is an unacceptable burden for both the atmosphere and the oceanic surface layers. So a costly direct disposal of carbon dioxide into the deep sea with all its unknown implications would indeed have to be envisaged [17].

If the prospects for an energy supply depending on either localized renewable sources or fossil energy resources are so dim, let us turn to the three alternatives that tap abundant sources: solar energy, nuclear fission, and nuclear fusion.

The energy flux from the sun to the earth, for example, is practically infinite compared to all possible human needs. Large-scale solar energy as one of the main global energy options leads to hard technologies. Such an option needs more than a few solar panels in the backyard. A few figures shall characterize this option. The key parameter is the average insolation at the earth's surface. In mid-latitudes, one

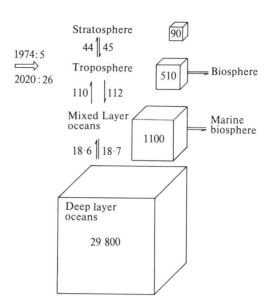

Fig. 12.10 Carbon dioxide reservoirs and carbon dioxide flows (figures are in units of 10^9 t).

square meter receives roughly 100 W on a yearly average. Present energy consumption densities in highly populated areas, such as Greater Calcutta, India, or the Rhine/Ruhr area, FRG, are about 10 W/m², and in rural areas, the densities are smaller by one to two orders of magnitude [18]. Despite this favourable ratio, solar energy is a dilute source that requires large areas of land.

The maximum-to-minimum ratio of the seasonal insolation cycle in mid-latitudes is close to 7:1; in southern desert regions it is nearly 2:1. These ratios point to the fundamental problem of energy storage, because, if solar energy is to be more than an auxiliary source, it must supply base load energy. This will most probably require the production of hydrogen, which would also alleviate the problem of long-distance energy transport [19]. With a most favourable siting in desert areas — which enjoy a low seasonal variability and an insolation of about 200 W/m² — it is optimistic to estimate an overall efficiency of 10 per cent of producing, storing, and transporting hydrogen from sunlight; energy requirements of 50 TW would translate into a land requirement of 2.5 million km². This 2.5 million km² of land covered with solar installations then compares with the 13 million km² used for present agricultural purposes and the built-up area of human settlements of 0.4 million km². Soft solar devices such as harvesting sunlight on the roof or in the backyard are confined to the last figure and can therefore supply only a few per cent of our future energy demand.

Thus solar energy (see Table 12.2) is clearly understood as a hard, large-scale, and remote technology, to be viewed perhaps as a new kind of agriculture. There will be supply and deficit regions, and — in the best tradition of the oil system — a world market for energy carriers that are derived from solar energy. We are certainly

Table 12.2 Options for a non-resource constrained energy supply

Source	Potential (TW year)	Technological maturity	Side-effects
Fission (Breeder)	10^8	Sufficient for power plants Not yet sufficient for large-scale fuel cycle	Storage of fission products Emission of radionuclides
Fusion (D-T)	10^8	To be developed	Storage of activated material Emission of radionuclides
Solar	(∞)	To be developed for large scale	Land and materials requirements Climatic disturbance? Storage and transportation

not accustomed to this dimension of solar technology, but it is this global context in which it is to be compared with the main alternative, nuclear power.

Nuclear power can be derived from either fission or fusion reactions. While fusion is still awaiting demonstration of its technological and commercial feasibility, fission is a commercial energy source already today. The light-water reactors (LWRs) depend on cheap uranium and will not solve the global resource dilemma. As LWRs can, on a net basis, use only one per cent of the uranium atoms, this technology adds an equivalent to the global energy resources that is comparable to that of crude-oil reserves. Uranium resources in conjunction with the LWR technology can only feed into the electricity submarket. For both reasons, uranium is not a direct economic substitute for oil. Nuclear fission becomes a rich and abundant source of energy only through the principle of breeding. The fast-breeder reactor technology increases the utilization factor of uranium by a factor of 60, or possibly more. Much thinner uranium deposits can be economically exploited, and the energy resources of 10^7 to 10^8 TW years accessible through fission breeders point to another way out of the long-term resource dilemma.

Fusion, the nuclear brother of fission, also depends on the breeding of tritium from lithium. Lithium resources promise to become an energy resource equivalent to 10^8 TW years once fusion can demonstrate its practical feasibility. Waste-disposal problems, containment of radioactivity, the use of liquid metals, and other features of the two breeder types are qualitatively similar in most respects, although they are quantitatively better for fusion [20]. When speaking of the nuclear option in general, we will refer to both technologies. In connection with an early substitution of fossil energy we will in the following only refer to nuclear fission, since, even if the optimistic target dates of present R and D programmes can be realized, fusion would not be able to significantly change the global energy balance within the 50-year time-horizon available for a transition. Here a similar argument holds as for localized renewable sources: it is vital to develop fusion, but its potential benefits have to be seen in the proper perspective.

This short analysis of the alternative energy sources clearly indicates that mankind will ultimately have to shift either to nuclear or to solar energy, or more likely, to a contribution of both energy forms. The only responsible question is not 'if' but 'how' and 'when'.

At this point it is important to look briefly at the environmental question. The problem of environmental pollution has arisen already at the present energy consumption level of 7.5 TW. Many resources are restricted in their use long before they are physically exhausted. High-sulphur coal is a case in point. From our present experience it is legitimate to ask whether the global environment is not bound to deteriorate as a consequence of producing 50 TW, the order of magnitude of our long-term demand scenarios in Fig. 12.5.

The environmental acceptability of single nuclear power plants providing 1 GW of electricity is a public issue these days. Producing 50 TW, i.e. 5000 times more, by fission breeders, for example, would force mankind to handle and dispose of some 10^4 t of fission products per year. An overwhelming degree of meticulousness is required to keep such a flow of potentially dangerous material separated from the environment. According to present maximum permissible dose rates, not more than some 10 g of fission products per year would be allowed to leak to the environment [21].

Harvesting 50 TW from solar energy on 2.5 million km^2 of land would lead to a similarly staggering effort. Many features, such as apparent waste heat as a consequence of albedo changes, or interference with natural vegetation, make it a new kind of agriculture. But there is at least one fundamental difference in that 'solar-culture' cannot rely on a natural recycling of its structural materials. With the solar tower concept as a reference technology and a lifetime of 50 years per installation, an annual flow of waste materials would result that is on the order of 2×10^{10} t; this is equal to three times the present material throughput of the world economy [4], excluding the hydrocarbon component of 'Demandite' in Fig. 12.3. By switching from fossil resources, i.e. energy stored in deposits, to a practically unlimited flux from the sun, we would merely shift the resource problem from energy resources to materials resources; a similar shift would occur with environmental problems. The investigation of this interplay is still very recent [22], but it is clear that energy strategies that serve to solve the resource dilemma must cover this dimension.

12.5 A closer look at energy demand

The difficulties that obviously pile up for an adequate future energy supply on a global level are fundamental. Therefore a broad range of energy demand projections has been chosen to illustrate the general situation. In the light of these difficulties and in view of the long-time horizon for adapting to this situation, it is now appropriate to have a closer look at energy demand. This is necessary if we want to learn more about the specific problems of the transition phase; and it is helpful in order to see the prospects of energy conservation in the proper perspective.

For a start, let us consider the unusual suggestion of 'supplying energy without consuming energy', the negentropy city of C. Marchetti [17]. The main elements of such a scheme are outlined in Fig. 12.11. The negentrophy city could, at least in principle, be built with today's technical components. It would provide all the services, mechanical power, heating and cooling, etc., of our present energy systems without tapping any energy source at all. Let us start in Fig. 12.11 with the use of the temperature differences in tropical oceans. A kind of steam engine produces mechanical power A by extracting heat Q_1 from the surface layer at temperature T_1, discharging a smaller amount of heat Q_2 to deeper ocean layers at temperature T_2. Instead of using the mechanical power to drive a generator and to transport electricity to the consumer, compressed air is produced. The compression heat, which equals A, is discharged to the deeper ocean layer at temperature T_2.

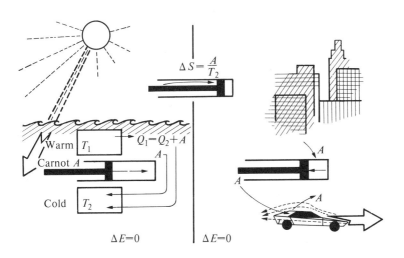

Fig. 12.11 An extreme case of energy conservation: the negentropy city of C. Marchetti [17]. (A work; E, energy; Q, heat; S, entropy; and T, temperature.)

On balance, no energy has been taken from the ocean. The net result is a simple mixing: The oceanic temperature difference has been reduced or, technically speaking, the entropy of the ocean has been increased. In turn a certain amount of compressed air at ambient temperature is obtained that does not contain more energy than before compression. It might be helpful to recall that no energy was taken from the ocean so that the energy of the compressed air cannot have increased. What has changed instead is the capability of the compressed air to produce work, and this is expressed quantitatively by the reduced entropy content of the compressed air ΔS. As it is easier to operate with a quantity that is used up instead of being increased in a process yielding a useful output, we will henceforth deal with the negative complement of entropy, simply called negentropy. Compressed air carrying negentropy can be stored and transported to the place where it is to be used, for instance, to propel a car. There the process is reversed: expanding in a cylinder, the air tends to cool off. The amount of energy that is extracted as heat from the ambient air to maintain the temperature of the expansion cylinder, is delivered as mechanical power. When the car is driven it is again quantitatively transformed into heat by friction, and the energy balance is closed. Using this principle it is possible to heat homes (for example, by combining a compressed-air expansion engine and a heat pump), or perform all the other services delivered by today's energy technology.

Our *Gedankenexperiment* is an extreme case in two ways. It can be interpreted as an unusual alternative energy supply, or as the ultimate case of energy conservation. As a supply alternative, the technological scheme of Fig. 12.11 would not entail classical environmental problems. There is no waste heat and without fuel combustion there are no chemical pollutants; but other problems appear. The low negentropy density of the ocean and the low negentropy carrying capacity of compressed air translate into enormous material requirements for the technological devices designed to handle the 'working fluids' ocean water, and compressed and ambient air. In terms of 'Demandite', the supply concept of Fig. 12.1 would dwarf the material problems of a global solar option. New indirect environmental problems would also arise. Understood as the most advanced case of energy conservation, our *Gedankenexperiment* points to the shift of problems to occur in our economies if energy were substituted by other economic values. A 'back-of-the-envelope' calculation yields prohibitive costs for the technologies deployed in the scheme of Fig. 12.11. So the main lesson of our *Gedankenexperiment* is that energy demand itself is a function of the type of supply technology chosen. This choice is obviously determined by economic and various other factors that, in any case, are exogenous to the energy system. Technological arguments therefore are not sufficient to estimate which ratio of energy input to economic output — generally called the energy coefficient of an economy — can finally be achieved.

For these principal reasons, forecasting energy demand is extremely difficult. And indeed, a rigorous scientific treatment of the complex role of energy in economic processes does not exist. Long-term energy demand is an open-ended question, but this does not mean that, given enough time and effort, any level of energy demand can be achieved.

For the purpose of orientation, we want to outline some of the mechanisms which will contribute to the decrease as well as to the increase of the overall energy coefficient in the foreseeable future. This will lead us to a more specific demand projection and reduce the broad range of energy needs given with the scenarios of Fig. 12.5.

In order to clarify traditional, but inconsistent and misleading concepts, Fig. 12.12 is shown, which details the various steps of conversion and transformation of energy all the way to those places where economic values are produced. Primary energy, derived in bulk quantities from natural deposits, is gradually adapted to the specific needs of final consumers. Final energy demand, depending on settlement patterns and economic sectors, aggregates into complex demand patterns in space and time [23]. The energy sector of an economy is specialized in converting primary energy into secondary energy, i.e. a group of energy forms with well-defined specifications. Secondary energy, i.e. light fuel oil, petrol, alternating current with a fixed frequency and voltage, hot water with a given temperature and pressure, etc., is then transported, transformed to appropriate power levels, and distributed to the final consumers. More than one-third of the heating value of primary energy

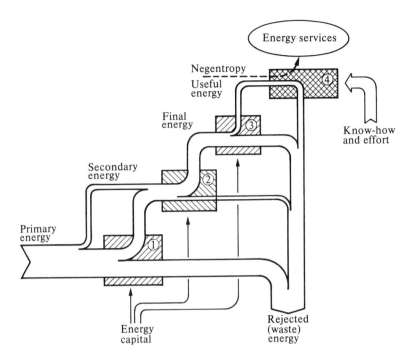

Fig. 12.12 Energy flow and demand for energy services.

is invested in this process of upgrading and supplying energy in modern economies such as the Federal Republic of Germany. Refineries and central power and district heating stations are part of a chain of technical operations that are decoupled from the final consumer, in order to utilize laws of scale and obtain higher load factors for capital-intensive conversion devices.

Substantial further 'losses' occur at the consumer end (box 3 in Fig. 12.12) when final energy is converted into useful energy by operation of productive tools and devices. Typical examples are a bulb, an automobile engine, or a stove. They provide light, motive power to overcome friction losses, or heat to compensate for insulation losses of a house, respectively. Useful energy in a modern economy, again measured in terms of its heat equivalent, amounts to approximately one-third of the heat equivalent originally supplied from primary energy [24]. It is important to note that useful energy is not an economic value as such. It is instead used to provide a particular service. Such a service at the same time requires capital and labour inputs and constitutes the economic result, traditionally measured in monetary units, that contributes to the gross national product (GNP).

Examples of energy services are a beam of light focused on a particular object, a running car on a free highway, or a preset temperature level in a room. In all cases useful energy is not actually consumed. In the process of provision of an energy service it is quantitatively returned to the environment, almost always directly in the form of heat. In the case of chemical reduction of mineral ores, useful energy is discharged after some time along with the depreciated devices. After all, there is no qualitative difference in the technical processes represented by box 4 in Fig. 12.12 and those processes in boxes 1 to 3 upstream. Whether useful energy is finally 'lost' or 'wasted', in the sense one generally speaks of 'waste heat' in electricity production, is a question of semantics. In reality energy cannot be consumed in an economic process. There is a strict law of conservation of energy in physics. But there is no such law for negentropy. This is the deeper reason why the 'negentropy city' of Fig. 12.11 could be operated 'without consuming energy at all'. Our *Gedankenexperiment* thus suggests that the only quality actually consumed in economic processes is negentropy [25]. Various efforts exist to substantiate this observation [26, 27]. Recalling Shannon's identity of negentropy and information, we find a quite natural transition from the technological to the economic stratum. In line with the physical explanation of energy consumption we must interpret energy as a proxy variable for negentropy that is derived from natural resources and that enters productive processes in economy. As a quantity, energy is proportional to information or negentropy as long as the spectrum of final energy forms or, more precisely, the technical energy infrastructure does not change. Different proportionality factors apply to electricity, coal, or oil, and it is in this context that economists speak of different economic efficiencies of various energy forms [28, 29]. Only on the level of information can total energy, capital, and labour substitute each other in the classical economic production functions, which relate the input of such factors to the economic output. This output is phenomenologically measured in terms of GNP. Empirically verified macro-economic production functions of the Cobb–Douglas type describe, in reasonable approximation, the interplay of various production factors, practically independent of the specific economic system considered [30]. These functions have the form

$$\text{GNP} = AE^{(1-\alpha-\beta)}C^{\alpha}L^{\beta},$$

where E stands for energy consumption, C for capital stock, and L for labour input; A, a, and β are 'constants'. A contains among other factors the coefficients of conversion into a common metric for E, C, and L. In the light of a more precise interpretation of energy consumption, this equation allows us to discuss the basic possibilities of decreasing the energy coefficient of an economy.

To begin with an apparently direct strategy, we shall assume that the energy infrastructure remains fixed and the energy throughput is reduced. In order to maintain the original economic output, a higher capital or labour input is needed to compensate for the deficiency in the production factor energy. Such a strategy will hardly provide a significant potential for energy saving. To make this obvious we will consider the quantitative relationship between these production factors. Energy input in relation to total capital stock turns out to be a practically fixed ratio for various world regions (Table 12.3). Whereas the capital stock *per capita* is roughly twenty times larger and differs in composition for developed market economies in relation to developing countries, the energy needs to operate one US dollar of capital stock are practically the same. Only the centrally planned economies consume nearly twice as much energy per capital stock, which is probably a

Table 12.3 Energy consumption and capital stock

	Capital energy input ratio (1963 US $/W)	Capital stock per capita* (1963 US $)
World	1.15	2000
Developed market economies	1.4	8500
Developing countries	1.3	380
Centrally planned economies (without China)	0.7	2700
	-	

*Data from [31].

consequence of the predominance of their heavy industry and their different resource situation. Considering the widely different contributions from the primary, secondary, and tertiary sectors to the GNP of developed and developing nations, it does not seem very promising that specific energy needs in a global context be reduced by a different choice of productive technologies. No doubt, some smaller countries could specialize in less energy-intensive products, but this would only shift the energy demand to other places in a region. It is also disillusioning to compare the economic values of human labour and energy. Table 12.4 gives the average wages per working hour for three different countries. (Since

Table 12.4 Energy value and human labour

	US	FRG	India
Average compensation of employees, US $/hour	4	3.9	0.05–0.10
Total energy consumption, kW *per capita*	11	5	0.6
Fraction of average working time (%) to pay for the energy bill*	8	4	37–19

*Primary energy, 11.5 $/bbl equivalent

a substantial fraction of the working force in India receives part of their compensation in non-monetary form, an estimated range was introduced for this country.) In order to pay the world market for the primary energy that is consumed per capita in each country, the average worker in the USA has to spend 8 per cent of his working time; the worker in the FRG 4 per cent; and the worker in India between 19 and 37 per cent. Although these figures change upon adjustment to the local energy prices, the orders of magnitude and the trend remain the same.

In order to save 10 per cent of the total primary energy consumption, the average worker in a developed country could, for economic reasons, spend only a few minutes a day; the worker in India might spend half an hour. Labour thus clearly cannot substitute significantly for energy as a production factor. The quantitative relations in Table 12.4 further demonstrate that economic development has led to a gradual substitution of energy for primitive human labour; even for India the tendency is clear to compensate a labourer for his know-how to operate the capital stock. Human 'labour' in the original sense of the word is no longer an important production factor, and a resubstitution of human labour for scarce energy is unlikely, even in developing countries. Instead, labour must become skilled labour and thus significantly contribute to the generation of negentropy by providing sophisticated know-how. In fact, it is a well-known feature of production functions, such as the Cobb–Douglas function above, that they only partly explain the GNP and its growth; high shares of GNP increases must be explained by the so-called 'technological progress'. Therefore, quantity A is often put as $\exp(\rho t)$, where ρ reflects productivity increases per year, or, in our language, additional negentropy inputs through sophistication.

So only the indirect way of improvement of our technical infrastructure, which is represented by capital stock, and quality of labour can lead to a reduction of the specific requirements. This necessitates research and development as well as education. But this would be only a first step. To change the energy needs, capital stock, and the organizational pattern of utilization of labour will have to be changed. And ultimately we have to be aware that the main goal of all efforts to improve our economic mechanisms is to increase total productivity, and not to minimize any one production factor in the above equation. In other words, energy conservation thus will always compete with capital and labour conservation.

Figure 12.13 quantifies a specific group of technologies for which efficiency improvements were possible in the past. The efficiencies of various prime movers, developed in the course of time and actually installed under economic constraints,

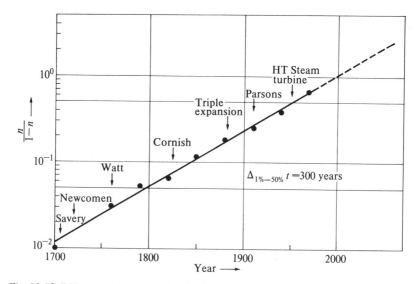

Fig. 12.13 Efficiency improvements of prime movers.

again follow a quite regular logistic pattern. It is important to note that higher efficiencies could have been and can now be achieved if the only purpose were to demonstrate the limits of technology; magneto-hydrodynamic energy conversion or multi-stage working fluid arrangements are present-day examples. Similar plots can be given for other groups of technologies that characterize the conversion and transformation steps in boxes 1 to 4 in Fig. 12.12 [24]. Fig. 12.13 shows that energy conservation is not a recent trend; already James Watt contributed to it. More important, however, are the extremely long time periods that characterize technoeconomic progress. A further doubling of the efficiency of prime movers would require another 50 years, if we extrapolate from Fig. 12.13.

Thus the hopes for a fast and thorough decrease in the energy coefficient seem quite low. Energy conservation most probably will not substantially influence the necessity of a transition to non-resource constrained energy sources, as discussed earlier.

Unfortunately, one cannot even expect a continuation of the past trend of energy efficiency improvements in Fig. 12.13, considering economies as a whole. The present generation of light-water reactors operates with saturated steam tur- bines and an overall energy efficiency of 32 per cent whereas the maximum of fossil power plants is now 42 per cent. According to present investment programmes, the line in Fig. 12.13 will drop until the year 2000 and not continue to rise as indicated. An even more dramatic reduction of energy efficiencies will go along with the gasification of coal in substitution of scarce natural gas. Generally speaking, the lesser quality of alternative energy resources and/or our insufficient techno- logies to harvest the negentropy content of new sources — whatever applies — ask for a higher primary energy input than that of today. Fig. 12.14 conceptualizes this new situation. The past substitution processes in the global energy system, which led from coal via oil to natural gas (Fig. 12.8), have undoubtedly helped decrease the overall energy growth rates with respect to the economic growth rates. This effect will fade out with the lowering of the oil and gas shares in the national energy balances. In analysing future energy demand it is absolutely necessary to clearly separate at least the few factors mentioned so far. Energy demand is a

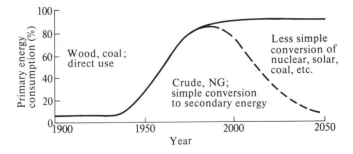

Fig. 12.14 A widening gap between primary and secondary energy.

poorly understood component of the present energy problem, and prudent planning obviously should not anticipate optimistically low energy coefficients for the future.

12.6 A global reference scenario

We will now focus on the transition of the energy system away from traditional and cheap fossil energy resources to non-resource constrained new sources. In doing so, we will separate the energy sector from the economy and consider those inter-actions which appear crucial.

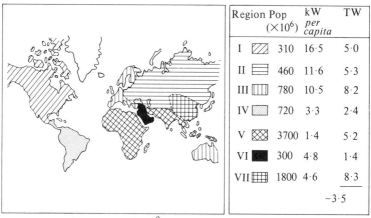

Region	Pop (×10⁶)	kW per capita	TW
I	310	16·5	5·0
II	460	11·6	5·3
III	780	10·5	8·2
IV	720	3·3	2·4
V	3700	1·4	5·2
VI	300	4·8	1·4
VII	1800	4·6	8·3
			~3·5

Population: 8×10⁹ Energy: 35 TW OR 4·4 kW *per capita*

Fig. 12.15 Reference demand scenario; world, 2030.

Fig. 12.15 summarizes the results of minimum-energy-demand estimates that are based on considerations of the preceding section. The world is disaggregated into seven regions that roughly correspond to: North America; the Soviet Union and Eastern European countries; Western Europe with Japan; Latin and South America; South-East Asia and Africa; the Middle East; and China. Population figures are taken from N. Keyfitz[32]. They are significantly lower than the UN data used for the set of scenarios in Fig. 12.5; the total world population projected for 2030 is not larger than 8 billion people. In many other respects we have also used moder-ate assumptions. Energy demand growth rates *per capita* in developed regions were assumed to decline steadily, approaching saturation in 50 years from now. For developing regions higher growth rates and a later saturation were fixed.

The projection of 16.5 kW per capita for the region of North America points to the high degree of conservation implicit in our estimates. The present US level is already 11 kW *per capita*. In 50 years time, Western Europe and Japan would ex-perience a saturation at twice the present consumption; South-East Asia would have slightly more than doubled its present consumption, which is 0.6 kW *per capita* including non-commercial energy forms, such as firewood, farm wastes, and dried dung. Our reference demand scenario arrives at a global energy demand of 35 TW and an average *per capita* level of 4.4 kW in 2030, which would not grow significantly thereafter. These figures refer to an energy input to the economy equivalent in quality to that of the present primary energy spectrum. Changes in this quality, which inevitably go along with the gradual introduction of alternative energy sources, will be accounted for in conjunction with the supply technologies.

Fig. 12.15 quantifies neither a most likely nor a particularly attractive development. Major catastrophic events that might further reduce energy demand are excluded. The main purpose of our reference demand scenario is to provide a lower, and yet consistent, minimum-demand projection. We purposefully neglect the effort and problems here related to the implementation of economic strategies that finally lead to the gradually decreasing energy coefficient defined in our reference demand scenario; instead we will briefly outline the minimum efforts necessary if the minimum demand is met by gradual recourse to the alternative energy sources described above.

Until 2030 the cumulative primary energy demand of the reference demand scenario (Fig. 12.15) amounts to 1000 TW years, whereas, according to present knowledge, economically recoverable fossil energy reserves only total 800 TW years. Only a small fraction of the 35 TW in 2030 can consequently be supplied in the form of conventional oil, gas, and coal. Because of the limited potential of natural renewable energy source (Table 12.1), nuclear and solar energy must cover a substantial share of the energy needs in 2030.

Solar energy, apart from some local uses, and nuclear energy cannot directly feed into the economic infrastructure, however. These primary energy forms of sunlight and particle radiation must be adapted to the specific requirements of our production devices, which requires a series of technical steps as indicated by Fig. 12.12. To simplify matters we assume that neither the settlement nor the economic production pattern of today would be changed to facilitate the technical deployment of either source. In this most likely case, nuclear and solar energy must provide a spectrum of secondary energy forms similar to those at present derived from fossil energy resources. In fact, this assumption does not constrain our further arguments; any other mutual adaptation of nuclear or solar energy and the economy would shift, but certainly not reduce, the efforts characterizing such a basic transition in the energy system.

Fig. 12.16 describes in more detail the principal step of adaptation. It corresponds to box 1 in Fig. 12.12, which today represents mainly oil refineries and conventional power plants. Natural gas now is bypassed and directly enters box 2. In order to use the downstream infrastructure, i.e. boxes 2 to 4 in Fig. 12.12, nuclear or solar energy will have to be mainly converted into synthetic hydrocarbons. A further extension of the present infrastructure within the coming two, or even three, decades on the bases of crude oil, natural gas, and partly also on coal, converted into synthetic gas [33], will hardly leave any other choice.

Hydrogen has in recent years attracted much attention as a substitute for hydrocarbons. Hydrogen can be produced by splitting the water molecule; yet progress in the area of thermochemical cycles is minor [34]. In addition, hydrogen would be difficult to penetrate those submarkets that are now covered by liquid fuels because of their easy distribution and storage characteristics. Thus a hydrogen economy

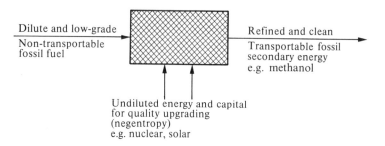

Fig. 12.16 Using energy and capital for the upgrading of residual fossil fuel.

would entail costly modifications in all boxes of Fig. 12.12.

Consequently, for the period of the next 50 years, the main product to be derived from nuclear and solar energy should be a liquid synthetic hydrocarbon fuel that is, to the largest possible extent, compatible with the specifications for petrol, light fuels, oil, etc. Methanol is a key substance in this respect, also because it would minimize the requirements of carbon input which still has to come from fossil resources, most probably from coal. If nuclear or solar energy were to provide all energy requirements for methanol synthesis from coal, global coal resources would support a long-term supply of liquid fuels. At the same time the carbon dioxide problem would be reduced by a factor of two [35]. For this or any other upgrading of residual fossil fuels, substantive quantities of both energy and capital must be invested. Although energy from nuclear or solar sources is not resource constrained it depends on the availability of capital. So capital, needed to secure an adequate long-term supply, must be considered the main economic substitute for scarce natural energy resources now in use. It is therefore equally important to check capital-formation capabilities of the global economy (or of certain parts of it) as to analyse the potential extension of oil and gas reserves.

Again we should first determine orders of magnitude. In order to quantify capital needs for nuclear and solar energy, reference technologies must be selected. As with the global energy demand projection, our point is not to predict a particular technological choice of the future but to arrive at reasonably indicative technoeconomic parameters. These parameters are then meant to approximately describe the average minimum economic efforts in order to tap nuclear and solar sources.

Table 12.5 summarizes a reference supply scenario for the year 2030. The underlying idea was to meet the 35-TW demand scenario of Fig. 12.15 by stretching traditional fossil energy supply and localized renewable resources (including soft solar) to their upper, and possibly unrealistic, limit. Nuclear energy combined with dirty and not directly usable residual fossil sources then was assumed to supply, if possible, the deficiency of liquid secondary energy, at present derived from oil. Hard solar energy, in view of its inherent economic and geographical disadvantages, finally was considered a complement that is designed to meet a still existing gap between the 35-TW oil equivalent demand and the aggregate supply of all other sources.

Table 12.5   Reference supply scenario world, 2030

Oil, gas, coal	5 (th)	Direct fossil
Localized renewable sources (biogas, wind, soft solar)	3 (th)	Fossil replacement
Hydropower 0.7 TW (e) Λ	2 (th)	Fossil replacement
LWR 3 TW (e) Λ	10 (th)	Fossil replacement
Methanol (nuclear and coal)	10 (meth)	Oil replacement
Methanol (hard solar and coal)	5 (meth)	Oil replacement
Total	**35 TW**	
Total fossil consumption	**12 TW**	

(e) electric, (th) primary thermal, (meth) methanol

The first entry in Table 12.5, 5 TW for direct and traditional uses of conventional oil, gas, and coal, appears a high estimate for 2030. But one must recall that already today 7.5 TW are drawn from these sources; and the values chosen for 2030 — the end of our 50-year transition period — should certainly be maintained for several decades thereafter in order to satisfy particular needs. Since substantial amounts of fossil carbon, mainly coal, will be needed for the production of synthetic fuels, our supply scenario implies vigorous efforts to extend present fossil reserves. Localized renewable resources were assigned a total of 3 TW, equivalent to the heat supply now derived from oil or gas. One out of the 3 TW was allotted to biogas, by which we mean all sorts of non-commercial rural sources including dung. The biogas technology would significantly enhance the efficiency of simple combustion of low-quality organic matter. Without such improvements the substitution potential of rural biomatter would certainly be much lower [36]. One may recall that in India about 0.3 kW *per capita* now come from such sources, already causing problems of deforestation. A replacement of 1 TW of fossil energy by wind power is a high figure, too; in coastal areas it would mean covering almost 0.5 million km^2 by wind mills [10]. Further, 1 TW for soft solar also appears a high figure. Roughly calculated more than 50 per cent of the suitable roof area of all houses in the year 2030 would have to be covered with solar panels to this end.

At present, electricity amounts to approximately 12 per cent of the secondary energy spectrum in developed countries. For our 35-TW reference demand scenario we assume, in line with the conservation efforts implied by this figure, a slight increase of the electricity share to not more than 15 per cent. This corresponds to an electricity consumption of roughly 4 TW(e) in 2030. To hydropower we have assigned 0.7 TW(e) or an oil equivalent of 2 TW(th), a figure that is not impossible but which implies rigorous use of practically all hydropower sources. More often than not this will be at considerable environmental expense. Another 0.3 TW(e) could probably be added by cogeneration of district heat and electricity, using part of the 5 TW traditional fossil energy mentioned above. The remaining 3 TW(e) would then have to come from nuclear power plants, and we assume that they will be produced in LWRs. Fossil energy, renewable sources, and nuclear electricity thus add up to 20 TW of primary energy, leaving a gap of 15 TW. As explained earlier, these 15 TW should be supplied in a quality economically equivalent to crude oil and natural gas used today. In our reference supply scenario we chose 15 TW of methanol production until 2030. Fifteen TW out of 35 TW roughly correspond to the present share of liquid and gaseous final energy carriers in the energy balance of developed countries.

The parameters of our reference supply scenario that have been determined so far now allow to quantify the general scheme of Fig. 12.16 for the introduction of large-scale non-resource constrained nuclear and solar sources.

12.7 The substitution of endowments for natural energy resources

Production of more than a few TW from nuclear energy requires breeding. It is common to assume that the first core plutonium inventory of fast breeder reactors (FBRs) comes from LWRs. According to our reference supply scenario, 3 TW(e) of LWRs built up until 2030, provide a plutonium output sufficient to support a parallel build-up of 5.6 TW(e) of nominal FBR capacity by the same time. According to Fig. 12.17, the electricity production for the consumer market by FBRs would

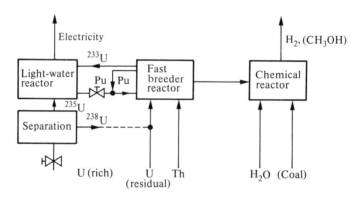

Maximum yield: Continuous 3 TW(e) +10 TW(meth) From 2030 Onward

Related endowment: 10×10^6 t U (rich) + 17×10^{12} $

Fig. 12.17 Transformation of fuel resources and capital into an endowment.

not be significant. A capacity of 5 to 6 TW(e) corresponds to 12 to 15 TW(th), which would be used partly as process heat and for electricity generation for electrolysis. Clusters of FBRs used in this way could be coupled to a chemical reactor for coal gasification. We refer here to a process by which coal is gasified by way of the iron bath technique and a subsequent catalytic synthesis of methanol. For stoichiometric and energetic requirements, electrolytic oxygen is added to the iron bath and electrolytic hydrogen to the synthesis, both of which are labelled 'chemical reactor' in Fig. 12.17.

With about 5 TW(th) of coal this scheme would produce methanol with a heating value of 10 TW. Methanol, like oil today, can be shipped over global distances, which is advantageous also because it may be difficult to realize a siting policy for accommodation of 12 to 15 TW(th) breeding capacity close to consuming centres. Methanol would allow the location of large conversion units in remote places, a major portion probably on some islands. This would facilitate both construction – 300 GW(e) per year otherwise appears too high even on a global level – and the disposal of waste heat without negatively affecting the transport of coal. In a way, large methanol-synthesizing complexes, operating in the range of hundreds of gigawatts up to 1 TW and more, could be seen as direct substitutes for present oil fields. In view of the difficulties encountered with nuclear energy today, 10 TW of methanol derived from such a scheme seem a highly optimistic figure. In any case it will not be more than 10 TW, and this still leaves a gap of 5 TW.

But the argument goes further. After 2030 the operation of 12 to 15 TW(th) of FBR capacity permits not only continuous plutonium refuelling of these FBRs but the surplus breeding gain can be used to produce ^{233}U from thorium, which in turn would constantly refuel the LWRs. We assume these LWRs to remain in the countries where the electrical load is, since it is expensive to transport electricity over large distances. After 2030 the configuration of Fig. 12.17 only needs ^{233}U and thorium, whose supply is no problem due to the inherent features of breeding. Operation of this configuration is *de facto* decoupled from the customary problems of fuel supply if we neglect the coal supply for a moment. It would constitute a permanent source of energy comparable in many respects to the negentropy city example of Fig. 12.11.

The build-up of the configuration by the year 2030, however, requires a total of roughly 10 million tonnes of cheap natural uranium. This is the present estimated order of magnitude of ultimately recoverable uranium reserves [37]. Should the FBR construction capacity be assumed to be smaller, the uranium requirements would be larger and the time of decoupling such a configuration from customary fuel supply would be later. Instead of 10 million tonnes, 20 to 30 million tonnes of natural uranium must so be envisaged, and the decoupling might be as late as 2050 or 2060 instead of 2030. What we face, therefore, is a trade-off between negentropy from natural resources on the one hand and from know-how on the other hand.

It is important to note that the 10 million tonnes of natural uranium in the scheme of Fig. 12.17 are not simply burnt. Part of this amount is transformed into plutonium which is then used in analogy to seeds in agriculture. Plutonium, together with the capital representing the economic value of the nuclear devices, constitutes an endowment, a source of electricity and methanol for an almost indefinite period of time.

In present-day prices the fixed capital to produce 3 TW(e) in LWRs is approximately US$ 2.1 × 10^{12}. This is significantly more than what would be needed to produce electricity from oil or gas. The capital costs for the FBRs and the chemical reactors, providing ^{233}U and 20 TW of methanol, would amount to approximately US$ 15 × 10^{12}, an unusually large figure. This, too, underlines our earlier observation that capital will be the main economic substitute for scarce natural energy resources.

A word of clarification must be added. Strictly speaking, the scheme of Fig. 12.17 is not a complete endowment concept. Coal is used very efficiently as a chemical substance, and yet it is consumed. The world's large coal resources might support such a concept for many decades of the next century. Significantly more capital would be needed to operate the chemical part of the scheme in the long run with complete recycling, either by recovery of atmospheric carbon dioxide and reduction to methanol or by use of hydrogen alone.

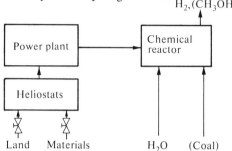

Reference yield: continuous 5 TW (methanol) from 2030 onward
Related endowment: 10×10^{10} t material$* + 0 \cdot 25 \times 10^6$ km$^2 + 15 \times 10^{12}$ \$

*World materials consumption (1975) 2×10^{10} t/year

Fig. 12.18 Transformation of material resources, land, and capital into an endowment.

Coal, a resource, thus obviously stretches the transition towards endowments. Coal properly used tends to decrease the overall capital costs of switching to nuclear energy. In this sense coal is introduced in a similar scheme, explained in Fig. 12.18, for harvesting large-scale solar energy. This scheme provides the missing 5 TW of liquid fuel for our reference supply scenario. Here the endowment arises from construction materials, land, and capital. Capital costs turn out to be twice as high per TW useful output for the most promising solar tower concept, compared to the nuclear endowment case.

The question of capital costs is fundamental with respect to endowments, and we will handle it as simply as is possible in order to bring out the salient points.

Table 12.6 gives capital-cost estimates in today's (constant) US dollars for the energy sector of a global economy. The estimates are based on the reference supply scenario of Table 12.5 and include primary to secondary energy conversion as well as transmission and distribution of secondary energy to the final consumer (boxes 1 and 2 in Fig. 12.12). As the spectrum of secondary energy was chosen to be approximately the same as today, transmission and distribution capital costs were taken from present system costs. We repeat here that the total primary energy consumption in 2030, depending on the choice of supply technologies, will exceed 35 TW. The case of solar methanol easily demonstrates that primary energy consumption will gradually become a void parameter. Strictly speaking, between 25 and 50 TW of primary solar flux will be 'consumed' to provide 5 TW of methanol, depending on the actual optimization of the mirror devices.

Table 12.6 Required energy capital stock: global reference scenario (35 TW), 2030 ($75)

Resource	Energy supply (TW)	Capital investments For production ($/kW)	(10^{12} \$)	For transmission and distribution ($/kW)	(10^{12}\$)
Oil, gas, coal	5 (th)	200 (th)	1	80 (th)	0.4
Localized renewable sources	3 (th)	100 (th)		20 (th)	
Biogas soft solar		1000 (th)	2.1	200 (th)*	0.42
Wind		1000 (th)		200 (th)*	
Hydropower	2 (th)	1000 (e)	0.7	700 (e)	0.5
LWR	3 (e)	700 (e)	2.1	700 (e)	2.1
Nuclear methanol	10 (th)	1500 (meth)	15	80 (th)	0.8
Solar methanol	5 (th)	3000 (meth)	15	80 (th)	0.4
Total			36		4.6
	Grand total		40×10^{12} \$ (Stock)		

*With storage

In contrast to primary energy, our 35 TW will still have an economic meaning. The expected slow progress in the economic productivity of an 'oil equivalent primary energy supply', discussed above, will gradually but not significantly reduce the energy coefficient of the global economy. This does not exclude an over-proportional growth of immaterial goods and services in the richer countries, which might gradually compensate for a declining growth rate of present material commodities. We are interested here in the interaction of the energy sector and the rest of the economy. As the energy sector within the coming 50 years will request, according to Fig. 12.16, typical heavy-investment goods rather than products of fine arts or sophisticated human services, we will put the interesting question of a changed composition of the future GNP aside and from now on operate with a fixed economic energy coefficient. In the same sense one has to interpret cost estimates of future technologies in today's prices: they characterize material efforts which will not change in time if some 'post-industrial' activities are added to our economic production spectrum.

After this clarification we can now ask for the economic growth necessary to support a build-up of capital stock of US\$ 40×10^{12} (the total in Table 12.6) in the energy sector until 2030. As shown in Table 12.7 the gross world product before the oil crisis was around \$ 3.6×10^{12} per year. The present state of the economy – to the extent it is of interest here – can crudely be characterized by two ratios: the capital output ratio (i.e. the number of years required for the gross product to equal the economic value of the capital stock of the economy), and the share of capital stock for energy purposes (i.e. the fixed capital needed to supply final energy to the consumer). The former is estimated to be approximately 2.5 years today, and the latter to be close to 25 per cent. Such an economy permits a consumption rate of roughly 60 per cent of the GNP. Forty per cent of the GNP is needed for investment and certain public expenses, e.g. defence efforts. Persistence of such values for these two ratios might be considered business as usual.

Table 12.7 Capital formation: business as usual?

Y_{70} (Gross world product, 1970)	$\approx$	3.6×10^{12} \$/Year
$\dfrac{\text{Capital stock } K_{70}}{\text{Gross world product } Y_{70}}$	$\approx$	2.5 (Years)
$\dfrac{\text{Energy capital stock}}{\text{Total capital stock}}$	$\approx$	0.25

In order to have an energy capital stock of 40×10^{12} \$ in 2030 with 'business as usual' (i.e. these two ratios remaining constant), the so-required Y_{2030} would be: 40×10^{12} \$/$(0.25 \times 2.5) = 64 \times 10^{12}$ \$ (2030).

If we assume this to be the case, a gross world product of $\$64 \times 10^{12}$ per year would be needed by the year 2030, or – compared with today – a growth factor of 18; but the growth factor of 4 available for energy is inconsistent. Thus it does not matter whether the constant energy coefficient assumed is approximately correct. Energy conservation over the next 50 years can by no means bridge the discrepancy between both growth factors. So we cannot expand on capital stock by business as usual; energy stocks have to have a larger share, which can only come from a reduced consumption share, as outlined in Table 12.8. The energy growth

Table 12.8 Strategic capital formation.

With energy growth to 2030 (35 TW) of 3%/year and assuming consistent economic growth 3%/year,
then $Y_{2030} = 17 \times 10^{12}$ \$.
In this case, 'business as usual' means energy capital stock in 2030 of
$17 \times 10^{12} \, (0.25 \times 2.5) = 10 \times 10^{12}$ \$ (stock).
But 35 TW in 2030 requires 40×10^{12} \$ energy capital stock, meaning–mostly for the endowments–
Excess stock required $= (40-10) \times 10^{12}$ $= 30 \times 10^{12}$ \$.

factor of 4 permits a 3 per cent overall economy growth rate, indeed a fairly likely figure, at least for the early decades of the 50-year transition period. This yields a gross world product of $\$17 \times 10^{12}$ per year, and a consistent stock for energy installations would be about $\$10 \times 10^{12}$ when business continues as usual; but $\$40 \times 10^{12}$ are required. A reduction of the share of GNP going into consumption from 60 to 52.5 per cent would allow accumulation of the required excess capital for energy stock of $\$30 \times 10^{12}$; but this means immediate action. When started only 15 years later, the necessary cut in the consumption rate is much more severe, i.e. from 60 to 45.5 per cent. Politically, institutionally, and socially this poses extreme problems, which can only be eased by starting strategic capital formation as early as possible.

A quick reaction appears mandatory as it would, after all, set the direction for the build-up of an economically tolerable energy system. A 5 per cent depreciation rate of the total energy capital of $\$40 \times 10^{12}$ in 2030 corresponds to an average of $\$12$/barrel of oil equivalent. When nuclear and solar methanol production are considered separately, the figure is $\$20$/barrel of oil equivalent. This makes up for the greatest part of the related total energy costs. After the year 2000 the cost of traditional oil is expected to be $20/barrel or even more, and thus the introduction of endowments indeed appears as an economically feasible strategy in order to overcome the problem of limited energy resources.

12.8 Conclusions and outlook

More details can and certainly must be added to the map of the technoeconomic energy future we have tried to outline so far. The terrain appears difficult, yet there is hope of finally solving the energy problem. In view of the minimum demand/ maximum supply situation, captured somehow in our 35-TW reference scenario, there are not many alternative technological and economic pathways that will allow the successful completion of the transition from scarce and cheap energy resources to technical endowments. These endowments, essentially nuclear and large-scale solar complexes with coal as a partner, exemplify the type of technology which would support a global population of eight and more billions for centuries. The only scarce resource appears to be time: any delay in the effort for choosing an appropriate energy strategy will magnify both technological and economic problems for an ever shorter transition phase.

The maximum supply scenario appears feasible in a global context, but it will create unprecedented regional and national difficulties. New political interdependencies will quickly develop. The oil crisis of 1973–4 must be seen as an early symptom which can also arise in conjunction with other energy resources, and in a different context. Coal is a case in point. Approximately 90 per cent of the global coal resources are located in three countries: the Soviet Union, the United States, and China. Present total production plans and projections for coal-mining up to 2020 [12] would support the coal requirements of the reference supply scenario (Table 12.5). Provision for coal export from these three countries to contribute to the build-up of a world coal market is too low, roughly by a factor of 10. In addition, transport technologies are not sufficiently developed to support an adequate global coal supply on the appropriate order of magnitude [35].

The general problem of balancing economic exchange most probably would not allow for an alternative solution either: that of siting nuclear conversion technologies close to coal deposits. According to Table 12.9, a country that imported 100 per cent of its energy needs in 1970 could have compensated this trade flow by exporting 3 per cent of its gross national product. At an energy price of $20 per barrel – and this is even too low a figure for nuclear methanol – the same country would find it extremely difficult to maintain its balance of payments. Solar energy, which is still

Table 12.9 Energy and economic exchange: allocation of GDP for energy imports

1970: 3 $/bbl   3% of GDP,   Total exchange between world regions 9% of GDP   Transition to large-scale nuclear, solar, and coal: > 20$/bbl Equiv.   > 20% of GDP

more expensive, as well as a selection of sites particularly suitable for the direct disposal of waste and/or waste heat, would suffer even more from the present terms of international economic exchange. This is also true for the deployment of wind power, geothermal energy, and other localized renewable sources. The US$ 30×10^{12} capital stock for endowments in 2030 (Table 12.6) points to a dimension other than the old question of a free and liberal flow of capital to the resources: not even small fractions of this sum of investment can be transferred between economic systems without negotiating terms of taxation, long-term price guarantees, and provisions for compensation transaction. Larger fractions inevitably raise the question of political leverage and, last but not least, of national interest and security.

Against this background we have to see the present criticism of energy technologies, mainly nuclear installations. Four points of concern have come to the forefront during the last years:

the possible destruction of such technologies by small groups or by accident, and consequent health and economic hazards [38];

the hardly controllable political power of monopolistic or oligopolistic supply organizations [39];

the destruction of human life-styles through the alienation of man from his natural basis [40, 41]; and

the deterioration of democratic civil rights.

A particular point with nuclear energy is the proliferation of nuclear military know-how as well as weapon-grade fissile material which have led to a certain political sensitivity and uneasiness. If we fully realize that the substitution of endowments for natural resources is an engineering solution and will not really solve the imminent political, social, and institutional problems of our present system, the energy debate of today appears a modest vanguard skirmish. Technology does not create these problems, it illuminates them.

The *Sachzwänge* which stem from the scarcity of traditional energy resources and engineering limitations largely predetermine the time for political action. We are convinced that a successful solution to the energy problem cannot be conceived on a national basis, however large a country might be. In addition to vigorous international R and D efforts, international co-operation in the actual deployment of new sources of energy appears mandatory. Only both components together would sustain confidence in the political ability to control and master the technologies demanded for tomorrow: to that end co-operation will have to pass a certain threshold of experience — as is the case with every technical concept. An internationally operated and controlled nuclear fuel cycle centre, if possible on a 'denationalized territory' would be one example, leasing of desert areas for harvesting solar energy in exchange for a certain fraction of the actually produced energy forms another. If such organizational schemes were implemented and tested already at the level of pilot and demonstration plants, this would help to address in time the actual institutional problems of a transition in the energy system. By the same token new technologies would not only be experienced as a tool of counterbalance and self-defence – an unavoidable result of a situation where prices for traditional energy forms tend to be fixed with respect to the minimum cost-level of alternative energy sources. Progress in the establishment of unusual joint technological ventures is more than a symbolic achievement. Along this road, technology could regain the more positive features of an integrating force.

If our political and social institutions do not anticipate and prepare well in advance for the related problems of a world that depends on powerful technologies, the crisis of science might well be followed by a crisis of institutions.

References

[1] Schuster, G. Crisis of science in the European Society? *Wirtschaft und Wissenschaft,* 25, 3 (1977). [See also contributions by J. Ravetz, J.J. Salomon, I. Prigogine, and P. Weingart in the same issue.]

[2] Hildebrandt, R., Schilling, H.-D. and Peters, W. Consumption of primary energy carriers in the world, the USA and the Federal Republic of Germany, *Reihe Rohstoffwirtschaft International,* Vol. 6. Verlag Glückauf Essen, FRG (1977).

[3] Wilson, C.L., *Energy: global prospects 1985-2000. Report of the Workshop on Alternative Energy Strategies,* McGraw-Hill, New York, (1977).

[4] Goeller, H.E., and Weinberg, A.M. The age of substitutability. *Proceedings of the 5th International Symposium of the UK Science Foundation: A strategy for resources.* Eindhoven, Netherlands, (18 September 1975).

[5] Meadows, D.L., *et al. The limits to growth.* Universe Books, New York (1972).

[6] Charpentier, J.-P. and Beaujean, J.-M. Toward a better understanding of energy consumption, *Energy,* 1 413-28 (1976).

[7] Recent population trends and future prospects: report of the Secretary General. In *The population debate: dimensions and perspectives. Proceedings of the World Population Conference, Bucharest, 1974,* vol. 1. ST/ESA/SER A/57, United Nations, New York (1975).

[8] Stolnitz, G.J. Some emerging interrelations between population change and economic development in low-income countries. *Fourth World Congress of Engineers and Architects.* Israel, 13-20 December, 1976.

[9] Desprairies P. *Report on oil resources, 1985-2020: Executive summary.* World Energy Conference, Conservation Commission, London (15 August 1977). [Similar reports exist on natural gas and oil.]

[10] Matthöfer, H. Energy sources for tomorrow? Non-nuclear, non-fossil sources of primary energy. *Reihe Forschung aktuell.* Umschauverlag Frankfurt, FRG (1976). [In German.]

[11] Partl, R. *Glacier power from Greenland.* RR-77-20, International Institute for Applied Systems Analysis, Laxenburg, Austria (1977).

[12] Peters, W., Schilling, H.D., Pickhardt, W., Weigand, D., and Hildebrandt. An

appraisal of world coal reserves and their future availability. *Discussion paper for the World Energy Conference, Istanbul, 1977.* Bergbauforschung, Essen, FRG (1977). [In German.]

[13] Marchetti, C. Primary energy substitution model. On the interaction between energy and society. *Chem. Econ. Eng. Rev.*, 7/8 9–15, (1975).

[14] Niehaus, F. and Williams, J. Studies of different energy strategies in terms of their effects on the atmospheric CO_2 level. Paper presented at *Joint IAGA/IAMAP Assembly*, Seattle, Washington, 31 August 1977. [To be published in *Tellus*.]

[15] Flohn, H. Großräumige Beeinflussung des Klimas durch menschlichen Eingriff? Historischer Uberblick und künftige Aussichten. *ENVITEC '77 – Internationaler Kongreß Energie und Umwelt*, Vulkan Verlag Essen, FRG (1977).

[16] Meyer, R.F. (ed.), *The future supply of nature-made petroleum and gas.* Pergamon Press, New York (1977).

[17] Marchetti, C. *On geoengineering and the CO_2 problem.* RM–76–17, International Institute for Applied Systems Analysis, Laxenburg, Austria (1976).

[18] Häfel, W. and Sassin, W. The global energy system. *Annual Review of Energy,* 2 1–30 (1977).

[19] Häfele, W. The contribution of solar energy to meet present and future energy demands. Paper presented at the *BMWF/ASA Symposium Sonnenenergieforschung,* Vienna. Austria (25 October 1977). [In German.]

[20] Häfele, W., Holdren, J.P., Kessler, G., and Kulcinski, G.L. *Fusion and fast breeder reactors.* RR–77–8, International Institute for Applied Systems Analysis, Laxenburg, Austria (1977).

[21] Avenhaus, R., Häfele, W. and McGrath, P.E. *Considerations on the large-scale deployment of the nuclear fuel cycle.* RR–75–36, International Institute for Applied Systems Analysis, Laxenburg, Austria (1975).

[22] Grenon, M. and Lapillonne, B. *The WELMM approach to energy strategies and options.* RR–76–19, International Institute for Applied Systems Analysis, Laxenburg, Austria (1976).

[23] Sassin, W. Secondary energy – today and tomorrow. In *Large-scale energy deployment and human environment: proceedings of a joint seminar.* Technical University of Vienna/International Institute for Applied Systems Analysis, Laxenburg, Austria (1977). [In German.]

[24] Häfele, W. On energy demand. Invited paper, *XXI. General Conference of the International Atomic Energy Agency.* Vienna, Austria (1977).

[25] Sassin, W. and Häfele, W. Energy and future economic growth. *Fifth World Congress of the International Economic Association,* Tokyo, 29 August– 3 September, 1977.

[26] Georgescu, N.R. *The entropy law and the economic process.* Harvard University Press (1971).

[27] Thoma, J. *Energy, entropy, and information.* RM–77–32, International Institute for Applied Systems Analysis, Laxenburg, Austria (1977).

[28] Schurr, S.H. and Darmstadter, J. Some observations on energy and economic growth. *Symposium on Future Strategies for Energy Development.* Oak Ridge, Tenn. (1977).

[29] Adams, F.G. and Miovic, P. On relative fuel efficiency and the output elasticity of energy consumption in Western Europe. *Journal of Industrial Economies* 17, 41–56 (1968).

[30] Evans, M.K. *Macroeconomic activity.* Harper & Row, New York (1969). [See, for example, Chapter 10.]

[31] Ströbele, W. *Untersuchungen zum Wachstum der Weltwirtschaft mit Hilfe eines regionalisierten Weltmodells.* Dissertation, University of Hannover, FRG (1975). [In German.]

[32] Keyfitz, N. *Population of the world and its regions.* International Institute for Applied Systems Analysis, Laxenburg, Austria (1977).

[33] Sassin, W., Hoffmann, F. and Sadnicki, M. (eds.) *Medium-term aspects of a coal revival: two case studies.* CP-77-5, International Institute for Applied Systems Analysis, Laxenburg, Austria (1977).

[34] Donat, G., Esteve, B. and Roncato, J.-P. Thermochemical production of hydrogen: myth or reality? *Revue del'Energie* 28 252–68, (1977).

[35] Sassin, W., and Häfele, W. The role of coal in the evolution of the global energy system: a reference case. *Third Conference of the International Institute for Applied Systems Analysis on Energy Resources.* Moscow, 28 November − 2 December (1977).

[36] Parikh, J.K. and Parikh, K.S. Potential of biogas plants for developing countries and how to realize it. *Proceedings of the Seminar on Microbiological Energy Conservation.* E. Goeltze Kg, Göttingen, FRG (1976).

[37] Foster, J.S., Dwet, M.F. *et al. Report on Nuclear Resources, Executive Summary,* World Energy Conference, Conservation Commission, London (15 August 1977).

[38] Adkins, B., *Public Attitudes Towards Nuclear Power, Colloquy on Energy and the Environment,* Council of Europe, AS/Coll/En-Env (29) PP8, 1977.

[39] Dahrendorf, R. Die Verfassung der Freiheit heute. In Otto Molden (ed.), *Zu den Grenzen der Freiheit.* F. Molden Verlag, Vienna, Austria (1977). [In German.]

[40] Schumacher, E.F. *Small is beautiful.* Blond & Briggs Ltd., London (1973).

[41] Lovins, A. Energy strategy: the road not taken? *Foreign Affairs,* 55, 65–96, (1976).

Commentary (I)
Umberto Colombo

I have read the contribution by Hafele and Sassin with great interest and have found its ideas highly innovative and stimulating. In my opinion, its main merit lies in having imagined a long-term future that contemplates an energy endowment for mankind without having recourse to massive energy sources whose feasibility is still to be demonstrated.

Most of the previous long-term studies on energy set out the problem more or less as follows: the oil era is destined to come to an end within a few decades, and in any event a progressively decreasing role of oil must be taken for granted. In the very long-term future (starting well into the twenty-first century), man will have one or two 'definitive' energy sources at his disposal: solar energy and nuclear fusion. It cannot be foreseen today when or exactly through which technologies these sources will become available, and it is thus necessary to direct a maximum research effort towards their development. In the meantime, vigorous energy-saving policies must be pursued, as well as increasing utilization of energy resources other than oil, whether fossil (coal, oil shales, tar sands, etc.) or non-fossil, among which nuclear fission energy is the most important. This approach has resulted in focusing the energy debate on problems of the next few decades and, in particular, on the development of nuclear fission and on its relevant risks and questions of social acceptability. We have thus ended up with an incomplete and even ill-defined discussion on long-term future energy.

Häfele and Sassin have envisaged an energy scenario up to the year 2030, and projections therefrom, that are compatible with a fairly consistent economic growth at world level over the coming decades, and which allows for a *per capita* energy consumption in the less developed countries, over a span of 70–100 years, equivalent to the present consumption of Europe and Japan. This scenario also provides for a continuous increase, even if not at the past rate, of the *per capita* energy consumption of the already industrialized countries, including the USA, Japan, and Europe.

Häfele and Sassin essentially assume a continuation of the basic trends of the past model of development of industrial countries, and even propose mechanisms that accentuate the trend towards the centralization of production and the structural rigidity deriving from it to which Dr. Danzin also refers in Chapter 14 when he mentions 'la cristallisation par gigantism'. This assumption is consistent with continuing urbanization and recourse to highly capital-intensive production technologies.

The amount of capital necessary to put the energy programme proposed by Häfele and Sassin into practice is enormous (US\$40 $\times$ 10^{12} that is a sum over 10 times greater than the gross world product today). According to the authors it is only by moving immediately in the direction proposed, that it is possible to avoid sacrificing private consumption beyond measure in the longer term.

To my mind, it is difficult to believe that in a world such as the present – dominated by the spectre of unemployment, a chronic shortage of capital for investment, a political power prevalently exerted at the level of single states rather than at the level of vast supranational organizations – it would be possible to mobilize the resources required for a programme of these dimensions. This programme, which is certainly well inspired from an ethical viewpoint, provides for an expenditure of some \$8-9 $\times$ 10^{12} just to accommodate the increases in energy consumption foreseen in Central and South America, Africa, and south and south-east Asia. Now, it is very difficult to imagine these countries obtaining such large capital which would have to be provided primarily by countries that are already industrialized, if the economic growth proposed for the less developed countries is to take place.

These prospects should be compared with the difficulties which even today are being encountered by far less ambitious aid programmes than the type provided for in the Häfele and Sassin proposal. One such project, in which I am personally involved on behalf of the Trilateral Commission, plans for the OECD countries to provide a sum smaller than \$2 billion per year for 15 years, just as a contribution to solving the food problem in one of the neediest regions of the world: south and south-east Asia (where by 1992 the population is expected to exceed 1.7 billion people); severe difficulties are being encountered precisely because of the size of the financing required.

In a more realistic appraisal, one is obliged to be somewhat sceptical about the likelihood that sufficient resources could be mobilized in favour of a Häfele–Sassin plan to supply the energy needs of the LDCs, and one fears that — contrary to the design of the authors — the plan they formulated might actually turn out as a proposal for pushing strong nuclear development for the sole benefit of the already industrialized countries, with the consequence of aggravating the already wide economic gap existing today between the North and the South.

Another criticism of the unquestionably stimulating contribution of Häfele and Sassin, relates to the provision for a considerable portion of the total 35-TW production capacity in 2030 to be dispersed in remote sites, probably on some islands in the oceans. Apart from the technical and organizational difficulties of implementing a programme of these dimensions, I should like to observe that it also goes against a natural preference to locate the strategic production of energy indispensable to the survival of economy in one's own national territory, whenever possible. In other words, it seems difficult to believe that the type of vast consent on a world basis required by the Häfele–Sassin plan to locate strategic energy production in remote areas could be reached in a relatively short time.

One original and highly interesting aspect of the Häfele–Sassin plan relates to nuclear-fission–coal and hard-solar–coal hybridization tied in to the production of methanol, and to the use of the latter as the principal energy carrier in place of petroleum. Hybrid methanol production from coal and nuclear heat is thermodynamically more efficient, compared with alternative solutions, is technologically feasible, and entails long-distance transportation of methanol, which is certainly simpler and more efficient than that of electricity or hydrogen. The whole idea, however, seems to me to be linked to the assumption of highly centralized and remote nuclear energy sites. Should nuclear power plants be installed in the consumer countries, it would probably make more sense to produce methanol (together with other liquid fuels) from coal at the pit head and to produce electric power or hydrogen via nuclear fission, using by-product heat for district heating or for other convenient purposes. This is, however, to be related to my estimate for global energy requirements in 2030, which is put at a much lower level than 35 TW. One should furthermore consider the negative aspects of the huge transportation system necessary to convey coal to the remote nuclear sites and methanol from such sites to the energy consuming areas.

The case of solar–coal hybridization is still more unrealistic, not strictly from a technical viewpoint but, rather, considering the social acceptability and the political aspects. Here the problem is not to avoid the hypothetical risks of a plutonium economy, but merely to exploit suitable climatic conditions and the thermodynamic advantages of hybridization. I believe that there exists enough free land surface available within and not too far from the consumer countries to avoid remote siting, and that solar energy is sufficiently flexible to produce, according to need, electric power, heat, and biological materials that are usable as convenient fossil-fuel substitutes.

Perhaps the most relevant objection to this aspect of the Häfele–Sassin project is, however, the degree of strategic vulnerability of such a centralized and remote form of energy production. The system advocated by Häfele and Sassin requires a stable world order, and an extent of control and regulation that seem to me to be incompatible with the present world structure. Suppose that the Western countries were to adopt this energy model, and that the Eastern bloc, not to speak of China, were to reject it; this would lead to an inconceivable vulnerability gap between the big powers.

Summing up the criticism to this remarkable contribution, I would submit that the vision behind it is technocratic, conservative, and also somewhat dogmatic. The authors, in fact, display a high degree of confidence, if not of certitude, in the virtues of our way of life and of the present technological and economic culture. The chapter seems to ignore the grave and as yet unsolved problem of social acceptability of technologies, and to propound the advent of a most efficient, almost perfect society, but of a society which by necessity would have to be super-authoritarian and quite repressive, one of its problems being that of reducing to silence those who contest or delay the realization of the energy programme. In other words, we should not bind future generations to an energy system requiring enormous resources and leading to patterns of life, which might be contrary to the values they might wish to express.

These criticisms might appear severe, but in my opinion they reveal one of the peculiar merits of Häfele and Sassin's contribution: that of having presented an energy scenario which is so provocative as to worry and stimulate the reader to suggest alternative hypotheses and solutions that are more humane, and more innovative, not only in technological, but also in social terms.

Let us, then, try to delineate a somewhat different, and perhaps more desirable energy future, and verify its soundness on techno-economic, as well as political grounds. A fundamental objection of the thesis expressed by Häfele and Sassin is that the capital/energy input ratio is nearly constant in time and does not depend on the shape of development of the economy. In fact, it is widely recognized that this ratio increases in advanced economies as a result of a change in the mix of economic activities and improvements in efficiency of energy use. Another factor of rather great importance, which is neglected in the Häfele–Sassin scenario, is the increasing price of energy, which can effectively lead to substitution of products and activities, as well as to considerable energy savings, through improvements in efficiency and technological substitution.

We should consider that the development of modern information technologies, the increasing substitution of software for hardware, as well as the advent of biology-based production systems, are in themselves representative of a trend towards lower specific energy requirements. It is surprising, in fact, that Häfele and Sassin, who have so clearly expressed the concept of negentropy in connection with large, rigid, and centralized energy systems, should ignore the value of the negentropy available from the development of an 'information society', and of technologies aimed at better exploiting the order of biological processes.

In a similar way, Häfele and Sassin have not ascribed any value to the rather solid arguments expressed on several occasions by Lovins, who has shown the feasibility for the United States being able to achieve drastic reductions in the *per capita* energy consumption over a sufficiently long time [1], while still maintaining satisfactory economic growth. This would be done by a mix of conservation policy, diversification of energy sources, and changes of life style.

Lovins has suggested both technical solutions to be implemented and institutional barriers to be removed. It should be emphasized that such measures require a more appropriate use of market concepts rather than rigid planning, which could hamper entrepreneurial spirit. Significant structural and value changes in work, leisure, agriculture, and industry are necessary in order to reach the target, and this requires strong social innovation, that is not compatible with a technocratic approach.

As far as the problems of the less developed countries are concerned, different patterns of development must be conceived, that are less centralized, more respectful of the interplay between urban and rural life, less wasteful of energy and materials resources. A careful scrutiny of wrong and wasteful paths to development undertaken in the past within the industrialized countries should be made jointly with the LDCs so that they may avoid ill-effects, which are all the more unsustainable in times of high and growing energy costs.

It is therefore possible to visualize an energy scenario for 2030 which, while retaining the most useful concept of endowment, takes into due account the effects of technological change within a feasible and more harmonious global economic environment. In doing this, let us be as provocative as Drs. Häfele and Sassin, in the sense of stretching the imagination towards an energy conservation — and social innovation-oriented future.

I have assumed a level of 15 (rather than 35) TW for the global energy requirement of an 8 billion world population in the year 2030. Such a system would have the same energy consumption *per capita* as today, but it should entail a much less unbalanced regional distribution. While the post-industrial society of most OECD countries will certainly be characterized by lower specific energy consumption as the structure of the economy moves from the production of goods to the supply of services, the different type of economic development advocated for the less developed countries will avoid the soaring energy trends that have occurred in the course of the development of most industrialized countries.

In the scenario proposed, I have endeavoured to suggest how the quantity of energy required (15 TW) could tentatively be distributed among different sources, naturally without the pretence of furnishing such a well-studied and defined model as the one proposed by Häfele and Sassin. Concerning the 'soft' sources, a considerably greater effort than that indicated by Häfele and Sassin (3 TW) may be possible, accepting that it would no longer be necessary to make huge efforts in the development of hybrid nuclear-coal and hard solar-coal systems. For that matter, there would also be more room for these energy technologies in relation to a different way of life — more decentralized and less urbanized — and to a different pattern of development for the LDCs with respect to the kind thus far experienced by the industrialized countries. I would hazard a guess that soft technologies could supply some 5 TW by the year 2030, which would correspond to 33 per cent of the total energy requirements, against the approximately 8 per cent resulting from the scenario proposed by Häfele and Sassin. Fossil resources (coal, oil shales, tar sands, etc.), especially through gasification and liquefaction, should provide a further high percentage of energy needs. Moreover, I feel that before 2030 it should be possible to develop hard solar systems using large orbiting plants that could by then make a substantial contribution to energy requirements. Taking account of other resources, like hydroelectric, geothermal, and nuclear fission energy (which in this scenario should represent a decreasing fraction of the whole), the proposed energy system would entail an alternative form of endowment. Again, by that time, the future of nuclear fusion should be clear, and therefore it would be possible to develop it, if it proves to be technically and economically feasible.

The 15 TW of energy consumption indicated in the proposed alternative scenario is based on the assumptions that, with a world population of 8 billion in 2030, economic growth will be more contained than that in the Häfele–Sassin plan, and that the nature of the long-term development will not be represented by an extrapolation of recent trends. A different mix of technologies is expected to arise in response to the need for a more balanced development between centralized and decentralized systems, transport and communication, traditional economic activities and new activities related, for example, to materials recycling and environmental control. In this context, there will be growing opportunities for appropriate technologies, which should be understood not only as 'soft', but with a high software and R and D content. They may help provide for a way of life and a society where the energy demand for services and goods would be considerably lower. With an advance of the world economy at an annual rate of 2.5 per cent for the next 50 years, and assuming a global energy/GNP elasticity of the order of 0.6, this still represents a considerable global development.

Such a scenario seems to me to be more realistic and sound from NATO's point of view, in that it avoids giving solutions to the energy supply problem that might have negative implications on the international scene. For instance, a gigantic fast-breeder-based nuclear development in a world of counterposed political blocks might create problems similar to those that are encountered as the present race for more and more powerful and abundant strategic arms. Instead, a scenario that is based on a more modest, but yet satisfactory, economic growth, on increased energy efficiency, and on less emphasis on centralized energy sources and urbanization will largely avoid these kinds of problems.

Reference

[1]   See, for example, Lovins, A. Energy strategy: the road not taken. *Foreign Affairs* 55, 65–96 (1976).

Commentary (II):
Robert J. Uffen

My colleague Dr. Columbo was asked to concentrate his attention on energy resources, and in particular the chapter by Häfele and Sassin, while I have been asked to comment primarily on other natural resources, renewable and non-renewable. In addition, we have all been invited to bear in mind not only the science and technology involved, but also the socio-political implications. Many of my observations will arise from the contributions by Ashby (Chapter 2), by my colleague Kenneth Hare (Chapter 5), and by Peccei and Mesarovic (Chapter 13).

Non-renewable resources

The supply–demand gap. First, I wish to make one or two observations on the contribution by Häfele and Sassin which bear on non-energy, natural resources. They present a very useful and valuable analysis of global energy demand in the next 15 to 50 years. In particular, the comments concerning the serious possibility of sudden gaps appearing between demand and supply of energy some time in the next decade (as foreseen by the Workshop on Alternative Energy Strategies) may apply also to certain other strategic materials, like scarce metals. Ashby has extended this method of analysis to other non-renewable resources in a manner that puts into clear perspective those mineral resources which are likely to be plentiful, those for which substitution can be made, and those which are likely to remain scarce and have great geopolitical implications.

Available materials. Ashby's treatment is a welcome advance over the somewhat over-simplified picture which arises from the use of the concept of 'Demandite' (the average mineral mined by man from non-renewable resource deposits). Following identification of its composition, Häfele and Sassin observed (p. 15):

Realizing that all other nonsubstitutable elements within deposits down to this depth [1000 m] could support man's present consumption rates for millions of years, *we may safely conclude that fossil energy resources are the only resources which pose a problem that demands a principal solution in the future* (my italics).

This statement warrants closer scrutiny. Non-substitutable, non-renewable resources are rarely abundant when and where they are needed. They first must be discovered in concentrations which justify recovery; they must be mined, concentrated, processed, transported, fabricated, and distributed to the user. All this uses energy, costs money, may have enormous environmental impact, and involves complicated questions of ownership and control. Conclusions can no longer be drawn safely in isolation from the effects of other interacting policy decisions and geopolitical realities.

Ashby makes a clear distinction between the likely prospects for the use of traditional metals and alloys and the growing use of polymers, ceramics, glasses, and composite materials. Starting with identification of the abundance of the elements in the earth's crust, oceans, and atmosphere, he identifies strategic metals such as copper, tungsten, tin, and mercury, whose workable deposits are so small and localized that many governments may wish to stockpile them.

Reserves versus resources. Ashby clarifies the distinction between ore *reserves* and *resources* by the use of a 'McKelvey' diagram (see Vol. 1, p. 24). Such a diagram plots, in a qualitative way, economic feasibility against geologic uncertainty, and shows how increased prospecting, improved mining technology, and higher prices may lower the minimum mineable grade, increase the rate of discovery and thus extend both the proven reserves and the resource base. Reserves are defined as known deposits that can be mined profitably at today's price using today's technology. Resources are always much larger than reserves.

The concept of resource half-life. Ashby goes further and introduces a more quantitative and informative way of assessing resources. He shows the quantity of ore available as a function of the minimum economically mineable grade. He then introduces the interesting concept of 'resource half-life', defined as the time when projected supply falls short abruptly from projected future demand (see Vol.1, p.28). He assumes that this will occur when approximately one-half of the exploitable resource base is exhausted.

Of course there are great uncertainties, and the identification of a fairly well-defined 'half-life' depends on how abruptly the gap between supply and demand opens up, which in turn depends on policies controlling rates of production, as in the analogous case of oil production of the OPEC countries. Nevertheless, Ashby concludes that the resource half-lives for iron, aluminium, and copper are a century or two, while for lead, mercury, silver, tin, tungsten, and zinc, they are more like half a century.

He then points out that some of these key metals are geographically very localized and long before resources are physically exhausted, they may become politically inaccessible, except on terms dictated by the producing countries.

Geo-political shifts. Massive shifts are taking place in the control and production of many resources. Countries like Canada (which has long thought of itself as one of the great warehouses of minerals for the world) are finding it difficult to adjust. For example, three-quarters of the nickel mined anywhere in the world in 1950 came from Canada, but by 1976 it was less than one-third and at present the country is experiencing very large unemployment in its nickel industry.

Most countries of the Third World regard their mineral resources as the key to industrial development, and the movement toward a new international economic order requires that the poorer countries acquire a larger share of the exploitation of the world's non-renewable resources. The 1976 conference of non-aligned nations reaffirmed the Third World's right to own and exploit its own natural resources.

The prospect of mining metal nodules from the sea-bed is an example of a new technology that may upset the established markets, and has already raised questions about international ownership. There seems to be general agreement that the economic viability of ocean-mining will depend on the nickel content of the nodules, although the manganese, cobalt, and copper content will have some bearing. If major countries decide to safeguard access to essential metals by ocean floor mining, they might also impose penalties on traditional sources of imports. In other words, they may decide to subsidize ocean mining in order 'to stake out a claim' in order to guarantee their own sources of supply.

Substitution for metals. I would like now to return to Ashby's contribution. He proceeds to compare the important physical properties of elastic moduli, yield strength, toughness (roughly the opposite to brittleness), and melting temperatures of metals, polymers, glasses, ceramics, and composite materials. In addition he analyses the availability and evolution of such engineering materials. The elements for making polymers, ceramics and glasses, namely carbon, hydrogen, oxygen, silicon, are far more readily available than are the ores of metals, and he asks (Vol. 1, p. 23):

Could we, over the next 20 years, shift the emphasis of use of materials away from the expensive and potentially scarce metals, substituting for them ubiquitous materials such as oxides and silicates which are almost as common as dirt; or materials like carbon-chain polymers which are plentiful and widespread as coal?

The answer, he says, is that it appears to be that we could, although a very substantial research and development effort, spanning the next 20 years, would be required. At the moment, most of the new materials are expensive, but potentially, many of them are very much cheaper than the metals they would replace.

Technological shifts. A shift away from the use of metals toward polymers and ceramics would require new design methods, and this could have a great impact on technological obsolescence and the nature of engineering education, and adult education and retraining.

This is an example of the massive technological shifts which Lord Vaizey refers to in Chapter 15. It is interesting to speculate as to which countries, or groups of countries, will recognize this likely shift in materials design and technology, and will move to exploit it. Very likely it will be those which lack indigenous metallic resources but which have educational systems and 'managerial styles which lead to acceptability of, and enthusiasm for, technological innovation'.

Energy-related problems

Nuclear and solar energy. The exploitation of mineral resources of the continents and of the oceans, and the development of new materials to substitute for them, will require a great deal of energy. It may well be that the limiting factor on the availability of engineering materials will be determined by the availability of the energy required to process them. Häfele and Sassin made only passing reference to this possibility. They concluded that only the use of large-scale solar complexes or fast-breeder nuclear reactors will be capable of supplying the energy needs of a global population of eight billion in the next 50 years or so. They state:

This short analysis of the alternative energy sources clearly indicates that mankind will ultimately have to shift either to nuclear or to solar energy, or more likely, to a contribution of both energy forms. The only reasonable question is not 'if' but 'how' and 'when' (p. 25).

A new kind of agriculture? Such a large-scale use of solar energy would require 'more than a few solar panels in the backyard' (p. 24) and would require 2.5 million km² of land. This would be roughly equivalent to 20 per cent of the present world land area used for agriculture, and should be viewed as 'a new kind of agriculture' (p. 24). In making these estimates, it was assumed that it might be possible to attain an overall efficiency of 10 per cent in producing, storing and transporting hydrogen or methanol using sunlight. This seems rather high, perhaps by a factor of ten or more, so the land area required could be very much larger – so large that the land required might have to compete with existing agricultural and urban areas.

It would seem highly desirable to pursue these estimates further to determine if, in fact, sufficient unoccupied areas would be available. It would also be desirable to estimate the amounts of metals, polymers, ceramics, glass, or composite materials that would be required for such a vast enterprise. Would there merely be a shift from an energy resource problem to a materials resource problem? What energy would be required to process the materials, what wastes would result, and what environmental effects might be anticipated? How many people would be employed?

Some observations by Lord Vaizey about large-scale agriculture are perhaps appropriate here. On p. 183 he says:

Agriculture, to be successful, requires highly motivated and well-trained small-businessmen. Large-scale employment of large numbers of people, requiring bureaucratic styles of management, is curiously unsuited to the varied needs of agriculture. No large-scale employment structure in agriculture has ever been successful.

Perhaps the analogue is not complete?

Nuclear wastes. Häfele and Sassin conclude that light-water nuclear reactors depend on cheap uranium and will not solve the global resource dilemma. According to them, nuclear fission becomes a rich, abundant source of energy only through the breeding principle. They also conclude that, even if optimistic targets of present research and development can be realized, fusion reactions involving lithium as a fuel would not be able to change the global energy balance significantly within the next 50 years.

They visualize clusters of fast breeder reactors coupled to a chemical reactor for coal gasification and the synthesis of methanol (see pp. 36–40). For Canadians, accustomed as we are to highly efficient thermal reactors using heavy water as moderator, we are often surprised at the apparent lack of interest in the potential use of thorium as a fuel in thermal near-breeders. Current estimates indicate that a 'self-sufficient thorium cycle' may be practicable in a heavy-water-moderated thermal reactor with only minimal modifications. At equilibrium, this cycle would require no further uranium, and only small amounts of thorium, which is at least three times as abundant globally as uranium [1].

Häfele and Sassin regard the plutonium generated in fast breeders as a desirable endowment, despite their own acknowledgement on p. 26 of growing public opposition:

The environmental acceptability of single nuclear power plants providing 1 GW of electricity is a public issue these days. Producing 50 TW, i.e. 5000 times more, by fission breeders, for example, would force mankind to handle and dispose of some 10^4 t of fission products per year. An overwhelming degree of meticulousness is required to keep such a flow of potentially dangerous material separated from the environment. According to present maximum permissible dose rates, not more than some 10 g of fission products per year would be allowed to leak into the environment.

Fortunately Hare, in Chapter 3, has presented a clear and rational account of the present state of plans for the management of highly radioactive used nuclear fuel, reprocessing wastes, and reactor wastes in his third case-history. He points out that nuclear engineers apparently saw the permanent disposal of the longer-lived actinides as a minor problem that could wait until more urgent matters had been dealt with. Quite suddenly, however, public anxiety was aroused, mainly by a series of small but embarrassing failures in radioactive waste isolation. Twenty-one countries now operate reactors producing spent fuel and all face the problem of disposal, but the complacency has been so universal that no country is known to have developed an adequate repository.

Both Hare and I have been involved in Canada in assessing such problems and (despite media coverage to the contrary) he and I are in agreement that deep geologic disposal in stable rock formations like granite can probably be achieved, but that a great deal of research and development is required in a very short time (5 to 10 years) in order to demonstrate to the concerned public the practical achievability of what are, at present, only scientifically feasible proposals.

Whether it is the high-level wastes from reprocessing plants, or unreprocessed fuel bundles themselves that require permanent disposal, large amounts of new ceramics or glass, and materials with good sorption properties, would be required to immobilize all the radioactive wastes from the very large nuclear generating plants envisaged.

In an earlier but similar paper, Häfele paid more attention to problems of thermal waste, transportation and distribution of energy, and risk evaluation [2]. For the purposes of this volume, it might have been quite valuable to have had more discussion of these.

Distribution systems. My own experience with the Ontario Hydro Electric Power Commission has left me acutely aware of the constraints imposed by our ability to distribute electrical energy. Distribution problems will probably arise for most natural resources, renewable or non-renewable. It can be excruciating to have thousands of MW of electrical power 'bottled up' through the lack of transmission lines. This can occur as a result of natural forces, such as sleet and wind storms, accidents, or overloading. It can also arise through inability to acquire ownership of the required land in time, because of public opposition or vested interests.

It is a moot point whether many acres of prime agricultural land should be taken out of production for power corridors, so our policy is to lease the expropriated land back to the farmer, with the exception of the small area required for the towers.

Similar problems arise with pipelines, railways, and storage facilities, whether the resource being transported and distributed is fuel or food.

Renewable resources

Fish and the law of the sea. It seems to me that more attention should be paid to problems of renewable resources such as forest products, fish, livestock, feed grains, and foods. Perhaps they have been neglected because there have been several related conferences on world food supplies recently. The potential for international stresses over traditional fishing rights vs. the imposition of the 200-mile national limits is a prime problem.

What has been happening in world fishing is something of a tragedy. A natural renewable resource that Huxley called 'inexhaustible' a century ago has been subject to massive exploitation to the point that catches of herring, cod, and mackerel in the North Atlantic have seriously declined.

Early in 1978, for example, the Canadian government announced a new fee system to be imposed upon foreign fishing vessels, or vessels supporting foreign fishing fleets, operating within the country's 200-mile limit. The fee will be based on the size of the vessel and the number of days fishing, not the value of the catch, and is expected to bring in Canadian $10 million per year in revenue.

Such actions are expected to allow the stocks to recover, and to increase our fish reserves several times. Similar actions can be expected from other coastal nations.

Food and desertification. The world's population will probably double in the next 35 years, yet there seems to be little prospect of doubling the productive agricultural land. In fact, as Hare points out in his second case study (see Vol. 1, pp. 57–61), at the UN Conference on Desertification, which was held in Nairobi, September 1977, the evidence was that, while on a world scale there appears to be no progressive desiccation in progress, the deserts are advancing in many areas. Overstocking, careless cultivation, excessive use of ground water, the use of heavy vehicles, and failure to 'rest' the land weaken the soil, ultimately to the point where it will be deflated and eroded: 'the desert margins have long been hazardous places in which to live. . . [and] , on countless occasions, the scene of tragedies and strife' (Vol. 1, p. 57). They will probably remain so. According to Kates *et al.*, 'A long-term chronic decline in productivity of the resource base is the most serious manifestation of desertification, yet its direct social and behavioural effects are poorly understood' [3] . The major response appears to be that people move away, if they can.

The Aspen Institute for Humanistic Studies is initiating a 'Food and Climate Program' which, unlike most other food and climate researches, will focus on social, economic, political, and ethical implications, rather than on the climatic theories themselves. The programme will look at such issues as the state of food reserves, the trend of productivity, progress in research on climate forecasting, development of new species of food plants, and new technologies for food manufacture and storage. Questions to be asked include: should there be food production incentives? should grains be stockpiled? should there be a world contingency plan for a food disaster? what are the implications of 'agripower' — the use of food as a political tool?

Conservation and the future

Whether in the long term humanity will find ways to manage the global system, to avoid undesirable environmental or social effects of continued industrialization and resource consumption, is a matter of considerable concern. While new technologies may yet provide the energy and materials necessary for growing world population, we do have to consider the possibility that it may be unachievable in time.

Present indications are that we face a period of exceptionally rapid change and transition, the extent and depth of which are not yet clear. The future will not be simply an extrapolation of the past!

Some of the changes that are foreseen as inevitable – in energy use, in materials, in ecology, and in social behaviour – are embodied in the concept of the 'conserver society', the subject of a Science Council of Canada report [4]. The concept has not arisen suddenly, but public awareness of the term seems to have grown rapidly since the publication of the *Limits to growth* in 1972. The term was first used in 1973 by the Science Council of Canada in an earlier report [5]:

It arises from a deep concern for the future, and the realization that decisions taken today, in such areas as energy and resources, may have irreversible and possibly destructive impacts in the medium to long term. The necessity for a 'Conserver Society' follows from our perception of the world as a finite host to humanity, and from our recognition of increasing global interdependence.

A conserver society is, on principle, against waste and pollution. It promotes economy of design, i.e. 'doing more with less', favours re-use or recycling and questions the ever-growing demand for consumer goods.

In industrialized countries, the flow of materials in the economy has been used almost as a measure of economic well-being – it is a component of Gross National Product. From a conserver point of view, however, we would be better off to pay more attention to materials 'in stock'. Despite the fact that some materials are ubiquitous, such as hydrogen, oxygen, aluminium, silicon, carbon, and magnesium, their excessive use still uses energy and has environmental impact, so there is still a strong case for conservation. The essence of the policy is 'to do more with less' through demand reduction, design, and substitution, and system management.

It should not be construed, however, as it sometimes has been, that such prescriptions are aimed at freezing the *status quo*. To the contrary, by changing the style and some technologies and being more far-sighted, the world should find room for continuing growth.

Conclusion

I would like to conclude my remarks with two observations. The first is related to the differences in the time scales involved in scientific discovery and technological planning on the one hand, and socio-political needs on the other. Scientific discovery may require many years before it is digested and incorporated to a significant degree into society. For example, the leadtime for a 5-GW electric power station to be placed in service is now close to 15 years, when allowance is made for public participation in the decision to proceed. Twenty-year technological leadtimes are commonplace. On the other hand, the economic and political cycles, with their frequent surprises, may have to be included in realistic planning. (Is it any wonder the public has difficulty in digesting and accepting proposals for the safe disposal of radioactive wastes over time-periods into the future that are measured in 'ice-ages'?). Consequently it is notoriously difficult to involve practising politicians in long-range planning. Secondly, it is remarkably difficult to remain optimistic about the future in the light of the current analyses and forecasts. However, we may take heart from Peccei and Mesanovic (p. 98): 'Looking at all these signs, it is reasonable to conclude that they are not simply glimmerings of hope but imply that a change of outlook is actually slowly taking place. Therefore, after having considered with great concern our world in disarray, even if it is at the apex of its knowledge and achievements, we are now able to perceive the first symptoms of a human renaissance.'

References

[1] Robertson, J.A.L. The CANDU reactor system: an appropriate technology. *Science* 199, (1978).

[2] Häfele, W. A systems approach to energy. *American Scientist* 62, (1974).

[3] Kates, R.W., Johnson, D.L., and Johnson Haring, K. *Population, society and desertification.* Component review, UN Conference on Desertification (1977). To be published in *Desertification, its causes and consequences.* Pergamon Press, London.

[4] *Canada as a conserver society: resource uncertainties and the need for new technologies.* Science Council of Canada Report No. 27. Science Council of Canada, Ottowa (1977).

[5] *Natural resource policy issues.* Science Council of Canada Report No. 19. Science Council of Canada, Ottowa (1973).

13

Dynamics of science, technology, and society: analysis and decision-making

Aurelio Peccei and Mihaljo Mesarovic

13.1 Introduction

This chapter has been prepared partly by a scientist and partly by a layman. In carrying out their assignment, the co-authors discovered that, despite their fundamental commonality of goals, and their mutual understanding and habit of reconciling any diverging viewpoints, many interstices and obscure, even contradictory, points remained in the final version. So the reader should know at the outset that the layman is responsible for §§ 13.1–13.3 and 13.5 and that § 13.4 has been contributed by the scientist.

It was, however, useful to realize the difficulty of synergistically blending science-inspired and more mundane thinking. It helped the authors to grasp just how difficult it will be to bring together in real life the much smaller, highly sophisticated, swiftly moving world of science and technology with the majestic but more complex, less prepared, and slower world of man — which in fact lags far behind. The task of harmonizing the two worlds is all the more difficult because, taking advantage of the fact that their motivations and aspirations were reputed to be so noble and beneficial as to be in the ultimate interest of society, science and technology have for some time been behaving as autonomous, self-centred, and self-justifying undertakings.

The freedom, independence, and unreined power inherent in modern science and technology have permitted them to acquire a strategic position in government, the universities, and industry, as well as the military establishment. From their lofty, privileged niche they are exercising an influence on society which dwarfs that of all other factors affecting society; and at the same time they raise unwarranted expectations and literally irradiate wave after wave of change before the attendant social and environmental costs, intermediate effects, and feedbacks can be appraised.

While the spirit of the entire exercise of commemorating the twentieth anniversary of the NATO Science Committee is clearly that of making science and technology more aware of, and responsible to, the overall concerns of society, a 'science-first' attitude is still perceivable. This attitude of the eminent scientists of the Committee, who have devoted their talent and inventiveness to the enhancement of human knowledge in their various disciplines, is fully understandable. It also goes without saying that the recognition of the immense benefits that mankind has reaped from its techno-scientific enterprise, and of the still greater ones it can expect from it henceforth, is a primary ingredient of any global reasoning about the future.

However, the stage has been reached when it is necessary to question whether the current situations and relationships, even with some possible corrections, are satisfactory and should be adopted as a model for the coming decades, and whether it is wise to let science and technology — which science? which technology? — ride roughshod over an unprepared society, or whether all this entails risks that we are not equipped to face.

One such risk is precisely that an exceedingly small minority, such as that represented by the scientific and technical communities, however enlightened it might be, will go on pouring out streams of inventions and innovations which 'wag' the immense body of mankind, now one way, then another, without anyone knowing what the overall effects can be, or whether the great majority of the world citizens is prepared to absorb them. In our opinion, this is a very real risk, one of the highest political and existential order.

Even now, thousands of millions of people are in a state of disarray, totally unable to keep abreast of things, and even decision-makers are at a loss to grasp what they should do and what the actual outcome of what they do is really going to be. The drama of our time is this *human gap* with respect to a runaway reality — which by tragic irony is a man-made and scientifically engendered reality. It is unacceptable that the human universe should be still further mutated by an *avant-garde* of scientists and technologists in ways that would widen this gap even further, because then it would probably be impossible for ordinary people and society to bridge it.

To stave off the mortal danger of being cut off from reality, it is most urgent for mankind to assess the human gap realistically, find ways and means to control and reduce it as rapidly as possible, and then see that it does not develop again. We believe that, given certain conditions, this fundamental objective is attainable. Therefore this chapter is devoted, not to discussing science and technology and their likely developments, but to presenting two arguments, titled *Rationalizing the decision-making process* (§13.4) and *The human revolution* (§13.5), which have to do with society and people generally and what is needed for them to live with their science and technology, without stifling them but acquiring instead the capacity of directing and using them intelligently.

Our prime concern must indeed be with the human being, both as an individual, and in his aggregations in groups and societies, and the improvement of them all, lest the revolutionary dynamics of the techno-scientific establishment eventually puts them irremediably out of step with their universe. On these bases, and considering the unprecedented exigencies which are emerging, we will examine the two ways whereby society at large can actually be stimulated and improved and made to develop higher-order capacities so that it can redress its situation.

The first way is eminently practical and operational. It concerns means and methods to rationalize human decisions and make them more efficient in the face of complexity, uncertainty, and change — which is precisely what society needs, *inter alia*, to guide and exploit its techno-scientific capabilities in an intelligent manner. An earlier project for The Club of Rome, led by M. Mesarovic and E. Pestel [1], proposed what has now been developed into a most useful tool aimed at extending man's natural orientative endowment by artificial means in ways similar to those whereby his natural capacities to move and to communicate have been developed a thousandfold by modern means of transport and communication. Section 13.4 reviews these possibilities.

The second way is that of a humanistic revival of man and his society – which in our view is both urgently necessary and entirely possible. Attention and action at this juncture should be focused on people and their individual or communal psychosocial adequacy and preparedness to continue the human venture with man as the protagonist on Earth. The question is nothing less than whether or not man, as a species, is capable of learning to survive and progress in the hectic real world that he himself is relentlessly shaping. A positive answer depends exclusively on him, for what is required is not only that he should develop his great innate but hitherto neglected capacity for adaptation, in order to match the waves of change that will occur, but also that he should guide these so as not to be caught unprepared, even by their indirect or intermediate effects. These exigencies stress the necessity of a revolution within man at least equal to those he brings to pass outside himself. We recognize that this is a yet untrodden path but believe that it should be thoroughly explored and hope that §13.5, which deals with it, may open a broad debate on this whole matter.

However, before dealing with these two topics, and as a complement to this introduction, two explanatory sections are necessary. Section 13.2 is devoted to elucidating the glossary used. Section 13.3 is aimed at preventing us from starting to speculate on or hypothesize about the shape of the future without foundation and framework. As a matter of fact, to be meaningful, any discussion about the future requires that the origin of it all, namely the state of the present, first be envisaged as realistically as possible.

13.2 An interpretation of basic concepts

Some explanation of the meaning attributed to various terms of the science–technology-society equation is necessary, for it is interpreted differently in different cultures or contexts and by different people.

For the key concepts Webster's *Third New International Dictionary* gives the following definitions:

Science: accumulated and accepted knowledge that has been systematized and formulated with reference to the discovery of general truth or the operation of general laws; knowledge classified and made available in work, life, or the search of truth; comprehensive, profound or philosophical knowledge.

Technology: the application of scientific knowledge to practical purposes.

Society: a community, nation or broad grouping of people having common traditions, institutions and collective activities and interests; an international social order or community of societies and institutions.

Although our analysis does not make specific reference to institutions, it is not out of order to record, too, what Webster says about them, because the concept of institutions is strictly linked with that of society:

Institution: something that is instituted as a significant and persistent element (as a practice, a relationship, an organization) in the life of a culture that centres on a fundamental human need, activity, or value, occupies an enduring and cardinal position within a society, and is usually maintained and stabilized through social regulatory agencies.

Let us consider first science and technology, which represent the cutting edge of world dynamics. We submit that, for the purpose of this chapter, they form a logical continuum which cannot be broken into two separate and distinct segments as current terminology or practice might suggest. While a demarcation between science and technology may be appropriate when they are analysed by themselves or for other purposes, such a separation would be arbitrary and scarcely significant, if not confusing, when they are examined in relation to society.

What is fundamental science today will probably evolve into applied science tomorrow, and this will in turn breed a spate of technology destined to influence society in many though unpredictable ways. The process is self-reinforcing because at every stage it stimulates more research, more development, more implementation, and so on. Marxists like to call this the scientific and technological revolution, and until recently seemed to bank everything on it. The time-scale and unfolding of the cycle will vary from case to case and from place to place but, however long the lead time and however different the paths for reaching the phase of practical applications, what counts for our broad review is the outcome. And this, predictably, is that scientific advance is sure to engender, in due course, technologies that are bound to have a greater or lesser impact on society. The important thing is thus to gain deeper and clearer insights into the mutual relationships and cross-influences between the techno-scientific enterprise as a whole and society, as well as the latter's capacity to direct the former to its overall ends.

To develop our equation, however, it is not enough to look at science and its corollary, technological developments, as forming an organic complex in what may be called a vertical perspective. It is at the same time necessary to look at the horizontal span of this complex: the whole range of the sciences and technologies which must be considered. Similarly, we must determine which different kinds or aspects of societies should be considered when we refer generically to society. The terms of the equation evidently cannot remain undefined from this aspect, and in our view their definition should be as all-embracing as possible rather than being restrictive, as we may be tempted to make them. As a matter of principle, we are convinced that, at this stage, in view of growing complexities and interdependencies of everything human, reductionist, fragmentary, or sectoral approaches to any major question are more likely to mislead than to clarify.

In the case under review, we believe that only by contemplating the entire panoply of sciences and their related technologies, as well as the assortment of societies that make up the present global system of states, is it possible to establish the overall framework required to get a better grasp of what occurs between them in the real world. Difficult as this exercise may be, and whatever our specific or proximate interests, holistic perspectives and approaches have become of primary importance, if not indispensable. We cannot in fact focus intelligently on the questions which are dearest to us unless we have first reasonably established the wider context in which they are embedded.

It would certainly be unwise in this vast and complicated matter to limit our reasoning on science by singling out a few disciplines, such as the exact sciences or the biological sciences alone, whatever the breadth of the boundaries we may wish to assign them. The eight fields selected by the NATO Science Committee for the retrospective/prospective review of scientific achievements cover a vast area indeed, and allow for highly interesting exemplifications and deductions, which, no doubt, is the Committee's intention.

But, when society is the main concern, these fields cannot by themselves represent science. The purpose of the Committee can be better served, in our view, by adopting a much wider concept of science. To leave out the human and moral sciences, or any other branches which are not usually included in the 'inner realm' of sciences through bias, expediency, or tradition would not only be unwarranted, but also likely to perpetuate and aggravate a state of affairs riddled by preconceptions and misconceptions which blur our vision. For instance, what are very often currently held to be 'problems' or 'solutions' do not correspond to reality, precisely because our conceptions are based on incomplete or improperly defined premises.

One reason advanced for keeping the 'soft' sciences in the background is that they cannot be, or have not yet been, systematized like the 'hard' sciences, and thus are more prone to subjective or controversial formulations. While this may be true, it is no less true that society is largely being destablized and confused owing to the gross imbalances existing between different branches of knowledge and our propensity to use some of them extensively and neglect others.

Meteoric breakthroughs made in some fields, while others have stagnated, have created social and cultural disorder. This negative feedback loop has been well expressed, although in different terms, in the Brooks Report: 'Science is in disarray because society itself is in disarray, partly for the very reason that the power of modern science has enabled society to reach goals which were formerly only vague aspirations, but whose achievement has revealed their shallowness and created expectations which outrun even the possibilities of modern technology' [2]. In our view, society is dangerously lopsided, mainly because its paramount motor, science, is ill-balanced, and now the two phenomena are acting as each other's booster.

The necessity of expanding the current concept and scope of science has been recognized time and again, particularly since science policy has become an important item of debate. Alexander King, in his overview of the state of the planet, prepared for the International Federation of Institutes for Advanced Study (IFIAS), indicates that 'if science is to serve humanity as a whole and if its applications through technology are to be developed so as to ensure the maintenance of an environment suitable for human life, there will have to be a major reorientation of research programmes and attitudes in terms of priorities very different from the present'[3]. He adds that 'so far, there have been three main objectives, namely defence, economic growth and national prestige, for example, placing men on the moon. The social and service sectors of the economy have attracted research resources of a lesser order of magnitude' [3].

For our part, we are firmly convinced that the techno-scientific enterprise must be radically re-oriented. The Brooks Report has already indicated this necessity: 'It is realized that the immense social benefits that have flowed from science and technology are sometimes accompanied by social disbenefits. Thus, policies concerned with science and technology in the next decade will have to take into account, much more explicitly than in the past, the benefits and disbenefits, actual and potential, that may result from the application of science or the deployment of technology' [2]. Nowadays, one can be blunter. In the years and decades to come, human progress and the quality of life, even survival, will depend much more on inventions and innovations of a social, political, institutional, and educational character than on any advance of science *sensu stricto*.

We suggest, as an example for meditation, the case of the country of one of the authors, Italy, which is in the grip of a multiple crisis. Her hopes no longer lie in more science in the conventional sense, but in more comprehension, civility, common sense, tolerance, decency, and justice, and in better institutions, structures, mechanisms, and laws. Probably not a single one of Italy's current problems – with the notable exception of the expected energy shortfalls – needs more or different technology, but rather a little more wisdom in applying that which exists. Italy's predicament, however, is not singular in this respect. Each of us, we feel, could quote similar examples in the case of his or her own country and region, and together we should recognize those which concern the world at large.

Far be it from us, though, to suggest that further advances, even in the sciences and technologies which have changed the world in many a surprising and sometimes uncomfortable manner, should be discouraged. On the contrary, we believe that our techno-scientific undertakings should be pursued most actively. They are most inspiring and hold out great promise. But, precisely because of the extraordinary thrust towards change which is inherent in their nature, they must be subject to two precise and absolute conditions. One is that such advances and the changes they are likely to engender should be matched by at least equally strong developments on all other relevant fronts of knowledge, so that society is not thrown even further off balance than it is at present but, instead, is put in a condition to re-establish its cultural, social, and political equilibria – we hope at a higher level of human organization. The second, and corollary, condition, is that all necessary measures should be taken to prevent the world population, or at least a large majority of it, from being caught unprepared by such further eruptions of changes and developments, and instead be made able to adjust to and absorb them – for without people's responsible participation, any new progress will be founded on sand or even be counterproductive.

Let us again quote from the Brooks Report, which puts this in another way when it cautions that science policy should be regarded 'as policy for the mastery of technical and social progress as well as for the generation of new knowledge required by man to increase his understanding of himself, his societies and the universe and also for the utilization of such knowledge in the management of change, complexity and uncertainty which are the characteristics of our times'[2].

A similar evolution of our concepts and reasoning must be made with regard to society. To be brief, we shall merely recall that to an increasing extent critical, even vital, factors interlink all peoples and nations one to the other in a planetary system, from which none can withdraw without grave, probably irreparable, damage for all. New imperatives in defence and security, resource management, 'earth care', and goal-setting inevitably bring mankind together, despite all its cultural, ideological, institutional, and economic heterogeneity. Diversity enriches society, and different identities will continue to play leading roles locally, but then they must be reconciled in coherent regional and global wholes if mankind is to continue. This process of globalization of human society – which receives its impulse from the very progress of science and knowledge – cannot be reversed; so, to be realistic, we must recognize the organic interdependence developing in the world-wide system and consider society first in its totality, before focusing attention on any one of its different components.

The concept of socio-political institutions must also be seen in a different light. With the holistic approach to a more and more integrated society, which seems to us necessary, this concept cannot be reduced, as is often the case, to that of the instrumentalities required to attain societal aims. For the purpose of our inquiry, the concept of institution can even – temporarily – be absorbed into that of society itself. Family and sexual mores, the Church and beliefs, government and political associations, inter-personal and state–citizen relationships, the role and organization of the economy, of work and leisure, of education, of justice, and of science and technology themselves, all these are institutions – the socio-political institutions which characterize our different societies and are their cultural backbone, their true essence. The identity of each society and its capacity for self-realization and self-reliance, as well as its relationship with science and technology, are rooted in these institutions and their possible evolution.

We submit then that, in order not to place ourselves in a position where we cannot see the wood for the trees, the interpretation of the terms of the equation to be resolved should be broadened. Instead of restricting the concept of science and technology, this should be expanded into that of what really counts: 'knowledge'. Only in this manner can one make an initial attempt at assessing the adequacy and harmony of what mankind knows and is able to make use of, with respect to the knowledge it may need or wish to have. By the same token, organized mankind itself is, naturally and appropriately, the other term of the equation. This should therefore express the dynamic relationships between

Knowledge (Webster): the sum total of what is known; the whole body of truth, fact, information, principles or other objects of cognition acquired by mankind. and

Global society (our own definition): the present loose system of states and governments, bound to evolve in yet unperceived ways towards a more integrated, though still diversified, global community of men, peoples, and nations.

The proposed escalation of the terms of reference from those initially laid down towards ever higher and more spacious horizons may appear as an exercise in quixotry and utopianism. A harder look at the human condition at this juncture, though, will show that the proposal is not altogether dreamlike, even if it requires a particular effort where perception and imagination are concerned. Although we are as yet unable to embrace and rationalize the whole of human knowledge and world community and their interactions in a comprehensive manner, it is indispensable to recognize immediately that this must eventually be done as a routine precondition to guide mankind's precipitous march ahead. Not to know where one can, must or should ultimately go is the surest manner of getting lost on the way. If we start considering this approach now, though we are still unprepared to follow it through completely, we can at least gain an idea of the direction and distances with which mankind must familiarize itself in order to climb out of the present dangerous impasse and attain new thresholds of self-fulfilment and security.

13.3 An interpretation of world situations and prospects

Irrespective of any definition one may wish to adopt for the various elements of the science-society complex, it is impossible to formulate a rational line of reasoning about its future without first assessing the current conditions and trends of the human system. Scenarios of the future cannot be pregnant with meaning unless they are consistent with the only reality we know or should know – that of the present. We shall therefore describe this dynamic reality as we see it, knowing, though, that any interpretation of the present state of human affairs and its prospects cannot but be subjective, tentative, cursory, and probably controversial. In making our assessment, we have of course not been able to take into account all the factors from different fields which combine to make the present so confused and baffling, but we have selected a number of the most relevant. We believe that the resulting picture provides a fair summary of the conditions prevailing at this turning-point in history. Our chief objective is in any case to establish and clarify the rationale of this chapter.

In brief, we think that, after a period of extraordinary achievements and crises, the end of the 1970s and the 1980s will almost certainly be years of unprecedented tests, trials, and dangers for mankind. What is euphemistically called the foreseeable future may, instead, hide the unexpected to an uncommon degree, and be a time of traumatic discontinuities and reversals, rather than the relatively surprise-free period of recovery and development we should all like, and many still hope, to see coming. It will probably also turn out to be a decisive phase in the human venture that will influence mankind's way of life, and perhaps even life itself all over the planet, for many decades, if not centuries to come.

Stripped of all rhetoric, the truth is that the alternatives facing humanity are extreme. At this stage, a new ascent, leading to ever greater accomplishments and the full realization of the human being, is probably within our reach. Yet, a dark age marked by the decline, suffering, and humiliation of our species is also quite possible, even probable. The extraordinary dynamics that are inherent in the human system make both outcomes plausible, but at the moment we must be aware that only dismal conclusions can be drawn from the current train of events, of which we are both witness and protagonist. Unless a supreme effort is made without delay to modify this course, civilization as now conceived will race ahead towards disaster.

In fact, much of what will actually occur will depend on what mankind as a whole – and particularly the liberal democracies – decides to do or leave undone during the next decade or so. The generations now active cannot escape the unparalleled responsibilities they must bear at this extraordinary juncture. They must rise to the challenge and redirect human developments in accordance with the new realities. If they fail to do so, they will widen still further what we have called the human gap, bequeathing to their successors a condition wellnigh beyond repair.

Blind human proliferation is the basic factor. Even the Soviets now agree with this. One of their foremost academicians recently said that 'the first global problem which has to be solved is how we can scientifically determine the policies we must adopt in order to control the size and quality of the world's population. Of all global problems this would appear to be the most important and difficult.' [4]

It will take just ten years or so for the world population to jump to five billions (5×10^9) and then it will still continue to multiply. Although fertility rates are decreasing, population equivalent to that existing at the time of the First World War will be added to the present one before the end of the century. This impressive demographic growth cannot, however, be viewed only as a quantitative phenomenon, for it will no doubt also provoke qualitative changes throughout society, the more so because the demand for food, goods, and services, as well as for space and mobility, is bound to increase still faster than population. Even if every other factor remained unchanged, the sheer presence of so many more people would cause all problems to grow quickly in magnitude and complexity, and eventually get completely out of hand.

Other elements are undergoing critical changes. This is certainly the case with four of the major components of the human condition: natural environment, basic resources, the state of society, and the violence complex.

The natural environment. The world's ecological systems are being steadily degraded on land, in the sea, and in the atmosphere. Their condition has already been seriously impaired, but the impact of larger populations and more intensive human activities will certainly make things substantially worse, to the point that it is doubtful whether remedial action can be taken before the damage is irreparable, even with respect to human life. At present, there are only a few countries with adequate conservation policies, and there is practically no valid international machinery to protect, let alone restore, the ecological equilibria. The few agreements in existence here and there are widely transgressed, and difficult to enforce.

With cut-throat competition between national or sectoral interests intent on grabbing or cornering whatever is useful or beautiful, nature is being overexploited, befouled, and wounded capriciously. Neither world public opinion, nor the powers which should really be responsible for conservation, show any concern for the steadfast deterioration of our planet's biophysical environment. Let us consider for one moment the much-vaunted prospective Law of the Seas, which should have permitted all nations to utilize and at the same time preserve intelligently what has been termed the 'common heritage of mankind'. The mammoth UN World Conference which was convened for this purpose is still stumbling ahead many years later, session after session, but the most likely result will probably be a division of spoils, rather than the far-sighted management of our common wealth.

There is not even a reliable scientific theory or assessment of 'the outer limits' of the Earth's life-support capacity. This is an inadmissible vacuum in our vast body of knowledge. The UN Environment Programme is about to start a comprehensive study aimed at identifying, analysing, and illuminating the changes that, according to available information, have occurred in various aspects of the environment and of environmental situations over the last ten to fifteen years. This project, to be completed for the decennial of the 1972 Stockholm Conference, should provide indications on the current evolutionary changes on the planet in this key front, so that appropriate goal-oriented designs can, it is hoped, be conceived in order to correct them whenever necessary. A summing up of some major questions on the world's physical environment is also being prepared as a part of the OECDs 'Interfutures' study.

For the time being, though, one can register only negative phenomena and an appalling state of ignorance. For one thing, the concentration of carbon dioxide in the atmosphere is regularly and appreciably increasing, and so is the release of other man-made effluents such as nitrogen dioxide and chlorofluorocarbons. The real effects on the climate and stratospheric ozone layer are still matters for conjecture and controversy, but they cannot be dismissed as if they were known to be negligible. At the same time, man's activities also result in a relentless increase in the emission of particles and thermal discharges which add to atmospheric turbidity and increase air temperatures. The intensity of these phenomena already exceeds the self-renewal capacity of natural cycles, at least in certain areas, and their persistent impact and interactions, in ways that we have still to discover, are likely to modify yet further the physical condition of the Earth. If adverse natural climatic trends are in the offing, as some symptoms seem to indicate, a most serious situation may emerge before it is generally realized just how dangerous it is becoming, and remedies can be found.

More generally, we must admit that we simply do not know the environmental consequences of disposing of the many millions of industrial waste products and by-products wherever it is easiest or least costly to get rid of them, considering only the short term. Some of our discards, such as heavy metals (mercury, lead, arsenic, cadmium, and chromium) and certain chemicals (a number of which have no parallel in the natural world) are highly toxic, and tend to find their way into the human body by means of the food chain. Others are solids, whose heaps fill the dumping grounds and litter the landscape because many of them cannot be recycled, are not biodegradable, and are uneconomic to transport. There is no wisdom in maintaining that pollution is after all only a marginal problem, because it can eventually be abated to acceptable standards at the cost of just a few percentage points of GNP. A much more realistic assumption is that pollution is going to increase and will entail costs much greater than those required for its theoretical elimination and risks which cannot be reduced to purely economic expressions. The situation will become substantially worse before it can be improved.

The planet's landmass is suffering visible degradation owing to the combination of sheer demographic pressure, anarchical urbanization, and rapacious or improvident agricultural practices. A few years ago the insidious growth of deserts became a matter of great concern, so much so that a UN Conference on Desertification was convened in September 1977 in Nairobi. Without counting Greenland and the Poles, there are already 8 million km^2 of hot deserts. Other vast regions are threatened with various degrees of desertification, the areas where the risk is 'great' or 'very great' being roughly one-tenth of South America, one-fifth of Asia and Africa, and one-quarter of Australia. We do not know whether this downward trend can be stopped, but it is abundantly clear that appreciable results against the advance of deserts can be obtained only at the cost of an enormous effort, by transnational if not global strategies, and integrated management of all pertinent resources.

What we know is that soil, besides being indispensable for agricultural production, is also an essential component of the biosphere, and that everywhere, in both developed and developing countries, it is being continually lost owing to various reasons other than desertification. It is estimated that under natural conditions, soil may form at the slow rate of one cm every 125 to 400 years, though good agricultural practices can reduce this period to 40 years. Each year in the United States only approximately 3.75 t of topsoil are formed per hectare, under normal conditions, while the average annual loss of topsoil from agricultural land is calculated to be 30 t per hectare, so it is being lost 8 times faster than it is being formed. [5]

These phenomena are little known and hardly considered in our global assessments. The world's agricultural and biological capacities will be reduced even further by the allocation of extensive plots of land — and in all likelihood lands which are at present under cultivation — to settle decently the new waves of population expected on the planet. Agriculture has an uphill fight to increase production, while its operational base is being reduced.

In the long run, however, the greatest menace to human life derives from impairment and degradation of the biosphere. One example of the ravage-cum-ignorance characteristic of our time is the destruction of the tropical rain forests, which had endured in a stable state for tens of millions of years. According to the International Union for the Conservation of Nature (IUCN) and the World Wildlife Fund (WWF), 40 per cent of their green expanse has disappeared already, and the rest is currently being felled and burned at the staggering rate of 20 hectares/min, day and night — corresponding, every year, to an area the size of Denmark, the Netherlands, and Belgium combined. If this massacre is not stopped, they will physically disappear in a few decades, together with all the animals for whom they are home: the orang-utan, the jaguars, the birds of paradise, and many, many others. The richest lowlands, such as those of South-East Asia, which have more species than any other plant community in the world, are the most vulnerable; in the Philippines and Malaysia, they are doomed to vanish completely within ten years [6]. We are afraid that this appalling, massive destruction of precious eco-systems will have ominous effects on humans, probably more than the expected drying up of oil fields in approximately the same time.

Unfortunately, this impressive example of spoilage of our habitat is not isolated. Let it suffice to mention the decimation of other species, animal and plant alike, which we are making in the name of progress. According to the WWF and IUCN, the extinction has been recorded of 75 bird and 27 mammal species during the last century and of 53 and 68 such species so far this century. In addition, some 345 bird species, 200 mammal species (and among them some of the noblest) 80 species of amphibia and reptilia and between 20 000 and 25 000 plant species are threatened with extinction. Other species are gradually depleted below levels from which they would be able to recover. What impact these devastations may have on human ecology has yet to be ascertained. Everybody can, however, judge the magnitude of the outrage we are perpetrating against nature and the irreparable cultural losses we are inflicting on ourselves and our successors.

Setting aside any spiritual, philosophical, ethical, and moral considerations — which, however, are primary and fundamental in any discourse about man and his future — it is quite evident that society cannot remain in the present limbo of ignorance and indifference about what happens in its already impoverished and ailing natural environment. In view also of the enormity of problems that the on-coming landslide of people and the stepped-up human activites will bring in their wake, a wait-and-see attitude would simply be suicidal.

Comprehensive long-term policies and plans for each large region and the world as a whole have become indispensable, if we are to accommodate these swelling populations and provide for their needs, and at the same time guarantee that in the process vital ecological balances are not further and definitely upset. There can be no doubt that most stringent safeguards to preserve and possibly restore a healthy state of the human habitat and what remains of the pristine wilderness all over the globe — which is the very heart of nature — are now essential, and must be given a high priority for the sake of one and all.

Up to this very moment, however, all peoples and nations are behaving as if Earth were large and generous enough to forever accept and absorb everything which individually and collectively they are doing to it — even at the new, exceptional levels of this age; so much so that each of them is racing and struggling to grab more than the others, quite unconcerned that the sum total of their exactions might be exhorbitant and crippling. Now — just to survive — we have all to change. Earthkeeping, a yet unknown necessity for our Earth-pilfering civilization, must become our common credo. The cynics among us may object that a shared concern for the planet is something altogether alien to human nature. We believe that they are wrong, but even in this case the time available for change is uncomfortably short.

Basic resources. The situation regarding the basic resources required by the world's society and its expanding economy seems somewhat less foreboding. With a few exceptions, such as gold, mercury, antimony, silver, and helium, the availability of non-renewable resources is today considered by and large more reassuring than it appeared a few years back. All the same, a number of scarcities and higher prices are to be expected sooner than we would like and are prepared to cope with. [7]

Very great concern is, on the contrary, felt about energy. It seems almost impossible that our societies can count on enough, reasonably cheap, clean, and safe energy in the future. Growing demand, heavy reliance on, and gross mismanagement of our main resource so far – hydrocarbons, discontinuation of R and D investments on coal as an alternative fuel, political rivalries, and the lack of any international co-operation to organize a smooth transition to non-conventional energy sources are likely to be creating world energy shortfalls, probably as soon as the first half of the 1980s. However, the situation may become very critical before people at large realize that it is serious. Of course, the central piece of the picture is nuclear energy, which in its various forms is looming, according to different viewpoints as the great saviour or the ultimate perverter. There is no need to say more about this, because the energy debate is well known, and some twenty major studies by groups of experts, oil companies, and national and international agencies have been made in the last two years.

Notable among these studies is the WAES Report [8], which shows convincingly the difficulties to be expected in the period 1985–2000, whatever measures of conservation, conversion, and diversification may be adopted in the meantime. However, we wish to record our opinion that, farther ahead, the energy crunch is likely to become even more dramatic and have even more explosive socio-political implications. We are afraid that, if there are no profound societal changes and a sweeping cultural evolution all over the world, no measure of global co-ordination will ever suffice to provide the amount of energy demanded by a teeming, voracious, and improvident mankind. The consequences for peace, stability, and what may be called civilization are difficult to imagine.

Apart from the problem of energy shortages, which darkens all future prospects, mankind finds itself, literally, at the crossroads on this front. The question is not just to try to find a solution most likely to provide the best, though perhaps incomplete answer to this crucial problem; it also involves consideration of the type of society which will eventually evolve from the adoption of the nuclear energy base compared with, say, the solar option which seems to be the other main alternative. The points now being scrutinized are indeed of great importance and difficulty. Are both these solutions, or others too, feasible, at what cost, and with what lead time, results, and impact in different regions? Going nuclear, can the various fission alternatives be planned and implemented reliably from all viewpoints? Are mixed or different solutions possible and advisable? Is the eventual transition to fusion reactors something to be relied upon in the more distant future?

However, the fundamental questions piling up upon us with tremendous implications are not these, namely they are not of an essentially techno-industrial and operational nature, and do not concern energy only, although energy scarcity may trigger them. They, too, imply a search for solutions which is no less delicate and difficult than that required by the material change-over to non-conventional forms of energy, but which concerns, instead, the laborious, even painful processes of cultural and socio-political innovation which must go on at the same time, so as to put society in a condition to guide or absorb other changes – including those in the field of energy. These fundamental questions have not yet been asked; they have not even been clearly formulated, for they touch the very core of the dilemmas confronting mankind, and their newness is disconcerting. But they are infinitely more important for human destiny, and very urgent, since they involve the basic issue of what kind of world we are going to bequeath our children and children's children by the decisions that we are hastily making at this time; something even more fundamental is the choice of ethics for our species consonant with its new status of power and domination in the world.

Another source of concern, again owing to a large extent to our poor management of resources in a crowded, globalizing society, is food. Even if this is man's primary need, and the question of world food security is of vital and continuing importance to all countries, the efficiency of the world's food producing system is appallingly low with respect to the Earth's theoretical capacity and the existing knowledge. 'While there have been some encouraging factors in the last two years, the world food situation is still fragile. The average annual increase in the food production of the developing countries since the beginning of the 1970s was only 2.6 per cent, well below the 4 per cent target set in the International Development Strategy for the Second United Nations Development Decade' [9]. The distribution system, however, is even more deficient.

Soaring population and demand make it imperative not only that both production and distribution be improved, but also that a better knowledge be attained of the climatic vagaries which might affect agricultural performance. It is by now evident that in this field, too, fundamental long-term policies and plans are required on a world-wide basis. A study made recently in Holland at the initiative of The Club of Rome showed that, short of an all-out undertaking to organize food production and distribution rationally and globally, hunger in the world is going to grow faster than population.

All in all, even if in most sectors mankind is not confronted immediately with dramatic resource problems, in other sectors serious crises are looming, also because resources are erratically distributed on the surface of the globe and the nations which 'own' them are determined to enforce their sovereign proprietory rights. Manifestly, a sector-by-sector approach will no longer suffice to satisfy over time the needs of a more and more demanding multi-billion society. In the postwar period, there have been five full-fledged international commodity agreements, for wheat, sugar, tin, coffee, and cocoa. However, the field to be covered is much vaster, for there are at least 10 'core' and some 25 more 'main' commodities which should be considered, if for no other reason because the less developed countries depend substantially on their export. But, with so many commodities and the world's fragmentation into 150-odd states, it is unthinkable that recourse to such agreements could represent a practical solution. Endless negotiations and entangled chaos would perforce be the result. Nevertheless, this is still the tendency, and it is imposed by a *de facto* 'balkanization' of natural resources along the obsolete pattern of political sovereignties.

This should therefore be the time for a fundamental change of outlook and approaches. Some form of resource pool for the benefit of all peoples and nations, today and tomorrow, should be established, together with mutually agreed rules for restrained use and conservation of the scarce non-renewable ones. It seems quite evident to us that only a comprehensive assessment of the ensemble of human needs and wants over the long term – well beyond the span of one generation – coupled with equally long-term programmes to meet the ensuing demand, can ensure that whatever resources are available will be used intelligently, in a framework of good management and foresight. However, as we all know, mankind is still a long way from having reached the degree of maturity that all this will require.

Complementary to resource management is waste management. Policies of saving and substitution of materials, durability and miniaturization of products, recycling, waste recovery, integral energy cycles, 'small is beautiful', 'lower grade can do', etc. do not require international co-ordination, for each country can apply them self-interestedly, with a view to making its own system more efficient or reducing physical constraints that may hamper economic expansion. Along with austerity in general, they should not be viewed as a curse, but rather as a challenge to our understanding and ingenuity. It is, however, sad to note that also on this apparently easiest front, abandonment of our profligate habits is slow to come.

There is no need to stress at this point that this stubbornness of mankind in not perceiving what is its real and permanent interest is a bad omen. Simple tenets of good management would require that material resources be handled differently. We want, however, to insist once again on the importance of ethical principles. The Earth's resources generally, and not only those in the oceans, are a 'common heritage of mankind' – at least, if we ignore the rights that other forms of life have on them. Each generation must consider itself as a trustee for those to come, a *pro tempore*, or income, beneficiary of these resources, while no human group can stake a claim as the absolute owner, with exclusive and indisputable proprietary rights, on some of them. While lip service is paid to the principle of the common heritage, with only a few exceptions, any question of sharing natural resources with others, living or yet to be born, is flatly and universally rejected in the name of urgent needs or superior national interests.

In reality, then, as already mentioned, all human groups hasten to get whatever advantage or benefit they can reap from the common good as quickly and fully as possible, unmindful of the damage this collective practice will cause to the overall system on which they themselves depend. Even if there were enough land or water or oil or uranium for all, along this path mankind can only race towards tensions and strife. Unfortunately, however, scarcities are already perceivable on the near horizon. Therefore, unless there is a mobilization of forces to put the man – resources relationships on a sustainable basis, what has come to be known as the 'tragedy of the commons' will be played out on planetary scale by gigantic protagonists ready to flay each other with ultimate weapons.

State of society. It is readily apparent that the institutions and structures of the human system have not kept pace with the exigencies of these days of high technology, great mobility, instant communications, and integrated, vitally inter-dependent societies in a 'shrinking' planet. In no time, the old- and present-world order has become antiquated and inadequate.

Even technically, it cannot be rated as efficient. Its performance is mediocre, wasteful, and erratic; again and again it falls into the grip of uncontrollable eco-nomic crises which rock its very foundations. Military or otherwise authoritarian regimes of the right or the left are mushrooming all over the place.

But the true ills which make the present system's deficiency critical, and its condition extremely precarious, have to do with its lack of concern for the human being and his dignity and resourcefulness. It seems to accept as normal and in keeping with the way it functions that people and nations be separated by wide gaps which are no longer tolerable either morally or politically, and that one-third of the world population should live under the poverty line and be in no position to contribute to the common good.

The number of men and women whom the system keeps at the margin of society is appallingly large. Although the figures are well known, it is useful to quote them once more. A recent reckoning estimated that 800 million adults are illiterate, 250 million children are not enrolled in school, 570 million people are undernourished, 1500 million lack effective medical care, 1300 million earn an income of less than 90 dollars a year, 1700 million have a life expectancy of less than 60 years, and 1030 lack adequate housing [10]. To complete the picture, it must be said that, according to the International Labour Organization, upwards of 300 million able-bodied people are currently unemployed and that one thousand million new jobs more are necessary between now and the year 2000 — which under present con-ditions it will be impossible to create. Under the shock of the future, such an in-firm, unjust system is destined to crumble — probably by internal disintegration or upheaval.

Three times during this century — besides the Marxist revolutions — attempts have been made to reorganize the world for good, but in vain, either because the reforms were not radical enough, or because new exigencies appeared too quickly. The last time was after the Second World War, in parallel with the creation of the United Nations and its agencies, but mankind is again confronted with the need to find a really adequate, equitable, and durable order. The less developed countries, asserting that they represent three-quarters of humanity, press urgently for it, stressing particularly the economic aspects. Their posture has been made clear and, although it may look extreme or unrealistic, it is on the table. The industrial demo-cracies, which have much more to lose by a breakdown of the world's fabric, never-theless seem unable to decide what is both feasible and desirable. They play for time — which however plays against them — and hope to hold the line with a policy inspired by *status quo*-ism and rearguard defence. The socialist group's main ob-jective appears to be that of not becoming involved in the debate; as if the in-dispensable and inevitable reform of the world system was none of its business.

The result is that there is presently no dialogue among peoples and nations on how to organize their coexistence and co-operation for the common good. To be sure, a number of proposals have been made from many quarters. We believe that the most complete and best structured one is that contained in the Report to the Club of Rome titled *RIO – reshaping the international order* [11]. None the less, only fragmentary negotiations on a myriad of subjects have gone on, inconclusively, with the result that a stalemate position has been reached, at a time when mankind instead should be busy forging its future and discussing its long-term goals [12]. Yet, even the intermittent North–South rounds, though inconsequential so far, and the on-and-off East–West attempts to build up a *détente*, though not daring to aim at an entente, could provide useful starting points, if a new spirit of realism and boldness could be injected into the conduct of world affairs.

The task of shifting the international mood from one of petty bargaining to one of earnest search for synergies and convergences is, in all evidence, incumbent on the strong and the free. In fact, only the liberal democracies can undertake it. They must resist the temptation to use the plethora of international meetings and conferences as a tactical expedient to gain time or save positions. The time to start building a more responsive economic and civil organization for modern society is now. If this occasion is missed, it will be like letting a 'cosmic window' pass unused in a space launching.

Within and among nations, a new deal or social contract based on mutual understanding and tolerance, social justice, shared responsibilities, and participation has become imperative. We will discover very soon, but in many instances only too late, that no peace, no security, no progress, and probably no really civilized life will be possible without it. With the dissemination of information and education as well as of the means of violence, the centres of power are multiplying throughout the world, and the very ideals of freedom and democracy must be reconceived in tune with the new context of human life in its changing environment.

This very evolution and emergence of planetary challenges, problems, and exigencies call for a thorough revision, too, as we have already hinted, of the time-worn, if still apparently untouchable, principle of sovereignty and the concept of national state that it underpins. With more than 150 sovereign states, the human system is absolutely ungovernable and destined to founder. No one is in charge or feels responsible. A rethinking of the very bases of both theory and practice, not only of the international economic system – and, by the way, the role of science in society – but of the world polity itself in all its aspects has therefore become a compelling necessity.

To imagine that without this profound renewal a responsible and mature society can emerge from the present chaos is like believing in miracles. It is high time for modern man to realize that his world is undergoing basic metamorphoses and that, if he is not to be overwhelmed, he has to adjust his vision, his modes, and his society to them, even by anticipating them. No doubt full adaptation can only be the culmination of a long process. Yet, under the present circumstances, this process cannot be too lengthy either, nor can its beginning be delayed, for temporizing in putting its own house in order is a luxury mankind can no longer afford.

The violence complex. In the present state of disarray, intolerance, frustration, and fear, modern society is becoming an armed camp, which is tantamount to seeking security by pursuing its mirage. The facts are well known. Although the nuclear arsenals have in store the equivalent of more than one million Hiroshima bombs packing a monstrous 'overkill' capacity – a dozen times larger than the world population – the military establishment still engages nearly 50 per cent of all scientists on 'defence' work and employs about fifty million people in armed forces and related activities. As President Carter noted in his recent address to the United Nations, 'last year the nations of the world spent 60 times as much equipping each soldier as we spent educating each child.' We have just seen that basic human needs are to a large extent unsatisfied; nevertheless, the military budgets have reached a new record of upwards of 330 billion dollars per year, which means that 7 per cent of what is produced by human toil and ingenuity in the world is siphoned away in order to keep the arms race and trade in full swing.

Despite peaceful declarations from all sides and endless disarmament negotiations, modern society seems unable to abandon this demented behaviour. Even public opinion, once much more alert, now appears paralysed, apathetic, or resigned in the face of a danger verging on the inhuman. Yet it is only too likely that, unless curbed, mankind's potential for self-destruction will one day be unleashed by accident, mistake, miscalculation, or even a fit of insanity breaking out in some corner of our deranged society. This weird possibility is enhanced by the apparently unarrestable horizontal proliferation of nuclear technologies largely convertible from peaceful to martial purposes – which, according to many, is going to make a nuclear conflict almost inevitable.

In parallel, there is an escalation in civil violence that is equally frightening, because it corrodes the very texture of collective life, destroying society from within. The boundaries between civil violence, torture, terrorism, guerrilla warfare, covert military operations, and violence sanctioned by a formal declaration of war are anyway wearing very thin in many places. Violence is a disease that is insidiously spreading everywhere, in thousands of ways, so that a frontal attack on it is virtually impossible. Displacing mutual understanding, forbearance, and serenity, it tends to become second nature to a crowded and angry society.

The present generations are to be blamed perhaps more than all the preceding ones because, in the pursuit of national, economic, or ideological interests, they have irresponsibly nurtured violence at its current peak levels. If they do not succeed now in attaining the triple goal of quenching the propensity for it thus engendered, of uprooting the causes that might justify it, and of controlling the means that can propagate it, the world situation can never be redressed, nor will the human family be able to enjoy peace or go very far.

The preceding review of world situations and trends confirms what the Club of Rome has termed *the predicament of mankind* at a time of immense opportunities – and sets the scene for our reflections. We do not believe that the overall picture that emerges from the preceding pages can be considered as too pessimistic, although we would very much like someone to prove it. We have tried to describe cold facts and how, under the impetus of an ever larger and more powerful, but by no means wiser, mankind, they are evolving before our very eyes. The factors examined interact among themselves and with others, too, which we have not listed, making the tangle of world problems ever more intricate and baffling.

There are, though, some encouraging signs. They come mainly from the grass-roots of society – from the people themselves. Almost everywhere in the world, public opinion is concerned and increasingly aware that something fundamental must be changed in contemporary society. It is unable to say in so many words that, since problems have changed in their nature, scale, and thrust, human affairs should no longer be guided by conventional wisdom, old-time methods, short-term expediency, or power politics – but it senses all this. It is likewise unable to say how and why the current ways and means to foster development, measure economic performance, manage the monetary system, administer the sprawling cities, husband natural resources, educate the young, and so on, are no more acceptable – but it feels this, and no longer trusts those who preside over these activities.

An indication of the keen public consciousness that changes must be made, even at the cost of sacrifices, is the formation of spontaneous groupings of citizens springing up all over the place – like antibodies in a sick organism. Usually called 'non-governmental organizations' (NGOs), they are and form the peace movements, the conservation and ecological groups, women's lib, the population policy associations, the defenders of minorities, of human rights, and civil liberties, the apostles of amnesty, the peaceful protesters, the dissidents, the social workers, the consumer advocates, the conscientious objectors, etc. A fountainhead of fresh, innovative ideas, usually with a simple and intelligible dialectic having the ring of a *vox populi*, they together form an incipient second-layer network, intent on watching and stimulating the official networks of government and intergovernmental agencies.

Whilst these ferments of renewal are a sign of society's **vitality** and must be reckoned with, they have yet to find a force capable of catalysing their potential and giving it the thrust necessary to influence events decisively; and even if people at large may feel that most NGOs somehow reflect their inner aspirations for a cleaner, fuller, and more humane life, they also feel that these movements need clarification, and some kind of visible legitimacy and support. This means, too, that ordinary people, while practically no longer trusting anybody or anything vested with authority, are desperately eager to find something or someone they can indeed trust and follow. This *yearning for leadership* is very strong in the world.

We are convinced that the leadership that could satisfy the profound, deep-seated expectations of today's world populations can no longer be leadership on a national scale. Even ordinary people, probably even those in the United States and the Soviet Union, have the feeling that true, long-lasting peace is no longer a national affair when all states depend on so many uncontrollable external factors, and in any case the hidden or overt hand of foreign powers and superpowers pulls the wires everywhere. They know that, if their currency is devalued, or their trade lags, or some domestic industries close down, this is probably due to events originating somewhere else in the world. They have heard too much about the multinational corporations, about OPEC and petrodollars, about balances of payments and the monetary system, about international corruption and terrorism, to believe for one moment that their own governments can do very much about key issues, even if they wanted to. They have seen on television the Soviet armies enter Budapest and Prague, the US airforce bomb Vietnam, wars waged by proxy, tragedies fanned from abroad in Cyprus, Lebanon, the Horn of Africa, and elsewhere, mammoth conferences convened on world problems, such as the environment, food, population, water, and human settlements – so much so that it has eventually dawned on them that it takes much more than national leadership to change what has to be changed in the world in order to change their own condition.

New leadership: who can take up this vital challenge to our generation – forcefully and credibly? It cannot be the institutions and their instrumentalities as they are now, or other similar ones which might be created for one purpose or another. It cannot be the Church, any church. There are too many denominations, and none of them has risen and faced up to the drama of modern man alone with his destiny. It cannot be a political party, because all of them have short views, solicit consensus rather than lead, and are ready to trade principles for votes whenever expedient. It cannot be the unions, too often entertaining parochial views or defending short-term sectoral interests, and having all but forgotten their pristine international inspiration. It cannot be the scientific community either, because science is no longer considered pure, having too often forsaken its search for truth to seek honour and privilege, or to serve one master or the other rather than the universal interest of man. Nor can it be counterculture, so insightful, but at the same time so uncreative, just as it cannot be the often correct but too mercurial youth movements. All these social strains and actors are indispensable to give support to a new leadership, but are in no condition to provide it themselves.

The problem is fundamentally an ethical and existential one, but it must be solved in political terms. The new leadership can only be a political force of great moral standing, capable of speaking for the world. But which of the large political human groups has, or can build up, the requisites to propose a new direction for mankind?

We prefer to leave the question open, but want to stay on record as confirming what has already been mentioned, in the sense that probably only the liberal democracies, and more specifically those of North America and Western Europe, can muster the formidable intellectual and organizational resources and the wealth of information, knowledge, and experience required to conceive and propose to the other human groups a grand design for mankind's future. This assertion should not be construed as an ethnocentric vision of the world. If, as we sincerely believe, only the West is at present in a position to mount such an effort, then this places it under a great obligation – the historic obligation to take the initiative of proposing to all the other members a new overall course in the interest of the entire human family, not its own interest alone.

Viewed in perspective, this obligation finds another justification in the fact that it was the West that opened the present path which held the promise of glorious developments now revealed to be so risky, and it was the West, too, which led other people on to it. There is, moreover, a potential sanction attached to this obligation. If the West proves to be unwilling to fulfil its function in this world emergency, when it alone can start the movement to redress the situation, it will have to answer one day for its forfeiture or deficiency.

However, the question is not who will be blamed or praised for what. The question to be faced is that the danger point for mankind, as we have tried to show, is just a few years ahead, and that before then, while its options are still slightly open, it must begin to steer a safer course – at the hands of those who can do the steering.

13.4 Rationalizing the decision-making process

The dynamics of science and the dynamics of society are intimately related. Science, as any other human activity, reflects the economic, social, and cultural conditions of the milieu in which it is being practised and, vice versa, the advances in science, or lack of them, influence the evolution of the society.

The world dilemma, as identified by the Club of Rome, poses a new challenge to science, and opens questions as to the very concept of what science is and how does it relate to other human activities. Several of the principal characteristics of this new challenge to science follow.

(1) *Science will increasingly have to focus its attention on its relationship with society as an object of inquiry.* According to Branscomb, 'Conventionally, science has been thought to be useful in proportion to the technological possibilities it creates. Today, science is more important and more frequently it is used as a means to understand the options and consequences associated with deployment of technology'. [13]

Science is expected to provide a basis for rational policy analysis and decision-making on national, international, or global levels. Society cannot afford the luxury of omitting science and scientists from the process of policy analysis; this is too important an activity for society to be left without the benefits that science can bring.

(2) *Science must become futures – rather than past-oriented.* Traditionally, science was concerned with the explanation of why certain events took place or with providing a causal description of some phenomena as observed. In futures research, instead of talking about explanation, the scientists must talk about anticipation; instead of stating what has happened, the scientist will have to state what might happen – reasoning from the logic of the past and the evolving conditions in the future. The very concept of the scientific method (which is viewed as synonymous with science itself or with any activity which is sufficiently mature to deserve that name) requires repeated experimentation and verification. However, if science is to become concerned with the future (i.e., with the objective of sorting out what might happen in the future), a new methodology – a new standard of excellence – and, indeed, a new mental and psychological outlook have to be developed.

Methodologically, instead of relying on observations and laboratory experiments under isolated conditions, the futures science has to rely on thought experiments, since the future, by definition, has not happened yet. In order for these experiments to be credible, they have to be based on all the data and observations derived from the past as well as on present data which are relevant for the future. But the solidity of analysis is based on the rigour of logical arguments, rather than on conformity to past observations. This brings a fundamental uncertainty into scientific inquiry. Science has to accept and internalize uncertainty in the basic scientific methodology. The methodological challenge is how to cope with this fundamental uncertainty and yet preserve excellence and integrity in the scientific enterprise.

It should be stressed that the uncertainty regarding the future does not result from a lack of knowledge at a particular stage in historic development or from a lack of data. It is in the last analysis the essential uncertainty involved in the freedom of choice by individuals or societies; i.e., the freedom to select any one out of a set of equally likely or beneficial (i.e., equally preferable) alternatives. In other words, the uncertainty in the futures analysis is fundamental. It is what one might call 'Heisenberg's uncertainty principle of future analysis'. It is in order to minimize this uncertainty that the logic and precision in analysis and reasoning has to be strengthened.

Quite obviously, a new mental attitude also has to be developed by the futures scientists. The ultimate fear of a scientist is to be proven wrong. A futures scientist will have to learn to live with the fact that he might be proven wrong and this is necessarily failure of his scientific ingenuity but might very well be intrinsic and, to a degree, inevitable in the given situation. The proof of excellence is whether he could have anticipated the outcome if he had taken an alternative – possibly more 'solid' and rigorous – approach.

(3) *Science inquiry has to be conducted in 'real time', i.e. concurrent with the evolution of the system under study.* The scientist has to understand the dynamics of the futures research endeavour. Since one is concerned with events in 'real time', i.e., in the events as they evolve and which change while being observed and analysed, one cannot wait for all of the uncertainties to be estimated and only then draw the conclusions. If one waits for the events to take place, then one is again going back to the past-oriented approach and in that way is retreating to the old concepts of scientific inquiry. A futures scientist has to make statements about his findings in time and on the basis of whatever evidence he has available at the moment. He has to learn how to live with the fact that he will have to update and modify his findings. The 'excellence' in science could not be measured by how much that finding has to be modified, but rather on the evidence as to whether the scientist has drawn the most defensible and most logical conclusions on the basis of whatever evidence he has at the time when his findings are reported.

(4) *The object of science has taken a relatively new spatial dimension.* The object cannot be confined to a laboratory nor can it be controlled. The object is society or indeed the globe. In other words, an 'experiment' always has to be conducted in a situation in which the scientist himself is part of the experiment. The experiment will influence the experimenter and, furthermore, the experiment

cannot be conducted exactly in the same form twice. We are confronted with yet another form of the uncertainty principle: the scientist – the experimenter – and the object of the experiment are intertwined and dependent on each other; i.e., the object of study is an 'open system' which is closed by the experimenter in the process of analysis.

If science is to undertake the challenge of developing a new methodology for policy research and futures investigation, a wholly new attitude toward its own enterprise has to emerge. The society that is providing impetus to science to develop in that direction in the first place must be willing to appreciate and accept the findings of the new scientific inquiry. In other words, the success of futures research must be measured by its relevance and ultimately on its impact on societal development. In spite of all academic pronouncements to the contrary, science and scientists will always be dependent on the appreciation and understanding of the public.

Bridging the gap between the new policy science futures research and society is the role of *technology*. The function of technology is to present proofs and results of futures research and policy analysis in a form useful for and usable by society. Just as the dynamics of the new scientific enterprise (i.e., engaging in scientific research on a system which evolves in time) requires a new attitude, the new technology which will make the futures research useful requires the opening of many new avenues. Technology ought to bridge the gap between science and scientific enterprise on one hand, and the arena of decision-making and public discussion on the other. If science is to be of any value, its findings must be communicated to the individuals and groups upon whose response the validity of futures findings is based. Science therefore has a stake in seeing the fruits of its endeavour implemented and used. To the extent that this is done, the scientific effort as such could be judged to be successful or not. To make this feasible is the role of technology. Technology therefore will have to make the scientific findings compatible with the users, i.e., with the humans involved. It should enable practical use of the scientific analysis while man or society are thinking about the future options and preferable alternatives. Apparently, what is necessary are new tools – new instruments – which can be used in the process of policy analysis, in the process of considering future alternatives, and thinking about the future in general.

The APT system – an attempt to test the practicality of using science in policy analysis.
In order to test the feasibility of developing systems which will link the futures-oriented science and policy analysis with the societal decision processes, a computer-based policy assessment tool (termed the APT System – Assessment of Policy Tool) has been constructed. Its usefulness has been tested in conjunction with a world system model by investigating the alternative paths for global development.

The APT system is envisaged as an instrument whose use can help in the assessment of likely consequences of proposed policies. Such a computer system is merely an aid to the decision-maker and is in no way a substitute either for policy formulation or for the policy selection; its role is restricted to the improvement of these processes by enabling formulation of a broader and more comprehensive set of policies, and by increasing the likelihood of success of a policy finally selected for implementation.

Any policy-making process has an intuitive component and a logical or rational component. Intuitive aspects include perceived goals, the interpretation of the acquired experience and knowledge, and the like, while the logical aspects have to do with the facts and data (numerical, quantitative information) and the logical sequences of events triggered by or affected by policy implementation. The role of the APT system is to add a new dimension to the second aspect and in so doing also to augment the domain of the first.

Some of the factors which provided motivation for the development of the APT concept and which guided its design and construction follow.

Complexity. National policy formulation and selection has become increasingly complex in the following two ways. First, *interrelationship* between domestic policies and foreign policies has increased considerably. As Harlan Cleveland aptly put it recently, 'every domestic issue is partly international and every international issue is partly domestic'. The energy, food, population, development gap, trade, and monetary crises provide ample examples to illustrate the point. Second, *inter-relatedness* has increased between policies traditionally considered, at least in the practice of decision-making, to lie in separate domains. To paraphrase Cleveland again, every foreign minister must be a little bit of an expert on the matters of energy and food – not to mention economics in general – while every minister of agriculture had better have a grasp of foreign affairs. While separation of policies into different domains was always a poor approximation of reality, it does not suffice any longer.

Rate and magnitude of change. In the past, many (although by no means all) policies could be based on a somewhat experimental approach. By implementing some aspects of policies, one could assess from the resulting trends what the policy would produce if fully implemented over the entire period of time. Such an approach, frequently called incrementalism or 'muddling through', appears both practical and prudent as long as the penalties involved are not unduly high. However, the rate and the magnitude of changes associated with various current policies make the approach increasingly risky. Technically speaking, one cannot rely on the real world 'feedback mechanism' for correction and adjustment, but must try in the best way possible to *anticipate* the consequences of intended options.

Lead time. The period between the time when the decision to implement a policy must be made, and the time when its beneficial (or harmful) consequences become apparent, has lengthened considerably, to the extent that in many vital areas it can stretch over a decade or more. For example, the return on government and private investment in the nuclear energy area can be expected only in a decade or so and depends on both the prevailing conditions at completion time and the ability to sustain a continuous effort (i.e., investment) over a rather long time period with associated uncertainties and changing conditions. Lead time concerns are related to the rate and magnitude of change concerns mentioned above, and together they indicate a need to take into account the 'dynamics of the system in an anticipatory manner'. The APT system was constructed to assist decision-makers in meeting such increasingly difficult decision situations.

Some factors which were considered in the construction, and continue to be relevant in the application and revisions of the APT system to assure its use in practice, are described below.

(i) *Reliability.* The construction of the tool has to be based on as many facts and data as are available and must use as much relevant theoretical and practical knowledge as possible. An important decision should not be based on incomplete information.

(ii) *Comprehensiveness.* A rather full spectrum of relevant factors must be covered. For example, there is no point in considering economic implications in minute detail while overlooking technical feasibilities and constraints; nor is there much to be gained from analysing technical possibilities without taking into account the economic and human support needed for such a technology. The task is further complicated by the need to comprehend the complex set of interrelationships among the large set of factors involved. This requires a balance in the detail with which each of the related domains (economics, technology, resources, demography, etc.) is to be represented.

(iii) *Efficiency in use.* This is a point too easily overlooked. Policy-making takes place under time constraints and in an organizational environment with established procedures and processes. It is useless to develop policy recommendations after the time has passed when the decision was needed. It is equally useless to make these recommendations in a manner that does not fit existing organizational patterns, or which escapes the human elements of decision-making.

(iv) *Realism in use.* The APT system, in stressing anticipatory policy making, involves a certain amount of 'crystal-ball gazing' and 'futures prediction'. This is clearly a risky business and one must be especially careful to distinguish between that which can be said about the future and that which cannot be said with any degree of certainty. To meet this challenge, the APT system is built on the concept of scenario analysis; i.e., on assessing the likely consequences of certain future events and policies. It is in such an 'if–then' context that credible and responsible use of the APT system is envisioned. In other words, the aim of the APT system should be to augment the power of reasoning as practised in prevailing decision-making processes, rather than as a substitute for that process.

Basic components or modules of the APT system are shown in Fig. 13.1. A retrieval data base, B, contains all available data that might be useful in analysis of a given set of policies or problem situations. A computer model, M, contains a representation of the relationships between policies (actions) and their evaluation indicators; i.e., the variables in terms of which different policies are compared. It might be useful to recall the often stated fact that it is in the very nature of any decision-making

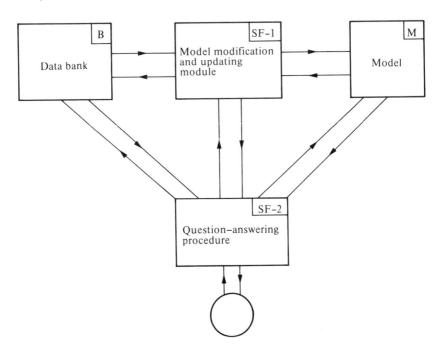

Fig. 13.1. Basic components of the APT system.

(or selection) process that a model must exist. It is only in reference to that model as an image of cause–effect relationships that any discrimination between different policies and selections can be made. A model, in effect, represents the understanding the decision-maker has (at the time when the choice is made) about plausible outcomes of alternative choices. Such a model must exist, even if only as a mental image. The success of the decision-making depends on how well such a model reflects many relevant aspects of the real situation. When the considerable amount of information needed for policy evaluation is in numerical form and when the model involves a large number of interconnected relationships, it is only natural to use a computer to untangle the sequence of events and the flows. Indeed, this is the only practical or even conceivable way in a situation of even moderate complexity. Model building therefore is not constraining in any way, but, if properly done and properly used, can be only illuminating and enlightening.

The third component is a software package, SF–1, which links the data base and the model and allows the updating of the model in the light of newly acquired data or, more generally, changes in the model parameters, equations or structure in order to keep the model up-to-date, as well as to allow the testing of alternative hypotheses of how the real system functions. The fourth module of the APT system, SF–2, is a question–answer software package which permits an efficient implementation of alternative scenarios and presents the results of comparative evaluations, according to some preselected criteria, in a form most effective for the policy analyst.

The APT system can be used as a tool for planning or policy analysis in a variety of ways. Some of the typical situations are shown in Figs. 13.2–13.5. First, past data can be retrieved and past trends can be investigated or identified from the stored data (Fig. 13.2). Second, the trends from the past can be projected into the

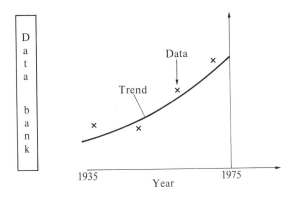

Fig. 13.2. Data and past trends analysis using data stored in the APT system.

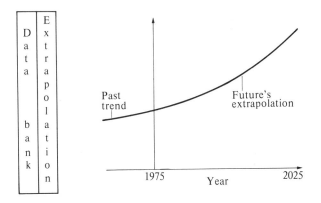

Fig. 13.3. Extrapolation of trends from the past into the future without any additional assumptions.

future as evolving from the past data and without additional assumptions as to the constraining relationships between them (Fig. 13.3). Third, a subset of data (and associated variables) — a submodel — can be analysed in order to get a better understanding of how a subsystem responds and functions under alternative conditions (Fig. 13.4). For example, an economic sector, agriculture, can be analysed in isolation in order to develop an understanding of how it would perform in alternative conditions; i.e., what is the maximal production potential in an ideal situation. Finally, the complete model can be analysed as it functions in its proper environment (Fig. 13.5); e.g., the alternative developments of a given nation or

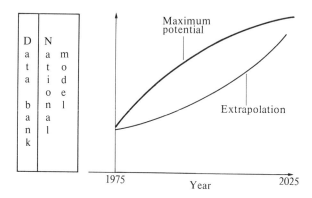

Fig. 13.4. Analysis of national potential arrived at by analysing the national model as to how it would respond and function under alternative conditions.

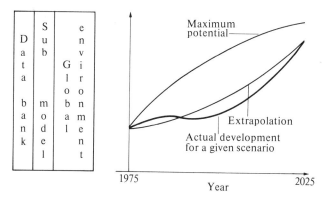

Fig. 13.5. Development in the global environment – the alternative developments of a given nation or region are analysed in the context of the total world system.

region, properly embedded in the context of the total world system.

The flexibility of the APT system as a tool for 'thinking about the future' should be apparent, even from this brief description. The system does not 'lock-in' an analyst within a preconceived structure of a fixed model. At one extreme, it allows the analyst to base his or her assessment purely on the basis of the past data and whatever mental image of the future one has; furthermore, it permits the use of a given model with a variety of alternative assumptions and premises, and permits comparisons of the results with a more direct assessment based only on the 'raw data'. At another extreme, not one but several models can be used in the course of policy analysis in order to test the importance of the consequences of alternative assumptions regarding the actual functioning of the real system. In summary, the APT system, as a tool, provides an opportunity for a maximal use of quantitative and logical information and reasoning, without introducing any additional conditions. In other words, the system does not tell the policy-makers what to do, but rather lets him or her find out what is the preference ordering among different policies under a variety of assumptions; ultimate decisions remain in the hands of the policy-maker.

The basic mode of using the APT system in the scenario analysis norm is shown in Fig. 13.6. A scenario is defined as a sequence of future policies together with a feasible sequence of concurrent future events. A scenario contains, in essence, assumptions regarding the uncertain future conditions in which the system will operate, as well as assumptions about poorly understood aspects of the system's behaviour. A scenario is the 'if' part of an 'if–then' type of futures analysis.

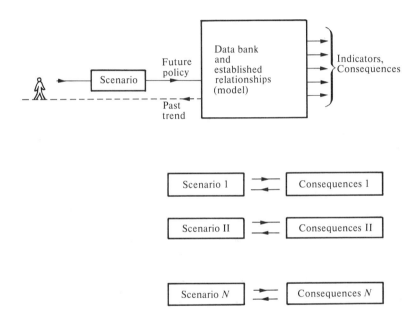

Fig. 13.6. The basic mode of using the APT system in the scenario analysis norm.

The crucial importance of efficiency in scenario analysis should be stressed here. In practice, one has to consider a rather large number of alternatives. Thus the consequences of implementing any one of these alternatives should be presented in a comprehensive manner, and within a sufficiently short span of time so that a policy-maker (or analyst) can compare the outcomes — not only in specific numerical terms, but also by comparing the general behaviour and response of the system. It is not unlikely that the final choice will be made on the basis of a feature whose importance becomes apparent only in the process of comparing the outcomes of alternative choices. Also it is not uncommon that, after the response of the system to a set of scenarios is analysed, new scenarios are conceived or triggered by some features observed in the system's response. It is through such a process that the policy analyst really develops a 'feel' for the dynamics of the real world system in question; such an improved understanding in itself can make a significant contribution, not only to policy selection, but to the policy formulation process as well. To facilitate the process of scenario analysis, video display terminals are used in practical implementation of APT.

So far, we have presented the APT system in its role for policy analysis. The system, however, can also be useful in the policy formulation stage; for example:

(1) It can help to decide which aspects of the 'adaptive' behaviour of the system should be taken into account (through interactive mode time analysis).

Two methods of scenario analysis are available in this respect.

Batch scenario analysis in which a scenario is formulated and implemented over the entire future time period, as specified by the user.

Interactive scenario analysis in which a scenario is implemented in steps; i.e., over a segment of the future time interval, shorter than the entire time horizon selected by the user. At the end of each partial time interval, assumptions are reassessed and, if necessary, policy changes are made designed to cope better with the newly developed insights. The results of such an interactive process is a scenario that reflects the perceptions of the analyst as to how the economic, social, and other processes will react to the changes resulting from the initial policy implementation as well as the structure of the model and data base (Fig. 13.7). The actual scenario is evolving during the process of analysis.

(2) It can help in translating general objectives — often formulated only verbally — into specific quantitative measures for policy implementation (through hierarchical mode of scenario analysis).

Two modes of operation are envisioned in this respect (Fig. 13.8).

The direct method in which a scenario is expressed in terms of specific data and values; e.g., the size of coal reserves.

The *hierarchical method* in which the analyst defines his goals, selects policies he believes will attain that goal, and finally decides specific data values by utilizing a tree structure of goals and policy options.

The easiest mode of using an APT system is in a batch direct scenario analysis. For that purpose, various forms of scenario sheets were developed for use with the APT system. The simplest scenario form is on the qualitative level, where a choice is made by indicating preferences among ordered alternative options. No numerical or technical choice is needed for qualitative level scenario formulation. As an illustration, qualitative scenario sheets for the analysis of energy issues are shown in Table 13.1. A set of increasingly detailed quantitative scenario forms has been developed for various applications of the APT system.

The feasibility of practical application of the APT system has been tested on various national and international policies and issues, using a computer model of the world system — referred to as the WIM system. It represents the second generation of the model used for the Second Report to the Club of Rome, *Mankind at the turning point* [1]. A full description of the WIM model is beyond the scope of

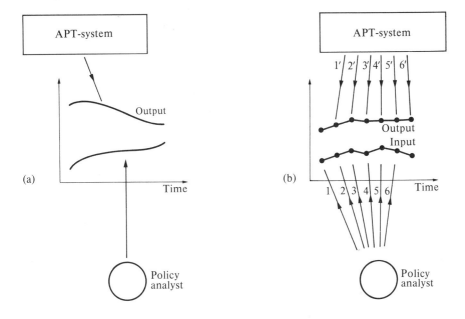

Fig. 13.7. The APT system used for (a) direct scenario analysis, and (b) interactive scenario analysis.

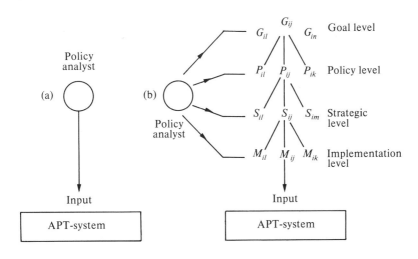

Fig. 13.8. Two modes of operation of the APT system for scenario analysis: (a) the direct method of scenario formulation; (b) the hierarchical method of scenario formulation.

Table 13.1 Oil analysis scenario form: qualitative level

Policies and parameters			Code for computer implementation
Potential resource estimates	high	(3000 bbl)	p11
	*medium	(2500 bbl)	p12
	low	(2000 bbl)	p13
Oil demand reduction with	high	(0.45)	p21
price increase (elasticity	*medium	(0.225)	p22
of demand)	low	(0.15)	p23
Oil supply increase with price	high	(1.0)	p31
increase (elasticity of	*medium	(0.75)	p32
supply)	low	(0.5)	p33
Annual increase in	high	(5 per cent)	p41
oil prices	medium	(3 per cent)	p42
	*none		p43
	decrease	(—per cent)	p44
	other	(exogenous)	p45
Limit on oil prices	high	($16.50)	p51
	*medium	($13.50)	p52
	low	($10.50)	p53
Oil consumption	substantial	(15 per cent)	p61
conservation	medium	(7.5 per cent,	p62
		10 per cent, WEUR)	p63
	*none		
Relationship between oil	full	(1.0)	p71
and investment good prices	partial	(0.5)	p72
	*none		p73
Economic growth target	fast	(5 per cent,	
		6 per cent, LDC)	p81
	*medium	(3 per cent,	
		4 per cent, LDC)	p82
	slow	(1 per cent,	
		2 per cent, LDC)	p83
Population policy	stringent	(20 yr)	p91
	*medium	(35 yr)	p92
	none		p93
Monetary recycling	efficient		p101
	*fair		p102
	poor		p103

*Standard values.

this discussion and can be found elsewhere. It suffices here to mention that the model represents long-term global world development processes. Its principal components are: population dynamics, the world economic system, physical transformation processings governed by man's activities (such as energy and materials processing, agricultural processes, etc.), resource development and exploitation, and ecological impacts. In addition, higher level socio-political and value factors are accounted for through scenario analysis procedures. All these components (which can also be referred to as levels, strata, or spheres) are fully inte-

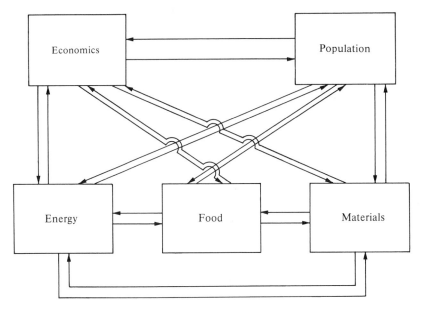

Fig. 13.9. The principal components of the WIM world system model.

grated into a total system, as they most certainly are in reality. In order to be able to account for the most crucial problem of today – namely, disparity in development in different parts of the world – the world system model is represented in terms of regions which in turn are fully integrated into the world system. The structure of WIM world system model is indicated in Figs. 13.9–13.11. For the sake of illustration, results of some scenario analyses are given in Appendix II.

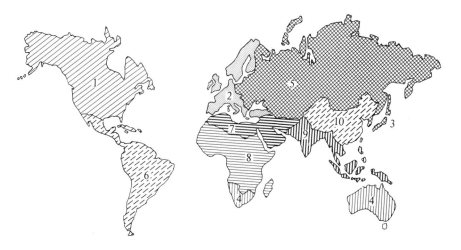

Fig. 13.10. The division of the world into regions for the purposes of the WIM world system model.

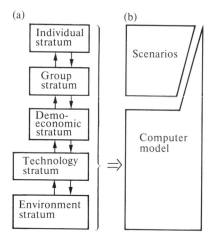

Fig. 13.11. Computerization of the WIM world system model. The behaviour of the world system is represented on five levels, termed strata (see a): individual, group demo-economic, technology, and environment. Each level provides a representation of the world system using knowledge as embodied in different sets of scientific disciplines, from psychology and nutrition to ecology and geophysics. All strata are interrelated in a total model.

Not all of the relationships and processes can be represented by a computer model based on the cause–effect kind of linkages. The subjective aspects of the individual and group strata are accounted for by designing a set of appropriate scenarios giving alternative sequences of plausible events and social and individual choices. For example, decisions and choices on the group stratum, regarding population policy would be represented by a corresponding set of scenario and time sequences of gradual changes in fertility rates.

An experiment for the assessment of the practicality of using the APT system and WIM model

In order to test the usefulness of the APT system and of the WIM world model, an international programme has been organized with the following objectives:

(1) To bring long-term and global considerations to the level of national decision-making where the really important decisions influencing global development are being made.

(2) To make the planning and policy assessment tool and a national policy-relevant global model available to national decision makers and international officials who are willing to test this procedure in practical policy considerations.

(3) To bring together policy analysts, decision makers, and professionals representing different economic, social, and cultural regions of the world in a joint assessment of the long-term future of mankind using the tools and models developed.

Feasibility of achieving stated objectives of the APT programme has been tested by a series of demonstrations, briefings, and seminars held in several countries in different parts of the world.

These meetings as a rule had the following format: description of the APT system and the world model, its structure, underlying assumption, and data base; presentation of the reference scenarios for the long-term development of the respective country and region in the global context by means of an on-line APT system; formulation of scenarios and alternative assumptions by the participants and analysis of the response of the system to these changed conditions using the on-line connection. In all these briefings, a central computer system based in Cleveland, Ohio, has been connected via satellite and telephone link with terminal and projector facilities at the respective location. Such an arrangement has demonstrated the flexibility and accessibility of the world system data base and world model for analysis from almost any point on the globe.

The list of demonstrations held so far is given in Appendix I.

13.5 The human revolution

The extraordinary upsurge of modern science and technology, coupled with the multiplying power of new production techniques, has driven events in directions and over thresholds that our forbears never imagined. In a matter of few decades, a pluri-millennial epoch of slow human development has ended and a new dynamic epoch begun. 'In reality, it is the *human condition* on Earth which has changed. Man, from having been one of the many creatures of the planet, has now cast over it his uncontested empire'.[14]

At the roots of these changes lie the *material revolutions*. The industrial revolution was followed by the scientific one, and this by the technological revolution, which is still in full swing. The combined thrust of them all ushered in a new age. The watershed between the two epochs was marked by the appearance of high technologies and complex artificial systems – in aerospace, defence, production, transport, communication, information, data storage and retrieval, etc. – which have radically transformed life on the planet. Likewise it did not take long for the man-made world to become not only gigantic and formidable, but even monstrous. Man found himself increasingly unable to adapt to the extraordinary transformations he was causing. And he was confronted by another unexpected and bewildering development. Though the increasingly more complex systems he was able to engineer interlocked with the ecosystem's delicate networks, they lacked the latter's self-regulating and self-healing qualities, and soon began to throw the ensemble out of balance; they thus required his constant supervision and intervention. At this point, man realized that he must take upon himself regulatory functions for the entire natural–artificial system complex which he previously thought were in the domain of nature or of providence. According to Julian Huxley, 'his role, whether he wants it or not, is to be the leader of the evolutionary process on Earth, and his job is to guide and direct it in the general direction of improvement'.

Man's agonizing discovery, at present, is that he is unable to fulfill this task. Instead of riding triumphantly, as he had deluded himself, the tigers of the material revolutions, he is perceiving that his power is treacherous and that he is, directly, in danger. In this plight, he is not supported by a strong ethic of the species – nay, he is utterly divided in groups and grouplets pitted one against the other – and he is more than ever uncertain about the meaning of his own existence. While, in the past, 'myths, religions, and philosophies did bring positive answers to the problem of meaning, and while it was believed for a long time that science would bring the final, definitive solution, we now realize, at last, that the problem of meaning is the one to which no scientific answer ever could be provided' [15]. Thus man, 'the only highly social animal whose code of conduct is largely transmitted culturally rather than genetically'[15] is in deep trouble precisely at the base of his strength – his cultural uniqueness.

However, this shocking discovery is salutary, for man now knows that he cannot race ahead blindly. The realization that he is up against changes he cannot rationalize and to which he is unable to adjust fully or quickly enough, and that he does not know how to control the very mechanisms of change that he has set in motion, is an eye-opener. It will help him to see that the current global crisis, in which old and new problems of population, food, education, resources, energy, crowding, inflation, poverty, alienation, militarization, disorder, eco-degradation, injustice, etc., are growing and intertwining beyond recognition, is closely related to the difficulty he has in achieving the level of understanding and sense of responsibility demanded by the new power which he derives from the material revolutions.

If this rough analysis of the predicament of mankind is fundamentally correct, as we believe, then *the problem is clearly within man*, not outside him; any possible solution entails, and can only stem from, a change of man's heart and mind. Such a conclusion is fraught with quite radical implications. For one thing, it means that everything that is important, from quality of life, to survival and human destiny generally, will depend essentially on the development of the human being, on his capacity to evolve so as to be in harmony with his changed and changing universe. For instance, as we have already noted, injecting more science and technology into the system will be good, but only to the point that people are prepared to absorb and make proper use of them; it will be bad, even destructive, if they are not.

Consequently, this means that man, both as an individual and as family, society, or mankind as a whole, should by definition be the prime objective of human concern and action. This is not the case nowadays, since we invariably put the accent on things that are certainly important and useful and have to do with people – such as the scientific and technological revolution, industrialization, agricultural development, plans, programmes, and projects, factories, dams, and canals, mechanisms, models, structures, systems, processes, automation, and more generally hardware and software and the attendant knowledge, information, and education – but in which people are taken for granted, or seen somehow as a peripheral element, a kind of accessory to the core issue or item. The primacy attributed to science and technology which we have denounced earlier is an example of this attitude. Now this order of priority must be inverted, letting people themselves regain the central place which rightfully belongs to them.

For another thing, the character of this human revival must be revolutionary. It cannot be just platonic if it has to be sufficiently strong and capable of regulating and steering the material revolutions. In other words, it must be a true *human revolution*, creative and persuasive enough to radically renew and even reverse principles and norms now considered unchangeable, and foster the emergence of new sets of values and motivations – spiritual, philosophical, ethical, social, aesthetic, artistic – in keeping with the imperatives of this age. Moreover, to become the watershed required to change the human course altogether, it cannot be either a movement restricted only to some elites or nations, but has to involve and be something pertaining to large segments of world population. It is mankind as a whole that must undergo a deep cultural evolution, and substantially improve its quality and capacity if the entire human system is to be put on a higher level of understanding and organization, based on a stable internal equilibrium and felicitous communion with nature.

The crucial questions then emerge. Can man engender his own revolution – as he has done with the material revolutions? And, if so, in what way? We believe that he can. 'We have succeeded in improving the quality of athletes, cosmonauts and astronauts, of chickens, pigs and maize, of machines, appliances and materials; we have succeeded in the case of man's productivity, in his ability to read fast, and his capacity to talk to computers. However, we have never tried in earnest to sharpen his perception of his new condition, to heighten his consciousness of the new strength he possesses, to develop his sense of global responsibility and his capacity to assess the effects of his actions. I am sure that, if we try, we will succeed – not least because every step made along this road will show more clearly that it is in our fundamental interest to go further in the same direction.' [14]

The encouraging ferments which, as noted, can be detected at the grass roots of society, concern also the place and role of people. The question of the *human condition* is coming to the forefront, as occurred a few years back in the case of the human environment. And this probably is the most important development of this decade. Its starting move was the revolt against the great disparities in living standards and opportunity which, if still accepted as inevitable as recently as the 1960s, are no longer tolerated either among or within nations. The imperious demand for *a more equitable international order* is in fact the most striking manifestation of this human awakening. But it is now followed by the growing concern about *basic human needs*, which for the first time extends to embrace all the world's inhabitants. Yet another aspect of the same phenomenon is the strong movement for *human rights*.

At the same time, two sets of corollary ideas are also on the move. First, the very recognition that the satisfaction of basic needs and rights has become a *sine qua non*, a prerequisite for mankind's progress, means that making such satisfaction possible is a *primary obligation* for one and all. The intrinsic interdependence between the two embodies the basic principle of all cultures that rights cannot be divorced from duties. Further steps should lead to the realization that mankind's future can indeed be more securely built on the recognition of obligations and duties than by assertion of needs and rights, and that individuals, groups, and nations whose capacities are greater also have proportionately greater obligations.

Second, there is the incipient perception that mankind's needs can, after all, be met only by mankind itself. This of course is a truism – though one which is all too often forgotten. The syllogism is inescapable; all problems, if not caused by climatic vagaries or unexpected natural events, are due to human behaviour; therefore, any solution to such problems depends essentially on appropriately modified human behaviour. Nevertheless, contrary to this plain logic, people have almost exclusively been viewed so far as problem-raisers, and attention has been riveted essentially on their needs, wants, and expectations, either as biological organisms or economic units. Very little consideration has been given instead to their complementary but equally essential and irreplaceable capacity and function as problem-solvers and need-purveyors. Now the concept is dawning that the future can be shaped as a rewarding venture much more by enhancing what individuals and groups can *contribute* to society and environment in proportion to their capacity, than by focusing primarily, as now, on what they should or actually do *demand* in proportion to their needs. Here we have then the recognition that human quality and capacity have become paramount factors.

Looking at all these signs, it is reasonable to conclude that they are not simply glimmerings of hope, but imply that a change of outlook is actually slowly taking place. Therefore, after having considered with great concern our world in disarray, even if it is at the apex of its knowledge and achievements, we are now able to perceive the first symptoms of a human renaissance. All this confirms our thesis that mankind has in its very fold the keys to its salvation and success and that, after having looked for them in vain everywhere – amazingly except in the only place where they can be found, the human being himself – it is now slowly realizing its mistake.

What is necessary, at this stage, is to strengthen this movement, making the re-valuation of the human being an irreversible process. What is also necessary is to understand not only that he represents mankind's most precious, ubiquitous, and permanent resource, which has been largely ignored, constantly neglected, and badly employed so far, but also that his brain has an innate potential yet to be developed. Nothing exists in the biosphere comparable to his humanness, mental capacity, and creativity. Therefore, the fundamental knot of questions can be made more specific. Can this marvellous but latent human potential be developed and deployed for the benefit of man himself? And how can this be engineered?

The answers to these questions are what we have to discover and learn. Some of our investigations must be addressed to the ways and means of ensuring the employment or activation of all the human capacities that exist. However, solution or alleviation of this most urgent and most difficult problem are not within the scope of this analysis, which is primarily directed at the other, and even more difficult question of whether the continuous updating, upgrading, and enhancing of these capacities is possible, and how it might be achieved.

Our point of departure has, in fact, been whether people and societies will be able to catch up with the torrential techno-scientific advances that have already occurred, and avoid lagging behind again in the future. The basic idea is that mankind must learn to be always on a par with reality, and even ahead of it, if possible. All this, in turn, requires that its members develop a new alertness and new qualities and capacities in order, on the one hand, to achieve rapid, satisfactory self-adjustment when change occurs (i.e. continuously) and, on the other hand, to ascertain that their own activity does not provoke change to which adaptation is too difficult or impossible.

The first exigency is easy to understand, for *genetic adaptation* to change is already recognized as the principle of survival and evolution for the other species. In the case of humans, however, this cannot be the answer, if for no other reason because their genetic evolution would certainly be too slow in the face of the tempo of change. What is instead required therefore is their adequate *psychosocial adaptation* to new speeds, dimensions, complexities, interdependencies, and uncertainties. But they cannot rely either on adaptation *ex-post*, for this can also turn out to be too slow, or insufficient, and even impossible. The present bad experience of the non-adaptation of billions of people to changed condition should not be repeated. This is the reason for the second and even more important exigency: that mankind should constantly monitor and regulate its change-provoking activities so as to make them compatible with its psychosocial *adaptability* – which needs to be carefully assessed and sensed *ex-ante*. In other words, since mankind itself is the main agent of change, it must prepare in advance, acquiring the capacity not just to react, but to proact. This brings us to the postulate, which we submit, that at this point of human ascent, the future will depend essentially on man's and mankind's *anticipatory cultural evolution*.

We prefer to stop here, not least because an ongoing project promoted by the Club of Rome, called in short *Learning* [16] addresses itself to the problem of learning, education, and preparation precisely from this angle. Postulating that, from now on, cultural anticipation, rather than adaptation, is the key to human survival and progress, it intends to explore what the natural *learnability* of men and societies is, and how it can be developed as a deliberate and conscious response in advance to opportunities and problems arising in a dynamic situation loaded with extreme alternatives. A report is expected to be ready by the end of 1978, and we hope it will lead to a deeper and broader consideration of the issues we have presented here.

Yet, even now, it would be unwise to deny that, in this period of torrential scientific and technological advance, mankind has failed to evolve culturally quickly and fully enough to keep abreast of events and maintain its options for the future wide open. Unfortunately, the stark reality is that mankind is indeed increasingly off balance, befuddled, and in danger, while its options are dwindling. It is now imperative for it to regain its senses and to attain a sound equilibrium throughout the planet – however colossal and Utopian this enterprise may seem.

We have tried to convey the idea that *there is no other way of doing this except by developing and employing the inner human potential consistently and intelligently*. The question of human destiny has never been seriously posed in such realistic terms. The time has come to do so.

Appendix I

September 1975: Tehran, Iran. A demonstration called *Presentation of the potential of the global model and its usefulness for national planning*, organized under the auspices of the Planning and Budget Organization and given to the economic council of Iran consisting of the Prime Minister and more than a dozen key ministers in the Cabinet. The meeting lasted five hours and a number of options and assumptions formulated and proposed by participants were examined.

October 1975: Cairo, Egypt. Analysis of alternative patterns of development for Egypt and Middle East countries, a seminar and demonstration. A special private demonstration was given to Prime Minister Salem and a group of ministers.

February 1976: Hanover, Germany. A three-day workshop organized by the European community was held in Hanover in February 1976. Attended by 12 parliamentarians from the European parliament representing countries from Ireland to Italy and selected because of their responsibilities for and interest in long-term European and world development.

February 1976: Caracas, Venezuela. Testing the use of APT-system and the world model for assessment of long-term trends in Latin America, a demonstration attended by the Deputy Chairman of the North/South Conference in Paris and the Minister for External Economic Relationships of the Government and attended also by two dozen policy makers and professionals from Venezuela.

May 1976: Philadelphia, Pennsylvania. Club of Rome meeting in connection with the U.S. Bicentennial. Several demonstrations to Club members and invited guests.

July 1976: Washington, D.C. A series of four demonstrations at the General Accounting Office attended altogether by more than 200 persons primarily drawn from congressional and senate committee staff members but also with some attendance by senators, congressmen, and public officials. The emphasis was on US food policy options and their impact on global development.

September 1976: Washington, D.C. A one-day briefing and demonstration at the Department of Agriculture attended by about 100 professionals from Economic Research Services, USDA, and a select group of personalities from Washington with interest in US food policy options.

October 1976: Algiers. A briefing and demonstration at the Club of Rome meeting, attended by an international group of public personalities attending the Club of Rome meeting with heavy emphasis on analysis of North Africa and Middle East using the APT system and the World Model.

November 1976: Jakarta, Indonesia. Seminar and demonstration at Indonesian Institute of Science and Technology on the global prospects for development. A meeting organized by the Indonesian Institute of Science and Technology and attended by about 60 people from the Planning Ministry Research Institutions and policy makers.

November 1976: Washington, D.C. Presentation of use of the global model for analysis of US policies. A demonstration given at the meeting of the Simulation Society, attended by about 200 professionals in the Computer Simulation field.

December 1976: Paris, France. A meeting in the French Senate organized at the invitation of the President of the French Senate and attended by a select group of French Senators and public officials.

January 1977: Honolulu, Hawaii. A three-day seminar at the East West Centre with the participation of teams from a selected group of Pacific countries, specifically from Japan, Australia, Philippines, India, Indonesia, Latin America, Germany Italy, France, and US. Typically the team consisted of three members: one economist, one systems analyst, and one person with experience in governmental affairs and decision-making.

April 1977: New Delhi, India. Seminar on *Futures research and international economic decision-making.* Attended by the Deputy Chairman of the National Planning Commission of the Government of India, the Director General of the Indian Council of Research and two dozen selected government officials and academic professionals. Emphasis on the long-term development options for India and the subcontinent.

April 1977: Vienna, Austria. A demonstration at IIASA of the world food system model and APT system. A meeting attended by about 100 scientists working in Vienna.

May 1977: Toronto, Canada. A demonstration to a group of ministers and senior officials.

June 1977: Helsinki, Finland. Prospects for long-term European development, a special seminar and demonstration presented under the auspices of the European Cultural Foundation. Attended by about 50 professionals and policy makers.

June 1977: Washington, D.C. Demonstration and analysis of scenarios from Herman Kahn's *The next 200 years* at the meeting of the US Association of The Club of Rome.

July 1977: Dakar, Senegal. A colloquium on the prospects for the formation of a West African Economic Common Market and the relationship of such an integrated unit with the Arab world and Western Europe. A meeting attended by professional staff and policy makers in Senegal and other West African countries such as Mali, Liberia, and some others.

July 1977: Barcelona, Spain. Presentation at The First World Conference on Mathematics at the Service of Man.

August 1977: Mexico City. Analysis of the interaction of family planning and population development in economic growth, a demonstration given at the General Conference of the International Union for the Scientific Study of Population.

August 1977: Moscow, USSR. Analysis of the impact of long term development in the environment, a special three-day meeting held at USSR Soviet Academy of Sciences. Attended by about 100 top scientists and planners from the U.S.S.R.

September 1977: Stockholm, Sweden. Demonstration at the Stockholm Club of Rome meeting to the Heads of State.

October 1977: Houston, Texas. A presentation at the Alternatives to Growth 1977 Conference.

Appendix II

Scenario 1: projection of historical trends
A scenario of interest in any analysis is a look into the future if the trends from the past continue (*historical or past trends scenario* Fig. 13.12). This scenario is not to be regarded as prediction in any sense. More appropriately, it should be regarded as providing an insight as to when some changes might have to be made if the course of development is to be changed.

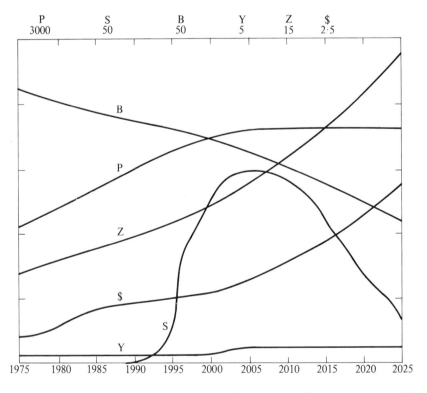

Fig. 13.12. Historical or past trends scenario: (P) population; (B) crude birth rate; (Z) income *per capita*, U.S.A.; (Y) income *per capita*, South and South-East Asia; (S) death due to starvation, South and South-East Asia; ($) price of food on the world market.

The *historical scenario* shown in Fig. 13.12 indicates two disturbing eventualities in the world food situation:

(a) A severalfold increase in world food prices (in real terms). That would most certainly drive domestic food prices unbearably high.

(b) A strong possibility of periods of food shortage in various regions of the world, which would lead to starvation on a disastrous scale.

Scenario 2: isolationist
Domestic food prices could be kept down by restricting food exports. Consequences of such an *isolationist scenario* (Fig. 13.13) would be truly disastrous. Additional tens of millions of people could starve (under unfavourable but feasible conditions of food supply) because of failure of the USA to produce or export food.

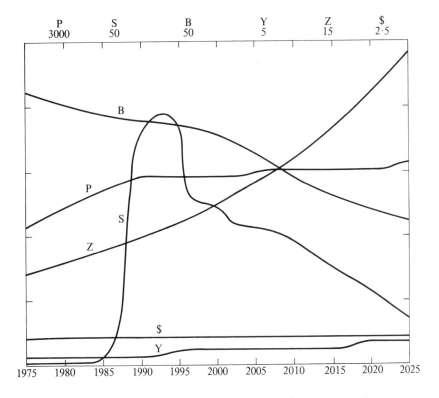

Fig. 13.13. Isolationist scenario. Symbols P, B, Z, Y, S, and $ as in Fig. 13.12.

Scenario 3: free world food market
The food need of the growing population could be satisfied if a truly free world food market is established with the USA as an active and full participant. In such a 'free world food market', scenario (Fig. 13.14) starvation anywhere in the world is avoided at the cost of:

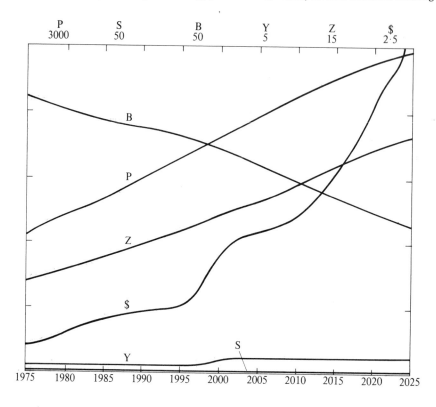

Fig. 13.14. Free world food market scenario. Symbols P, B, Z, Y, S, and $ as in Fig. 13.12.

(a) More than tenfold increase in world (and domestic) food prices (in real terms).

(b) Subsidies to the needy nations amounting to hundreds of billions of dollars (under unfavourable but feasible conditions) in order to maintain the market demand and stimulate production to the degree necessary to avoid starvation.

Scenario 4: intense family planning

An *'emergency family planning scenario'* (Fig. 13.15) in which an equilibrium crude birth rate is reached in fifteen years, would avoid extreme developments in the world and domestic situation. Any less radical approach to population will necessitate deliberate USA policies aimed at avoiding the extremes.

References

[1] Mesarovic, M. and Pestel, E. *Mankind at the turning point.* E.P. Dutton, New York (1974).

[2] *Science, growth and society — a new perspective,* by the Ad Hoc Committee on New Concepts of Science Policy of OECD, led by Harvey Brooks (1971).

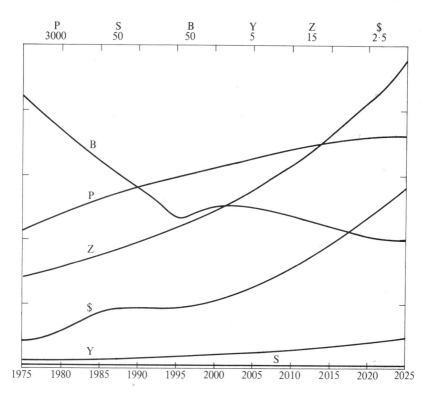

P	S	B	Y	Z	$
3000	50	50	5	15	2·5

Fig. 13.15. Emergency family planning scenario. Symbols P, B, Z, Y, S, and $ as in Fig. 13.12.

[3] King, A. *The state of the planet.* Pergamon Press, Oxford (1978).

[4] Kapitza, P.L. of the Institute for Physical Problems of Moscow, from the *Bernal Lecture*, delivered in London (7 October 1976).

[5] From Biswas, A.K. *Loss of productive soil*, a paper prepared for the IFIAS Soil Project (March 1977).

[6] From various documents presented during international meetings. See e.g., Vollmar, F. *Conserving one earth.* Paper presented at Kruger National Park (13 September 1976).

[7] See Meadows, D.H., Meadows, D.L., Randers, J., and Behrens III, W. *The Limits to growth.* Universe Books, New York (1972), and Gabor, D., Colombo, U., King, A., and Galli, R. *Beyond the age of waste.* Pergamon Press, Oxford (1977).

[8] *Energy–global prospects 1985–2000.* Report of the Workshop on Alternative Energy Strategies, directed by Carroll L. Wilson, McGraw-Hill, New York (1977).

[9] *The Director-General's programme of work and budget for 1978–79.* Food and Agriculture Organization of the United Nations, Rome.

[10] McHale, J. and McHale, M.C. *Basic human needs: A framework for action.* Centre for Integrative Studies of the University of Houston (April 1977).

[11] *Reshaping the international order–RIO.* (co-ordinated by Jan Tinbergen). E.P. Dutton, New York, (1976).

[12] See Lazlo, E., *Goals for mankind*. E.P. Dutton, New York, (1977).
[13] Branscomb, L. Science in the White House: a new start. *Science*, **196**, 848–52 (1977.)
[14] Peccei, A. *The human quality*. Pergamon Press, Oxford (1977).
[15] Monod, J. In *The place of values in a world of facts*. Nobel Symposium, Almqvist and Wiksell, Stockholm (1970).
[16] The provisional title is *Innovative and prospective learning for man and society*, and its leaders are M. Malitza, J. Botkin, and M. Elmandjra.

Commentaire (I)
Edmond Malinvaud

C'est un honneur d'intervenir après deux auteurs qui ont tant fait, dans le cadre du Club de Rome, pour sensibiliser nos concitoyens aux problèmes futurs de l'humanité et qui viennent de nous présenter un message d'une grande hauteur de vue. Devant les questions cosmiques qu'ils ont abordées chacun se sent humble et hésite à parler. Je ne l'aurais certes pas fait si les organisateurs de la conférence ne me l'avaient pas demandé. Je vais cependant tenter d'exprimer ce que beaucoup d'entre vous doivent penser.

Le développement de l'humanité s'est toujours fait d'une part grâce aux progrés de la connaissance et de la technologie, d'autre part grâce à l'adaptation des sociétés à ces progrès. L'adaptation a rarement été aisée ; chacun connaît le désarroi des populations rurales forcées d'émigrer dans le grandes agglomérations urbaines au moment de la révolution industrielle ; chacun connaît la sévérité du choc de la civilisation occidentale avec les civilisations traditionnelles d'Amérique et d'Asie au moment de la colonisation. Mais s'il n'est pas nouveau de constater un désarroi, il est vrai que le désarroi actuel revêt des formes particulières et doit retenir notre attention, car il constitue le problème majeur de notre génération.

Les deux conférenciers insistent à juste titre sur le fait que l'avenir peut réserver à l'humanité un degré tout-à-fait inhabituel de discontinuités traumatisantes et irréversibles. Peut-être exagèrent-ils la probabilité des éventualités défavorables. Mais peu importe. Chacun comprend que les incertitudes ont actuellement une ampleur considérable et un caractère mondial qu'elles n'ont jamais eus dans le passé.

On doit aussi manifester son accord aux auteurs quand ils nous montrent éloquemment qu'en face de ces risques une stratégie planétaire s'impose, que les décisions à prendre sont urgentes car leurs effets seront longs à se manifester, que les réalisations actuelles sont décevantes, mais qu'il ne faut pas sous-estimer le rôle des mouvements spontanés en faveur de l'écologie, du désarmement ou des droits de l'homme.

Peccei et Mesarovic nous disent que les démocraties libérales ont aujourd'hui une responsabilité historique pour concevoir, proposer et faire accepter une voie nouvelle, plus sensée, pour l'humanité. C'est sans doute exact : l'humanisme y a de profondes racines et le savoir scientifique y est très largement concentré. Mais les démocraties libérales n'ont plus le pouvoir incontesté qu'elles eurent pendant longtemps.

La difficulté des problèmes actuels tient non seulement à ce qu'ils requièrent un accord entre des nations nombreuses, regroupées en deux, trois ou quatre camps, mais surtout en ce que chacun des pays et chacun des camps surestiment ce qu'ils peuvent obtenir de la confrontation internationale. L'étude des situations conflictuelles montre bien qu'aucune issue ne peut être trouvée tant qu'il en est ainsi. Les peuples du Nord sous-estiment l'importance des sacrifices qu'ils doivent fournir pour assurer l'avenir de l'humanité; les peuples du Sud surestiment l'amélioration qu'ils peuvent obtenir pour leur niveau économique. Tant qu'il en sera ainsi, il faudra que les négociateurs fassent preuve de beaucoup de patience et que les mouvements non gouvernementaux comme le Club de Rome développent inlassablement leurs efforts de persuasion.

Bien que j'accepte ainsi très largement l'inspiration et les objectifs des auteurs, je dois exprimer un certain malaise devant ce qui est présenté comme deux voies d'approche pour résoudre, ou au moins améliorer, l'état actuel des choses.

Certes, les processus actuels de décision ne sont pas satisfaisants. Mais un logiciel pour l'étude des diverses éventualités, un logiciel tel que le système APT, peut-il jouer un rôle essentiel dans l'amélioration des processus décisionnels? Evidemment non. Un rôle secondaire oui, car bien utilisé il peut sans doute traduire en termes simples le résultat de recherches scientifiques sur l'énergie, l'alimentation ou l'environnement, recherches qui sont essentielles. Un rôle secondaire aussi, car des négociateurs pourraient y avoir recours pour arrêter certains détails d'accords internationaux une fois qu'ils se seraient entendus sur l'essentiel. Mais l'amélioration des processus décisionnels suppose surtout la volonté politique d'aboutir et de privilégier des considérations autrefois négligées.

Faut-il par ailleurs penser que toute solution des problèmes actuels doit passer par la transformation de l'homme? En un sens, oui. Ce sont de nouvelles conduites collectives qu'il faut susciter. On devrait reconnaître que le désarroi actuel provient en partie du mépris dans lequel la morale a été tenue depuis un siècle par la plupart des hommes de science. Les populations de cette fin de siècle ont besoin d'une morale sociale qui soit adaptée à la science et à la technologie modernes. Les auteurs ont raison de dire que les chercheurs en sciences humaines consacrent trop peu leurs efforts à la découverte de solutions positives aux problèmes actuels. Mais c'est la transformation des conduites collectives, spontanées ou réglementées, qu'il faut rechercher, plutôt que la transformation de l'homme en tant qu'individu.

Pour conclure et résumer cette brève intervention, je voudrais marquer mon profond accord avec les préoccupations des auteurs. Justement parce que ces préoccupations sont fondamentales, nous devrions y répondre avec encore plus de rigueur dans la recherche du vrai que Peccei et Mesarovic en ont déjà manifesté dans leur communication.

Commentary (II)
Denos Gazis

This is an interesting mixture of an evangelical Peccei and a mechanistic Mesarovic.

But, as the authors themselves admit in the introduction, the chapter is rather a layered suspension than a homogenized mixture. Nevertheless, it makes fascinating reading. It also displays the brilliance and the biases of both authors. I shall comment on the contributions of both in turn.

Aurelio Peccei is a humanist and an internationalist, and his contribution is an apology of these two points of view. He drives forcefully and convincingly the often-made point that our scientific and technological progress has outrun man's ability to 'relate to his surroundings'. I agree with this position, as well as with two other of Peccei's precepts: that more understanding of the effect of our actions and better social organization than we have had in the past are needed; and that participation of the public in decision-making is needed urgently, as is its prerequisite — information. I am also in sympathy with two specific suggestions: that a resource pool be created for the benefit of all peoples and nations, today and tomorrow; and that we strive to find something other than national sovereignty as the dominant concept of world systems organization.

I find myself diverging from Peccei's view in reading his emotional condemnation of man's assault on the environment. To be sure, we are guilty of a lot of insults on the environment. But Peccei's condemnation is too sweeping, and thus it loses some of its potential effectiveness. Peccei makes the eminently correct observation that we *do not know* exactly what we are doing to the environment. But in the next breath he acts as if he *does* know by asserting, for example, that phenomena caused by our actions exceed the self-renewal capability of our environment. I do not believe that Nature, without man's intervention, was distinguished in its 'self-healing quality' as it is claimed. In fact, it has heaped a lot of misery through the ages on all species including man. I am not prepared to argue that on balance we have improved on Nature's ways, but I *am* prepared to say that we can. I also believe that environment can be viewed as capital which can be exploited for the benefit of present as well as future generations, provided that any man-made changes either are reversible or can be adequately corrected and compensated by the benefit accrued.

Returning to the good points of Peccei, I can only applaud his focus, at the end of the chapter, on improving man's nature as the key element for the long-term salvation of mankind. I am also in perfect agreement with his message on p. 52, where he echoes John Kennedy on a world scale by saying, in effect: 'Ask not what the World can do for you, ask what you can do for the World'.

Section 13.4 of the paper is Mesarovic's apology for his brand of futures research. There are many good points in the contribution of Pestel and Mesarovic to futures research. I would single out their admirable attention to building a good data base as the foundation of their computer system, APT. The weakness of APT lies in the degree to which it attempts to model some ill-structured situations, such as the variation of world food prices with widely varying supplies, and the resulting deaths from starvation. Mesarovic correctly points out that some model underlies any decision-maker's reasoning. But the difficulty in spelling-out such a model and integrating it in a computerized 'what if' exercise is that a computer programme requires a clear statement. There seems to be some sort of uncertainty principle associated with modelling of ill-structured situations: the accuracy of modelling seems to be inversely proportional to the clarity of the statement. My own preference in the use of computers for a look into the future is to confine them to a task that they can do well; namely, sorting out data, doing some numerical digesting of the data, and presenting it to the decision-maker who does all the modelling, however imperfectly. To the extent that APT allows interactive modelling, it is moving in the right direction.

I would raise some questions about certain statements of Merarovic's concerning the desirable direction of science. He says that 'science must become futures rather than past oriented'. He goes on to describe futures research as, in effect, the only sensible form of science. I would accept futures research as an addition to, rather than substitution for, all other science, to the extent that such an addition is possible. I would also subscribe to a different interpretation of the futures orientation of science; namely, the requirement that all science be done with more attention than in the past on the possible consequences from its use, and more selectivity aimed at wide public benefits.

Peccei and Mesarovic are somewhat integrated on p. 98, where Mesarovic's tools are claimed as ones that enhance human adaptability, or rather replace it by some form of 'cultural anticipation'. To the extent that Mesarovic's model, and Peccei's exhortations, can raise the level of consciousness of people concerning such major world problems as food, population, conservation, and disarmament, theirs is indeed an important contribution toward attainment of this cultural anticipation.

14

L'impact sociétal des technologies de l'information

André Danzin

Le Professeur Porat [1] a donné une image saisissante de la mutation intervenue récemment dans l'histoire de l'homme en matière d'information. Séparant ce qui est traitement des symboles de ce qui est manipulations de denrées ou actions sur des objets concrets, il parvient à la représentation suivante de la population active aux États-Unis (Fig. 14.1).

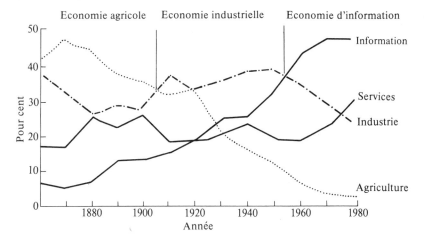

Fig. 14.1. Distribution de la population active entre les grands domaines d'activité, aux États-Unis de 1860 à 1980.

Bien sûr, il est quelque peu arbitraire de mettre dans une catégorie unique, les chercheurs, les enseignants, les notaires, les agents d'assurance, les banquiers et beaucoup d'autres agents qualifiés 'd'actifs de l'Information'; mais ne chicanons pas, il y a bien là la révélation d'un phénomène profond, le signe que nous parvenons, dans les sociétés post-industrielles à une 'économie de l'information' où l'essentiel de l'activité humaine sera consacré, non plus à la production de biens, mais à la communication et au traitement de signes, de paroles, d'écrits, de graphes, de statistiques, de symboles correspondant à des objets parfois réels, souvent abstraits.

Cette montée de l'information peut s'interpréter comme une marche triomphale de l'homme vers une «civilisation de l'esprit», mais aussi bien comme une asphyxie par la «complexité», comme l'enfoncement dans les «sables mouvants de la bureaucratie».

Ce mouvement vers la complexité coincide avec la brutale apparition d'outils nouveaux que l'on désigne sous le terme de «technologies de l'information». Est-ce la nécessité qui a fait éclore ces technologies? ou bien, au contraire, leur seule existence a-t-elle permis le développement des phénomènes de complexité? C'est là un problème qui n'a pas plus de réponse, ni d'intérêt, que celui de l'oeuf et de la poule. Il y a coexistence du besoin et des moyens de le satisfaire, l'un soutient alors l'autre dans sa croissance, d'où naît un mouvement impétueux de la création d'information, de sa communication et de son traitement qui nous entraîne vers un équilibre dont les termes nous sont, aujourd'hui, pour l'essentiel, encore inconnus.

14.1. Étrangeté et violence du courant d'innovation technologique

Remarquons au passage que cette poussée vers toujours plus de complexité et toujours davantage de psychisme, se trouve dans le droit fil de l'Évolution biologique. De l'être monocellulaire à l'Homme en passant par les reptiles et les mammifères, c'est à une progression constante de l'Information que l'on assiste, stockée dans la mémoire génétique ou diffusée dans les sociétés animales et humaines. Mais nous avons le privilège, en cette fin du XXème siècle, d'assister à une sorte d'accélération du phénomène sans l'avoir véritablement voulue ni peut-être même désirée. Car le courant d'innovation technologique reste caractérisé par son étrangeté, en même temps que par sa violence.

Étrangeté, le mot n'est pas trop fort pour dépeindre le caractère imprévisible de l'explosion des techniques. Vers 1945, l'électronique base ses progrès sur les phénomènes analogiques sur la circulation des électrons dans le vide en présence de champs électriques et magnétiques complexes.

L'apparition de la digitalisation des signaux est, à l'origine, mal comprise dans son importance essentielle. Elle est, cependant, appelée à jouer, en techniques électroniques, un rôle révolutionnaire que l'on peut comparer à l'introduction des quanta en physique théorique. Mais pour digitaliser à bon compte il faut pouvoir traiter avec une extrême rapidité un nombre considérable de signaux, ce qui suppose une localisation des phénomènes de mémoire ou de logique dans de très petits volumes et l'emploi, pour chaque commutation, d'une quantité infime d'énergie. Rien, vers 1945, n'annonce la disponibilité d'une technologie adéquate.

Vers la fin de 1947, Bardeen, Brattain et Shockley annoncent la découverte de l'effet «transistor», mais le dispositif proposé, pointe et germanium, est fragile et peu fiable. Cependant la voie est ouverte et les coups de théâtre vont se succéder dans les 15 années qui vont suivre par l'emploi du silicium, la maîtrise des surfaces du semi-conducteur (1952), les traitements de diffusion, d'épitaxie et d'oxydation, l'exploitation de l'effect de champ, l'intégration de plusieurs dispositifs logique ou mémoire sur un substrat unique de silicium, la réalisation de circuits complexes à très haut degré d'intégration (vers 1970), puis des microprocesseurs (vers 1975).

C'est en travaillant sur des ferrites pour la télévision que deux équipes de chercheurs trouvent, en Nouvelle Angleterre, vers 1951 les propriétés des tores à cycle rectangulaire qui, pendant 20 ans, fourniront les solutions nécessaires aux mémoires rapides de calculateurs.

L'apparition du concept de logiciel n'est pas moins accidentel. En voulant accroître la puissance de calcul et la rapidité de leurs machines, les constructeurs de calculateurs vers la fin des années 1950 se trouvent aux prises avec un être immatériel nouveau qu'ils désignent sous le terme de «software», sans se douter qu'ils assistent à la naissance d'une nouvelle industrie, celle des produits programmes et des logiciels d'application de toutes natures.

L'intention des ingénieurs soviétiques, par le coup de théâtre du premier Spoutnik, en Octobre 1957, n'était certainement pas de faire progresser les télécommunications. Et cependant, l'une des retombées les plus prodigieuses de la conquête spatiale est un phénomèe nouveau qui peut avoir la plus grande importance dans les relations internationales : grâce aux satellites, la communication des informations peut se faire aisément sur toute la planète et le coût de la communication est presque indépendant de la distance. . .

C'est encore au cours des 30 dernières années que sont devenues disponibles des techniques dont nous ne commençons qu'à apercevoir les impacts d'applications : lasers; hologrammes; fibres de verre; optoélectronique; mémoires à bulles; cristaux liquides; reconnaissance des formes, de la parole et de l'écrit; traitement des graphes; capteurs, senseurs et actionneurs de la robotique; multiplication des langages des machines; techniques de programmation; microprocesseurs.

La violence du phénomène d'innovation se traduit par la rapidité de l'évolution. Tout se passe comme si la création technologique continuait de répondre à des règles d'essais et de sélection de caractère darwinien. Le hasard est sollicité dans nos laboratoires; la recherche scientifique l'interroge systématiquement jusqu'à ce que sorte un produit chargé de promesses. C'est ensuite l'épreuve du marché qui fait la sélection. Ainsi naissent et meurent les technologies, l'une remplaçant l'autre ou laissant subsister des formes latérales. Lorsqu'une «espèce» est née, puis confirmée par l'effort de développement et par la réussite commerciale, elle se ramifie en sous-espèces nouvelles toujours plus performantes, toujours mieux adaptées à la complexité. Et les conséquences sont parfois stupéfiantes comme l'atteste la brutale redistribution de la division internationale du travail en horlogerie sous l'effet de l'usage des quartz et des microcircuits.

Cette remarque impose la prudence lorsque l'on essaie d'imaginer ce qui se passera au cours des 20 prochaines années. Car les armes technologiques à pied d'œuvre pour de nouveaux progrès sont plus nombreuses que par le passé mais, comme nous allons le voir, leur pénétration sur le marché va commencer à se heurter à des inerties d'ordre sociétal et à l'obstacle des besoins en capitaux.

Néanmoins on peut d'ores et déjà annoncer, sans craindre l'erreur, la poussée de nombreuses technologies nouvelles dans différents domaines dont nous allons examiner maintenant les impacts sur la Société.

14.2. Informatique et télécommunications

L'informatique s'est, jusqu'à ses dernières années, accommodée des télécommunications telles qu'elles étaient organisées par les monopoles étatiques ou par les grandes sociétés spécialisées dans la téléphonie. L'heure approche où l'ensemble des rôles sera remis en question car un grand nombre de techniques nouvelles vont apparaître qui s'accommoderont mal des positions monopolistiques. Le télé-courrier, la téléreprographie, les systèmes de téléconférence, la distribution de la télévision en circuit fermé, et peut-être, la généralisation du visiophone qui associe image et son avec toutes les facilités que donne la téléphonie, le radiotéléphone sur fréquence privée, toutes ces nouvelles techniques vont entraîner une profonde modification des missions de la poste et des télécommunications et de leur rôle en qualité de fournisseur de services au delà du seul service de la communication. Les réseaux de calculateurs à partage de ressources dont ARPANET aux États-Unis est un exemple, introduisent encore davantage la notion de service par l'accès à des banques de données spécialisées et par la quantité, la qualité et la rapidité des informations transmises. Quant aux réseaux privés, nationaux et internationaux, dont disposent les banques, les assurances, les sociétés de transport pour la réservation et la tarification des places, ne constituent-ils pas d'ores et déjà des entorses aux monopoles du courrier ou de la téléphonie? Et que dire de l'intervention des satellites de télécommunications susceptibles d'être reliés directement aux usagers?

La situation des États-Unis diffère ici très notablement de celle des pays européens et du Japon caractérisée par l'existence du monopole d'État des Postes et Télécommunications. Le problème est majeur politiquement et économiquement car il s'agit des degrés de liberté accordés au système de communications, c'est-à-dire au système nerveux de notre Société. Ce système doit-il évoluer par le libre jeu de l'essai et de la sélection sur le marché ou doit-il être contrôlé étroitement par les gouvernements, ou se dirige-t-on vers des solutions intermédiaires nouvelles? Le problème est brutalement posé par l'avènement des techniques dont la force interne ne pourra pas être entièrement contenue par des politiques volontaristes. Nous sommes ici, comme dans beaucoup d'autres domaines, bousculés par l'étrangeté et la violence du courant technologique.

14.3. Technologies de l'information et industrie

Lorsque Jean Monnet et ceux qui l'on accompagné dans son projet voulurent «faire l'Europe» en agissant sur une force concrète, ils choisirent le terrain du charbon et de l'acier où se localisait la puissance. Si l'on partait aujourd'hui des mêmes prémisses, à savoir s'unir là où se cristallise le pouvoir économique, nul doute que l'on serait amené à choisir les industries de base de l'acquisition, de la transmission et du traitement de l'information.

Plus explicitement il s'agit de l'immense domaine, encore en pleine croissance malgré la crise économique, des composants électroniques et des capteurs, de l'électronique professionnelle, du téléphone et des télécommunications, de l'informatique et de l'automatique, de la reprographie, des activités de service et de

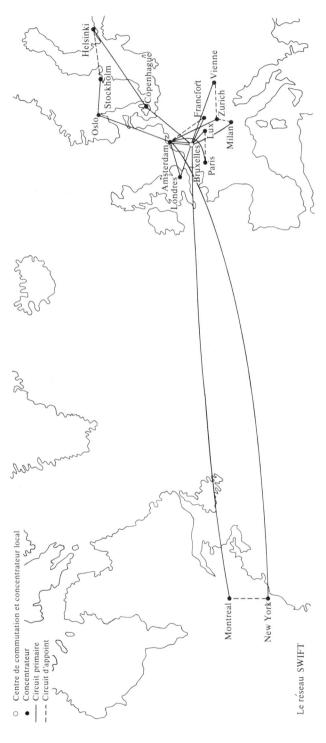

Fig. 14.2. Exemple de grand réseau de communication privé: le réseau inter-banques SWIFT (Society for Worldwide Interbank Financial Telecommunication). [*Source:* OCDE – DSTI/ICCP [7736.]

conseils en informatique, des centres de calcul des entreprises et de la fourniture d'informations scientifiques, techniques, économiques ou financières.

A partir d'une situation très modeste au sortir du dernier conflit mondial, inférieure à une contribution de 1% du PNB cet ensemble d'activités industrielles et de services représente d'ores et déjà plus de 5% du PNB dans les pays les plus industrialisés. Nous verrons plus tard les conséquences sur l'emploi. La place tenue dans ces industries et l'usage qui est fait des produits des technologies de l'information constitue une sorte «d'indice de développement». Les figures 14.3, 14.4, 14.5, et 14.6 suivantes montrent respectivement les surfaces (en millions d'hectares), la

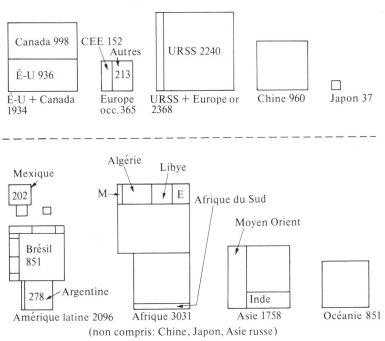

Fig. 14.3. La surface dans les 5 régions dominantes et dans les 114 autres pays (millions d'hectares).

population (en millions d'habitants), les PNB (en milliard de $) et la puissance installée en calculateurs (en milliards de $) pour les 5 régions dominantes et le reste du monde [2]

La cristallisation par gigantisme de l'industrie – problème d'éthique pour la société libérale

Un phénomène unique dans l'histoire de la libre entreprise, accompagne l'avènement de ces industries. Dans le domaine essentiel du traitement de l'information, un groupe multinational d'origine américaine, IBM, a en effet, réussi à conquérir une position étonnante de supériorité. Contrôlant plus de 60% du parc des ordinateurs installés sur les cinq continents, IBM annonçait pour 1976 un chiffre d'affaires

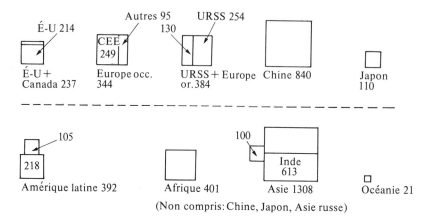

Fig. 14.4. La population dans les 5 régions dominantes et dans les 114 autres pays (millions)

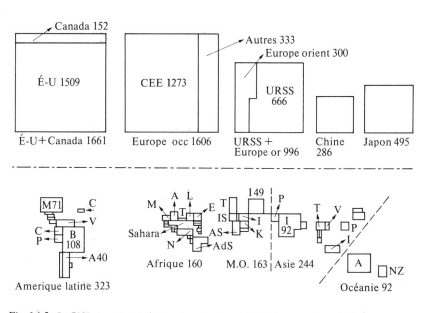

Fig. 14.5. Le PNB dans les 5 régions dominantes et les 114 autres pays (10^9 $).

consolidé mondial de 16,3 milliards de $, soit 8 fois plus que son plus proche concurrent Burroughs, et des profits nets de 2,4 milliards de $ après plus d'un milliard de $ de moyens financiers consacrés à la Recherche-Développement au sein du Groupe. IBM est en outre partiellement propriétaire du parc des machines installées, car la règle de ce métier n'est pas la vente mais la location ou le leasing; de ce fait, le cash flow atteint des montants considérables. En 1976 le cash flow d'IBM a représenté 4,04 milliards de $, soit plus de 2 fois le chiffre d'affaires de

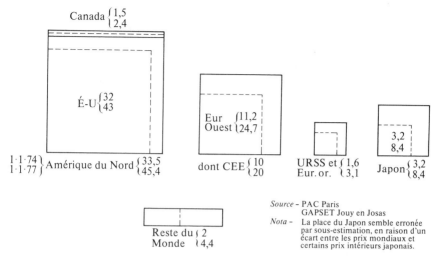

Fig. 14.6. Valeur approximative (minisystèmes exclus) du parc des calculateurs installés exprimée en milliards de \$ respectivement au 1.1.74 et au 1.1.77. *Nota*: la place du Japon semble erronée par sous-estimation, en raison d'un écart entre les prix mondiaux et certains prix intérieurs japonais' [*Source*: PAC Paris; GAPSET Jouy en Josas.]

Tableau 14.1 Les principales firmes mondiales d'informatique—Chiffres d'affaires en millions de \$ réalisés en 1976

		Chiffre d'affaires
Etat-Unis	IBM	16 300
	Burroughs	1 630
	Univac	1 430
	HIS	910[1]
	CDC	1 330
	NCR	1 100
	DEC	740[2]
Japan	Fujitsu–Hitachi	1 410
	Nec–Toshiba	640
	Mitsubishi–[3] Oki	290
Europe	CII–HB	660[1]
	ICL	540
	Siemens	530

[1] Le chiffre d'affaires consolidé du Groupe HIS–CII–HB est voisin de 1 430 millions de \$.

[2] La croissance de DEC, comme d'ailleurs celle des sociétés japonaises est supérieure à 30% l'an au cours des 3 dernières années alors qu'elle est inférieure à 15% pour les autres.

[3] Cette présentation ne tient pas compte d'une restructuration en cours au Japon, dans laquelle Oki quitterait la grande informatique et Mitsubishi rejoindrait Fujitsu–Hitachi.

son principal concurrent. Le poids économique de la société peut se comparer avantageusement à celui d'une nation comme la Finlande ou l'Egypte (40 millions d'habitants) si l'on retient pour critères représentatifs respectivement le chiffre d'affaires et la Production Nationale Brute, mais l'étalement géographique de la Compagnie lui confère un poids politique à certains égards plus important que celui de la plupart des États de taille moyenne.

Si l'on essaie d'imaginer les péripéties qui peuvent intervenir dans les 10 prochaines années, rien n'annonce que cette position de force puisse être menacée par le jeu de la concurrence commerciale normale. Au contraire, IBM est elle en train de conquérir de nouvelles places dans les domaines pleins de promesses des télécommunications et des minisystèmes.

Ce phénomène de cristallisation par gigantisme est l'une des caractéristiques de l'avènement des «technologies de l'information». Nous retrouvons la même tendance si nous analysons la structuration de l'industrie des circuits intégrés et des microprocesseurs où tout annonce une concentration des productions entre 5 à 8 producteurs mondiaux au plus et nous assistons peut-être à la naissance d'une tendance identique en matière de fournitures d'informations de nature économique, scientifique ou technique à partir de grandes banques de données internationales.

Pour les partisans de la libre entreprise, il y a là matière à amples débats. Est-il conforme à l'éthique du libéralisme de laisser se constituer des positions de forces invulnérables à la concurrence? Nous avons connu, entre les deux guerres mondiales, la recherche de solutions de puissance au sein des États-Nations et leur prédilection pour les formules autarciques; cette tendance caractérise encore les économies des pays socialistes. Assistons-nous à la naissance de nouvelles forces supra-nationales assez puissantes et organisées pour faire pièce aux gouvernements, en tout cas assez fortes pour briser les lois de l'essai et de la sélection par le marché? Nul domaine plus que les technologies de l'information ne pose ce problème avec une telle acuité.

Une ample littérature est disponible sur ce sujet, notamment pour commenter les différents procès intentés à IBM au titre de la loi anti-trust. Il semble que le gouvernement américain soit lui-même quelque peu désemparé par ce problème et aucune doctrine n'a été réellement énoncée. On peut supposer qu'au moins implicitement, le raisonnement suivant est avancé: d'une part, l'industrie informatique américaine, par sa potition dominante, ne concourt-elle pas à consolider la supériorité économique des États-Unis? d'autre part, couper en morceaux IBM ne conduirait-il pas à des situations de concurrence encore plus difficiles pour les autres compétiteurs?

On aurait tort toutefois de focaliser l'attention sur IBM. La même structuration en oligopoles mondiaux apparaît en microprocesseurs et en télécommunications.

Les stratégies concurrentielles – bases de la construction de la future division internationale du travail.

Résumons ce qui vient d'être dit: des courants violents d'innovations, de nature souvent imprévisible, font naître des activités industrielles et commerciales nouvelles dans l'immense domaine des «technologies de l'information». Ces courants pourraient engendrer une sorte d'anarchie mais ils se heurtent à des forces structurantes que sont les grandes sociétés de production et de commercialisation dont la puissance est assez grande pour exercer un contrôle efficace du marché.

Quelles sont alors les stratégies des concurrents et comment de nouveaux venus, entreprises créées à l'initative de leurs propriétaires ou par la volonté des gouvernements, peuvent-ils trouver leur place?

En dépit de sa simplification abusive, on peut classer les stratégies en 4 catégories:

— les concurrents qui tendent à exploiter l'étrangeté et la violence du courant d'innovation;
— ceux qui accompagnent ou profitent des positions acquises par IBM;
— les soutiens volontaristes des gouvernements en vue de créer des industries nationales indépendantes;
— les politiques globales d'avenir.

Pour bien comprendre la distribution géographique des efforts, il faut avoir présent à l'esprit le fait que l'informatique, c'est encore aujourd'hui, essentiellement les États-Unis comme le montre le tableau suivant:

1950	Apparition du premier ordinateur aux États-Unis.
1960	95% des 800 millions de $ d'ordinateurs livrés dans le Monde ont une origine américaine.
1975	10 milliards de $ livrés dont 85% américains.

Cette étude n'a pas pour objet les stratégies industrielles; nous devrons donc être très brefs, mais les conséquences sociologiques ne doivent pas être sous-estimées car il s'agit d'un facteur essentiel dans la division internationale du travail.

1. La création et le développement d'entreprises basés sur l'explosion innovatrice est un phénomène particulièrement significatif de la fécondité américaine.[1] De nouvelles technologies ou de nouvelles familles d'applications ouvrent des créneaux que ne peuvent pas couvrir, en raison de leur inertie ou par le jeu même de leurs priorités stratégiques, les grands oligopoles. Une équipe de chercheurs animée par la foi dans la validité de ses résultats, pénètre dans ces créneaux et en exploite toutes les richesses. Parfois la réussite est considérable et s'étend au plan mondial.[2]

2. L'allonifagerement d'IBM par l'adoption de fournitures identiques ou de services complimentaires adaptés est l'équivalent des phenomènes classiques en biologie de mionetisme et de symbiose[3].

3. Le soutien des gouvernements à leur industrie informatique en vue de créer un pôle de production capable d'équilibrer la puissance d'IBM ou, au moins, de donner des paramètres supplémentaires d'indépendance, est un phénomène spécifique des nations les plus puissantes de la Communauté Européenne : Royaume Uni, République Fédérale d'Allemagne, France. Chose curieuse pour des nations qui décidaient par ailleurs de se lier par le Traité de Rome, les concentrations industrielles se sont

---

[1] La fécondité du milieu nord-américain est le fruit d'une circonstance doublement favorable, la réceptivité et la sélectivité du marché immense et avide de novation d'une part, la structure mobile de la recherche scientifique et technique aux États-Unis d'autre part. Par opposition, l'Europe apparaît comme peu réceptive, exagérément divisée dans ses cloisonnements nationaux et dotée de structures de recherches mal adaptées dans leur rigidité.

[2] Tel est le cas, notamment, de Digital Equipment C°dont la progression annuelle a dépassé 30% en moyenne dans la période 1974–6 alors qu'IBM et les constructeurs d'ordinateurs de gestion ne dépassaient pas 15%.

faites par cristallisations à l'intérieur de chacun des États. Beaucoup de tentatives ont été entamées ou poursuivies pour élever les expériences à l'échelle de la Communauté, mais elles ont toutes tourné court, soit en cours de négociation, soit après une courte période de lancement. Lorsque pourra être faite avec la sérénité qu'autorise le recul du temps, l'histoire de l'évolution technologique de l'Europe pour la période que nous vivons, il est probable que l'on s'interrogera sur les raisons qui ont poussé les particularismes nationaux à s'enfermer dans des dimensions de marché intérieur, évidemment trop étroit, sur le pourquoi d'une politique industrielle qui n'a guère favorisé la stratégie des créneaux d'innovation et, plus encore, sur l'absence d'un pôle européen pour la conception et la production de grande série des circuits intégrés et des microprocesseurs qui constituent la pierre angulaire de toutes ces industries.

4. L'exemple d'une stratégie basée sur une politique globale d'avenir nous est fourni par le Japon.

Une étude de JACUDI [4] montrait dès 1973 que le Japon, apuvre en énergie et en matières premières, devait porter son effort principal sur les «technologies de l'information» faiblement consommatrices en produits de base, caractérisées par une forte valeur ajoutée en main d'œuvre qualifiée et catalyseurs d'efficacité pour la productivité des industries et des services. Bien que les proposition de financement du JACUDI n'aient pas été retenues dans leur totalité par le gouvernement japonais, l'industrie nationale s'est développée dans le sens préconiseé, notamment en ouvrant des marchés de consommation de masse pour les composants électroniques et les microprocesseurs par la conquête de positions prédominants en horlogerie électronique et en calculatrices de poche. Simultanément, s'est mise en place au Japon, à l'abri d'un protectionnisme de fait singulièrement efficace, une industrie des ordinateurs moyens et gros, prête aujourd'hui à exporter.

Dans une certaine mesure, le Plan Calcul français a été, lui aussi, l'expression d'une stratégie globale et l'une de ses retombées les plus efficaces a été la naissance d'une industrie de logiciel qui, par l'importance numérique de ses agents et par son chiffre d'affaires, se place au 2ème rang dans le monde.

6. Dans la division internationale du travail, un phénomène nouveau est apparu au cours des 15 dernières années : l'utilisation de la main d'œuvre du Sud-Est asiatique pour la fabrication des composants et pour le montage d'appareils électroniques de très grande diffusion. Hong-Kong, Taiwan, Singapour, la Coree du Sud fournissent une main d'œuvre féminine particulièrement habile et un encadrement intelligent et rapidement qualifié à des salaires et pour des coûts sociaux extraordinairement faibles. En règle générale, la conception des produits fabriqués, le savoir-faire et les machines sont importés à partir des États-Unis ou du Japon qui fournissent également les produits de base très élaborés, tel que le silicium traité par diffusions, épitaxies et oxydation. En outre les productions sont distribuées par des chaînes de commercialisation contrôlées par les pays les plus industrialisés. Cet ensemble d'activités de sous-traitance établit de nouveaux liens d'interdépendance dont la nature dissymétrique a fait parler de néo-colonialisme. Il est difficile de savoir où conduisent de tels rapports de forces. Il est certain que la proximité culturelle et linguistique de cette source de connaissances et l'expérience du succès

Tableau 14.2 Division internationale du travail—Emplois dans 4 pays du Sud-Est Asiatique dans les industries manufacturières de l'électronique (1975)

Pays	Population en millions d'habitants	Population active en millions	Actifs dans l'industrie	Actifs dans l'électronique et l'informatique [5]
Corée du Sud	34	12	2 500 000	100 000
Taiwan	16	5,7	1 900 000	130 000
Hong-Kong	4,3	1,9	600 000	70 000
Singapour	2,2	0,85	220 000	50 000
Total	56,5	20,45	5 220 000	350 000

A titre de comparaison, la France avait en 1975 à peu près le meme nombre d'habitants (52,7 Millions) et d'actifs (22,3 millions) dans l'industrie (5 900 000). Les industries manufacturières de l'électronique, de l'informatique et de la téléphonie, employaient 270 000 personnes dont 90 000 étaient affectées à des productions analogues (composants et petits materiels) que les 350 000 ci-dessus.

ne seront pas sans influencer la Chine Continentale dont la nouvelle politique semble être d'aller droit au but, vers l'acquisition de la maîtrise des techniques les plus avancées. La Corée du Sud, si elle était assurée de sa stabilité politique, pourrait devenir un nouveau Japon en tant que pôle de développement industriel sophistiqué. Inversement la tendance actuelle à confier à des automates perfectionnés les fabrications très délicates des circuits à très haut degré d'intégration qui sont les formules technologiques de l'avenir, fait perdre à la main d'œuvre une partie de son poids économique et peut conduire à rapatrier au sein des nations-mères une partie des sous-traitances précédemment exportées. Sur le plan de l'indépendance stratégique, le moment où la consommation américaine était dépendante des opérations de montage en Extrême Orient est maintenant dépassé.

Ainsi, l'explosion des technologies de l'information a provoqué une sorte de clivage entre les nations. Loin devant les autres, en toutes catégories, les États-Unis, ensuite les japonais, puissants par la continuité politique, le choix de productions à très grande diffusion commerciale et la protection de leur marché intérieur, enfin les européens de l'Ouest beaucoup plus importants par leur puissance de consommation que par leurs facultés innovatrices et le rang de leurs industries. Pour la main d'œuvre employée, une part importante en est située dans l'Asie du Sud-Est dont ces nouvelles industries sont l'un des moteurs du développement.

Quant à l'Union Soviétique et à l'Europe de l'Est, leur participation au concert international est faible et ils se comportent beaucoup plus comme des importateurs de savoir-faire et d'équipements que comme des concurrents potentiels.

Un phénomène économique unique : la réduction constante du rapport prix-performances.

La vie des produits industriels de grande série est marquée depuis le début de ce siécle par l'abaissement constant du rapport prix-performances, mais en aucun domaine la rapidité d'évolution et sa permanence sur une longue durée n'ont été aussi remarquables. Rien ne paraît devoir arrêter cette course dans l'avenir immédiat comme le montrent, à titre indicatif, les prévisions de la figure 14.7 ci-dessous.

Hausses
dépenses en personnel: +6%/an
Baisses:
communications: −11%/an
logique d'ordinateurs: −25%/an
mémoire d'ordinateurs: −40%/an

Fig. 14.7. Évolution previsible des coûts pendant la prochaine décennie. [*Source: Revue Datamation,* avril 1977, p. 62.]

Il y a là un gros phénomène sociologique car bien des applications s'ouvrent dès que devient rapidement récupérable le coût d'investissement d'une machine.

14.4. Les phénomènes sociologiques directement engendrés par les applications

Les domaines d'application des technologies nouvelles de l'information sont innombrables et l'on a dit, avec raison, que l'informatique était un fluide qui imprégnait toutes les activités humaines.

L'automatisation des industries

Les gains de productivité des industries manufacturières obtenus par approches successives grâce à l'automatisation sont connus. Aux solutions purement mécaniques, par engrenages ou dispositifs pneumatiques, se sont ajoutés des relais électriques puis, avec l'apparition des premiers calculateurs en temps réel, des conduites automatiques de procédé. Il faut savoir que l'introduction des microprocesseurs, à la fois extraordinairement économes et puissants, relance les possibilités d'automatisation, notamment en s'appuyant sur les récents progrès de la *robotique* qui met à la disposition de l'homme toute un série de capteurs nouveaux (peau artificielle, détecteurs visuels par laser et fibres de verre, reconnaissance de formes, etc.) dont les signaux peuvent être intelligemment combinés pour se substituer à l'intervention de l'homme. A l'IRIA, à côté de Versailles, nous expérimentons actuellement, dans le cadre d'un projet pilote désigné sous le terme de Spartacus, un ensemble de techniques qui, à leur aboutissement, devraient ouvrir la voie à la réinsertion sociale, dans le travail actif, de grands hanicapés physiques, ou encore, au travail par des robots isolés en milieux industriels hostiles tels que la présence de gaz chimiques dangereux, de rayonnements nucléaires, de hautes températures ou en milieu marin à grande profondeur.[1] A terme, on peut imaginer que des machines accompliront les tâches dangereuses, pénibles ou fastidieuses et feront ainsi disparaître la plupart des emplois d'ouvriers spécialisés sur montage en chaîne auxquels répugnent les habitants des pays développés. Cet objectif de l'âge d'or qui

[1] Sans parler des domaines spatial et militaire non cités ici car non directement rattachables à l'économie concurrentielle.

bannirait le travail manuel répétitif a été trop souvent annoncé pour qu'on le croit prochain. Microprocesseurs, miniordinateurs organisés en réseau, capteurs sensitifs, actionneurs en sont cependant les pièces disponibles qu'il s'agit seulement d'assembler judicieusement, sans nier cependant les difficultés qui restent à vaincre pour que cet assemblage soit opérationnel. Mais les 15 années à venir nous apporteront des progrès décisifs dans ce domaine. Trois conséquences sociales importantes pourraient en découler : le rapatriement d'industries récemment émigrées vers le Tiers Monde, pour des raisons de prix de main d'oeuvre (par exemple le textile); la diminution du nombre des travailleurs immigrés employés dans les pays industrialisés; la diminution globale des emplois manuels dans l'industrie.

La gestion des entreprises, des administrations et des services
La gestion automatique des multiples opérations d'écriture comptable qu'exigent la vie courante des entreprises, des administrations et des services est le domaine de prédilection pour l'application des ordinateurs. Au point où il en est arrivé, le développement de la machine appelle plusieurs réflexions d'ordre sociologique car autant de problèmes sont peut-être nés qu'il en a été résolus.

En premier lieu, l'homme serviteur d'un système informatique est-il heureux dans son travail? Dans bien des cas, banques, assurances, gros centres de comptabilité, la réponse est médiocre : la relation homme–machine a été traitée en privilégiant les facteurs mécaniques de la productivité et la technicité des solutions, mais les facteurs de fatigue et d'intérêt psychologique du personnel ont été, à l'origine au moins, insuffisamment pris en compte. Cette «déshumanisation du travail par la machine» est venue alimenter la protestation des syndicats d'ouvriers et d'employés.

Cette sensibilité aux conditions de travail paraît plus développée en Europe qu'aux États-Unis, particulièrement dans les pays latins. Elle appelle une correction et deux lignes de progrès se développent rapidement.

L'une est du ressort de l'ergonomie dans son sens classique; elle prévoit des dispositions des places de travail et des perfectionnements des matériels propres à diminuer la fatigue physique, notamment visuelle. L'autre est plus spécifique et l'on pourrait la désigner sous le terme *«d'ergonomie administrative»*. Elle consiste à mettre en application les principes de la psychologie cognitive. Le mécanisme mental des opérateurs est décrit et expliqué dans ses aspects observables et dans ses mécanismes intériorisés. L'équipe d'étude comprend au moins un ingénieur informaticien, un spécialiste de la psychologie cognitive et un groupe représentatif des personnels au profit desquels on recherche un enrichissement des tâches. Ce groupe de réflexion procède à la description systématique de l'interaction des sujets avec leur environnement, dans lequel le système informatique intervient comme élément de communication et, à partir de cette compréhension des phénomènes psychologie-machine, propose des solutions qui associent efficacité du travail et enrichissement des tâches. Par tâtonnement et en associant dès l'origine le personnel à la recherche des solutions, on parvient à des situations parfaitement tolérées qui peuvent même parfois être à l'origine d'une promotion culturelle des personnels. La méthode expérimentée en France pour réduire la fatigue et les risques d'erreur des contrôleurs du trafic aérien s'étend de proche en proche à de nombreux secteurs d'activité.

Mais l'automatisation des tâches de gestion est venue poser un second problème beaucoup plus fondamental que le précédent, celui *du partage de l'information au sein d'une entreprise ou d'une administration.* Rendue disponible par l'accès facile aux mémoires du système informatique, l'information doit-elle être *concentrée ou décentralisée*, doit-elle être l'apanage du pouvoir ou la matière première d'une large participation du personnel aux décisions mineures et majeures qui conditionnent l'avenir? La machine ici intervient comme révélateur de problèmes très anciens mais sa puissance joue à la manière d'un amplificateur pour rendre aiguë et visible au grand jour une question qui, en son absence, aurait pu rester dissimulée.

Différentes thèses s'opposent à ce sujet. Pour les uns, la mise en place d'un système automatisé de gestion conduit inexorablement à un renforcement du pouvoir hiérarchique, quelles que soient les techniques adoptées. Pour les autres, le choix de minisystèmes organisés en réseaux et dotés d'une autonomie locale large, assure la permanence des solutions décentralisées et participatives. Ce qu'il faut souligner, c'est que la technique ne comporte en elle-même aucun élément de fatalisme, mais qu'elle amplifie puissamment les effets des solutions retenues; l'information donc, si on en a la volonté, peut devenir l'instrument d'un pouvoir hiérarchique centralisé quasi dictatorial ou, si on en a le désir, peut fournir les élements techniques d'une politique de participation, largement décentralisée, en contribuant à l'apprentissage social du personnel. Les machines durcissent les conséquences des choix sociaux; les hommes restent libres et responsables devant ces choix.

La productivité des services

Alors que la productivité de l'agriculture et de l'industrie a fait des progrès prodigieux au cours des 30 dernières années, la productivité des services semble avoir quelque peu marqué le pas. Cette analyse ne rend compte que des apparences car, en fait, l'automatisation du traitement de l'information a combattu l'extraordinaire poussée de complexité qui s'est manifestée au cours de la même période; c'est l'équilibre entre les deux phénomènes qui conduit à un diagnostic pessimiste sur les progrès en matière de productivité des services.

Le développement des nouvelles technologies parmi lesquelles, notamment, les microprocesseurs, le télé-courrier, la télé-reprographie et le développement de nouveaux logiciels pour le traitement des textes et la classification automatique, la mémorisation sur microfilms ou sur mémoire magnétique, la généralisation de l'usage des cartes de crédits et de bien d'autres dispositifs à lecture automatique annoncent, comme très probable un gain, cette fois mesurable, de la productivité des services. Il n'est pas utopique de parler pour les 10 années à venir d'un objectif équivalent à la suppression de 20% des postes affectés au travail de bureau[1] ce qui revient à poser d'ores et déjà le problème des conséquences sociales sur le volume global des emplois.

[1] On rappelle que dans une société développée telle que la société américaine, 50% des emplois actifs sont affectés à des tâches de traitement d'information sur lesquels environ la moitié correspond à du travail de bureau. Un gain de productivité de 20% par la «bureautique» («office automation») équivaudrait donc à la suppression de 5% des emplois actifs.

Technologies de l'information et emploi

Le chômage est l'un des plus graves problèmes que rencontrent les pays industrialisés et les progrès de productivité qu'annoncent la robotique et la «bureautique» («office-automation») ne sont pas de nature à simplifier la donnée sociale. L'inquiétude provient surtout du degré d'incertitude dans lequel nous nous trouvons quant à l'étalement dans le temps et à l'étendue des gains de productivité. Car, s'il reste beaucoup d'inconnu sur la maîtrise des techniques et sur l'étrangeté de l'innovation technologique à venir, il en demeure davantage encore dans le domaine fondamental de la capacité à financer les investissements nouveaux et leurs frais de développement et dans le domaine très sensible du comportement psychologique des personnels touchés dans leurs habitudes de travail et menacés dans leurs carrières.

Si nous jugeons par extrapolation, en prolongeant les attitudes observées au cours des 20 dernières années, il est probable que rien aux États-Unis n'arrêtera l'esprit d'initiative, ni au Japon la politique d'une construction volontariste ardemment tournée vers l'avenir, mais on peut craindre que l'Europe, à la fois s'empêtre dans ses contradictions sociales, et connaisse des difficultés de financement.

Il paraît impossible de freiner ce mouvement vers la réduction de la main d'œuvre par gain de productivité dans les services, même si l'on considérait qu'il est contraire à une heureuse politique du plein emploi par exemple en Europe. Car il n'y a pas d'exemple qu'une technique donnant des armes pour un progrès de compétitivé soit arrêtée dans sa course. Ces gains de productivité conditionneront d'une manière très aigüe la capacité concurrentielle des différentes économies en position de rivalité; toute avance ou tout retard retentira puissamment sur les équilibres commerciaux.

Il serait, par ailleurs, illusoire d'attendre de la création des emplois dans les industries constructives – électronique, informatique, télécommunications – la compensation aux suppressions d'activités entraînées par ces gains de productivité. Pour les pays industrialisés, l'avenir apportera donc, soit la relance du chômage, soit la naissance de nouveaux modèles de consommation, soit la ressource d'activités nouvelles dans l'accompagnement de décollage économique de certains Pays en Voie de Développement. Nous reviendrons sur ces questions dans la conclusion de cette étude.

14.5. Les conséquences indirectes du développement de l'Informatique : l'évolution des concepts

L'informatique n'agit pas seulement à la manière d'un outil de productivité. Dans de nombreux cas, elle intervient comme un modificateur de concepts et, à ce titre, peut se comporter à la manière d'un *catalyseur de changements sociaux*.

La *médecine* nous fournit un exemple particulièrement démonstratif. Lorsque l'ordinateur y est utilisé comme un outil d'aide au diagnostic, il oblige le médecin à distinguer ce qui, dans l'exercice de sa profession, est art par rapport à ce qui est science. L'approche scientifique s'appuie sur les analyses chimiques, physiques et biologiques que l'on peut faire sur le patient et résulte d'une claire description des effets et des causes, de la mesure des certitudes et des doutes. Cette approche est complétée par l'intuition du praticien faite d'éléments irrationnels et informulés accumulés dans son cerveau par l'expérience, mais non aisément descriptibles. L'intervention de l'ordinateur oblige à prendre une conscience claire de l'ensemble de ces opérations mentales autrefois intériorisées dans l'esprit du médecin. Un tel effort de description ne peut pas se faire seul, mais en équipe. Le développement d'un logiciel d'aide au diagnostic est très onéreux; il n'est économiquement viable que s'il peut être amorti sur des séries répétitives et donc s'il est étendu par généralisation à plusieurs équipes médicales. Les analyses elles-mêmes se transforment; d'indices simples pris en compte par une lecture directe comme la température ou la fréquence du pouls, les appareils dotés de procédés automatiques pour le dépouillement des résultats permettent des investigations sophistiquées. Ainsi le traitement automatique de l'information clinique conduit-il la médecine vers une profonde transformation. L'intuition demeure, mais elle confie à la logique tout ce qu'elle peut lui sous-traiter, la démarche scientifique s'impose, l'individualisme du chercheur isolé cède le pas au travail en équipe, la généralisation des solutions propage la connaissance et normalise les méthodes, les techniques d'apprentissage du métier se perfectionnent. On comprend les obstacles psychologiques qui s'opposent à un tel mouvement, mais il s'agit d'une sorte de résistance d'arrière-garde qui cédera peu à peu à la force du courant.

On peut dire qu'aucune activité humaine où intervient un processus intellectuel n'échappe à ce ferment de mutation qu'est l'informatique et nous devons limiter notre description à quelques cas particuliers.

Le progrès des *mathématiques* est fondamentalement relancé. En amont, elles fournissent les bases théoriques du traitement de l'information; latéralement, elles développent des procédés de calcul qui resteraient sans objet si les machines ne permettaient pas d'exécuter avec une rapidité stupéfiante des quantités prodigieuses d'opérations; en aval, elles interviennent par la modélisation dans l'étude des systèmes complexes économiques et sociaux. L'analyse automatique des données, la solution de problèmes d'ingénieurs par la méthode dite des éléments finis, les calculs d'optimisation, le concept généralisable de boucle de réaction, l'application des probabilités, la notion d'ensembles flous, la simulation d'événements instables, catastrophiques ou fluctuants, les prolongements vers la thermodynamique, ont des retombées philosophiques de portée considérable sur les idées de liberté, de relativité, de hiérarchie, d'instabilité. Et nous ne sommes qu'au début d'un phénomène qui ira s'amplifiant.

L'informatique apporte, par ailleurs, aux *sciences humaines et sociales*, restées longtemps du domaine de l'observation ou du discours, une technique intellectuelle qui les rapprochent des sciences exactes.

Les premiers bénéficiaires en sont évidemment les recherches en *statistiques* et en *économie*. Les langages de commande et de programmation nécessaires aux ordinateurs et les méthodes d'analyse que permet la puissance des machines ont relancé la *linguistique*. Les études historiques peuvent s'appuyer sur l'accumulation des renseignements jusqu'ici inexploitables, archivés chez les notaires ou dans les paroisses. Les comportements sociaux peuvent s'étudier par les sondages et comme on l'a vu précédemment, la psychologie confine à une technique d'ingénieur lorsqu'il s'agit d'étudier les relations intellectuelles et affectives qui s'établissent entre l'homme, la machine et l'environnement auquel s'appliquent leurs efforts.

14.6. Les conséquences sociales indirectes : information et pouvoir

Les nouvelles technologies de télécommunication, de concentration et de traitement amplifient l'acuité de l'antique problème des relations entre le pouvoir et l'information.

Le mythe de l'ordinateur

Il faut s'arrêter quelques instants sur le caractère subjectif des réactions de la plupart de nos contemporains. La machine est assimilée à un cerveau; elle possèderait une intelligence artificielle et apparaît, de ce fait, dotée d'un pouvoir magique. Cette attitude irrationnelle est d'autant plus développée que l'instrument est moins familier, qu'il n'a pas été fréquenté dans l'adolescence et les techniciens eux-mêmes convertis tardivement à l'informatique n'échappent pas à un certain sentiment de mystère devant les possibilités des systèmes informatiques; ils perdent une partie de leur sens critique devant certains modèles de prévisions démographiques, économiques et sociaux. C'est en jouant sur la dramatisation théâtrale que conférait le recours à un ordinateur que le Club de Rome a pu faire passer son premier message; le contenu revenait à dire qu'aucune croissance exponentielle ne pourrait se prolonger longtemps sans conséquence catastrophique – ce qui pour un observateur disposant d'une modeste culture mathématique est évident – mais seul le recours à un modèle traité sur machine pouvait faire prendre l'avertissement au sérieux par les mass média et retentir sur l'opinion publique . . .

Ce parfum de magie explique la place importante qu'occupent les calculateurs et les engins électroniques dans les romans de science-fiction. En quelques années s'est constituée une littérature abondante dans laquelle le calculateur est capable de dépasser les limites humaines; le plus souvent la relation entre la machine et la société est conflictuelle; très rarement elle est volorisante. Déjà Georges Orwell dans *1984* avait imaginé une société sans liberté, basée sur la surveillance constante des individus par micros et caméras. Ses successeurs ont fait mieux dans la peur en décrivant des situations dans lesquelles l'homme est soumis aux caprices possessifs de la machine.

Lorsque l'on parle des relations de l'Informatique et du pouvoir, il ne faut jamais oublier que l'on cotoie constamment le psychodrame. Aux éléments rationnels, les interlocuteurs ajoutent inconsciemment les projections passionnelles de leurs angoisses au sein d'une société trop riche de moyens techniques nouveaux qu'ils ont mal assimilés dans leur culture profonde.

Informatique et libertés individuelles

Dans la plupart des pays occidentaux d'importants débats sont engagés sur les dangers que serait susceptible d'apporter à la vie privée l'existence des fichiers de renseignements sur les personnes; il est impossible d'en résumer en quelques lignes la substance, mais il est nécessaire d'en souligner le caractère immature car les

interlocuteurs mélangent constamment constats de fait et produit d'une imagination angoissée. D'où naissent de nombreuses contradictions:[1]

— exigences simultanées d'un droit à l'accès aux informations et d'un droit à
conserver confidentiel ce qui vous concerne en tant que personne.
— souhait d'une meilleure justice fiscale ou sociale et refus d'être soumis à un
système de surveillance qui exclurait la fraude.
— volonté d'être pris en considération dans son identité personnelle et simultanément, souhait de rester dans un complet anonymat.
— désir de connaître facilement tout ce qui est inscrit dans les mémoires des
ordinateurs concernant sa personne et refus d'être marqué par un numéro
d'identification unique.

On ne peut que plaindre les législateurs qui auront à trouver la ligne de meilleur
passage entre ces différentes contradictions et on doit espérer qu'ils n'introduiront
pas à l'excès des rigidités artificielles. Dans cet état immature du problème, il est, en
effet, difficile de discerner entre les contraintes qui auraient des conséquences
bénéfiques et celles qui se comporteraient involontairement comme des freins à
l'assimilation culturelle du nouvel outil technologique.

Informatique et libertés collectives
Plus grave est probablement le problème de l'amplification de pouvoir que con-
féreraient les systèmes informatiques à ceux qui en detiendraient exclusivement le
contrôle. Pour essayer de comprendre cette question, il faut en examiner successive-
ment divers aspects.

(a) *La société rigide, le pouvoir des informaticiens*
La conception et la mise en place des grands systèmes de gestion mobilisent des
spécialistes nombreux et de haut niveau, employés sur une longue période de temps
et représentent des investissements très lourds en capitaux dans le logiciel et dans la
saisie des données. Il est possible, par exemple, de concevoir que l'ensemble de la
gestion des prestations sociales soit, dans un pays, placé sur un système unique
d'ordinateurs, organisé en réseau ou concentré, peu importe. Un tel système gérerait
les dossiers de quelques dizaines de millions d'administrés et accomplirait plusieurs
milliards d'opérations par an.

Supposons alors que le législateur, pour des raisons de politique sociale, veuille
modifier profondément le mécanisme des prestations, par exemple passer d'un
système de prélèvement au niveau des salaires et des entreprises à un système de
prélèvement fiscal lié à l'impôt sur le revenu afin de moduler les aides sociales selon
les ressources. Il est probable que les informaticiens répondront qu'ils ne savent pas
faire, qu'en tout cas la transformation coûtera un prix exhorbitant et exigera des
délais très longs. Il y aura donc transfert, *de facto*, du pouvoir d'évolution du niveau
politique au niveau opérationnel. Toutes proportions gardées, la même mésaventure

---

[1] Nous verrons plus loin d'autres contradictions notamment le désir d'être protege contre
la violence et le terrorisme et de s'opposer à la constitution de fichiers de police, le désir
d'observer des situations dans la durée et le «droit à l'oubli», etc.

peut arriver à une entreprise dont l'ensemble des mouvements comptables et la conduite des procédés seraient confiés à un système informatique très perfectionné; il serait difficile d'y introduire des modifications de structure qui bouleverseraient les circuits d'information.

Plus généralement, le passage sur ordinateur tend à rendre rigides les jeux entre les informations. Alors que le langage humain, accompagné de mimiques et d'intonations, laisse une grande marge d'interprétation aux interlocuteurs et ouvre donc constamment des voies implicites à des conflits qui devront être résolus, mais qui peuvent s'exprimer d'une manière feutrée, la quantification de l'information qu'impose l'ordinateur, peut conduire à supprimer tous les jeux dans les engrenages sociaux. Le problème est probablement plus grave pour les pays de droit romano-germanique, notamment latins, habitués à se référer à la lettre de la loi et à s'en remettre à une logique des événements, que pour les pays de common law où le pragmatisme est la règle dominante.

(b) *La concentration du pouvoir de prévoir*

S'il est vrai que «gouverner c'est prévoir», la combinaison de la capacité d'être informé et de modéliser l'avenir donnera à ceux qui la possèderont, une singulière supériorité de pouvoir. Or, bien que ces techniques soient encore dans l'enfance et marquées de nombreuses infirmités, il devient possible grâce à la concentration des données économiques, sociales, démographiques, scientifiques et techniques dans des «banques» et grâce à la modélisation mathématique placée sur ordinateur, d'interroger l'avenir en imaginant différents scénarios concernant l'allure des variations des phénomènes de base.

Les infirmités tiennent au fait que la plupart des statistiques internationales sont entachées d'erreurs et que les indicateurs, notamment pour la mesure des phéno-mènes économico-sociaux, ne sont pas pertinents. L'insuffisante maturité retentit sur la pauvreté des modèles mathématiques qui ignorent certaines interactions majeures et procèdent par simplifications abusives. Mais tout cela est perfectible et l'approche de certains problèmes aisément isolables conduit à des résultats, d'ores et déjà, utilisables.

On imagine quels facteurs de supériorité détiendront ceux qui concentreront de tels moyens et à quel point leurs interlocuteurs qui n'y auront pas accès, seront démunis d'arguments.

Or, les dernières années ont vu le développement d'un nouveau phénomène de concentration et de gigantisme au profit des États-Unis en matière de banques de données économiques ou encore scientifiques et techniques. Du colloque organisé en Septembre 1977 par l'OCDE à Vienne, sur les flux de données transfrontières, nous tirons les renseignements suivants:

— à partir de leurs bases nationales et grâce à leurs filiales à l'étranger, les États-Unis assurent déjà au moins 70% des flux d'information transfrontières et disposent, notamment, de deux réseaux très puissants et très commodes d'accès: Lockheed–Dialog et SDC, sans équivalents dans le reste du monde.[1]

[1] — à l'exception, mais seulement dans certaines matières, scientifiques, du réseau de l'Agence Spatiale Européenne.
— sans parler des grands réseaux d'informations économiques, tous américains notamment Data Resources Inc., DRI (Boston) et Chase Econometrics (filiale de la Chase Manhatan Bank – New York).

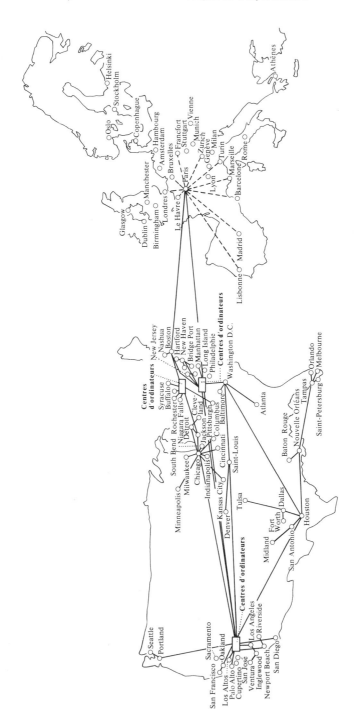

Source: La Documentation Française No. 16-1977

Fig. 14.8. Exemple de grand réseau international basé aux États-Unis : le réseau TYMSHARE par lequel sont accessibles depuis l'Europe les informations économiques distribuées par DRI (Boston). [*Source: La Documentation Française*, No. 16, 1977.]

- la part des États-Unis dans la production de bases de données de toute nature est déjà de l'ordre des 2/3. Le reste, en provenance des organismes nationaux et internationaux, y est collecté et transformé en termes informatiques très faciles d'accès.

- les États-Unis disposent dès maintenant de 600 banques de données accessibles (environ 100 milliards de caractères qui exigeraient 10 000 bandes magnétiques); l'Europe a accès à 60% d'entre elles.

- les facilités accordées par les systèmes de mémorisation et de télécommunication sont telles que certains européens préfèrent confier aux États-Unis la gestion des informations de leurs propres affaires. A titre anecdotique, la base de données des pompiers de Malmoe (Suède) capable de les renseigner, en temps réel, sur leurs sites d'intervention est à Cleveland (Ohio).

Tableau 14.3 Aide a la decision parlementaire–L'informatique au Capitole–Crédits et effectifs budgétaires du Congressional Research Service–CRS–entre 1971 et 1976.

Année	Crédits $	Emplois budgétaires
1971	5 653 000	363
1972	7 238 000	438
1973	9 155 000	524
1974	11 391 000	618
1975	13 367 000	703
1976	16 606 000	778
		(en fait 809)

Source: La Documentation Francaise n° 321 du 14 Octobre 1977 *La Révolution Documentaire aux Etats-Unis.*

Cela laisse présumer que l'information, aussi, appartient à cette catégorie de produits dont le prix est très rapidement décroissant lorsque la quantité traitée s'accroît. On sait que ces produits ont une tendance naturelle à se concentrer au sein d'oligopoles situés dans les sites géographiques où les marchés sont les plus importants. L'Europe découvre, une fois de plus, que ses divisions en États de trop petites dimensions économiques, la rendent inape à participer à la naissance de ces oligopoles. Lorsqu'il s'agissait d'énergie et de matières premières, la dépendance qui en résultait était essentiellement d'ordre économique; lorsqu'il s'agit d'information, la dépendance devient politique et culturelle.

(c) *Les relations entre le pouvoir exécutif et le parlement*
La poussée de la complexité des sociétés post-industrielles a bouleversé toutes les données de la vie politique. Le titulaire d'un mandat parlementaire, élu au suffrage universel, connaissait autrefois la plupart des problèmes de sa circonscription et il était capable de porter un jugement sur l'intérêt national des lois dont les projets étaient soumis à son vote par l'exécutif. Il était, par ses propres moyens, capable de proposer lui-même des projets de loi.

Aujourd'hui la complexité ne peut être abordée que par la spécialisation et en faisant appel à des études préalables dont la technicité décourage le non initié. *De facto*, le pouvoir de proposer les lois passe aux cabinets ministériels et aux groupes de réflexion qui leur sont associés. Et dans des cas de plus en plus nombreux, ces études seront faites sur ordinateur avec l'aide des banques de données dont le gouvernement pourrait avoir le contrôle exclusif.

De cette situation naissent deux revendications. L'une est la demande, par les parlementaires, d'avoir accès aux systèmes d'information utilisés par l'exécutif; mais qu'adviendrait-il alors du secret d'État, du droit d'initiative du gouvernement et des interprétations erronées, car un modèle n'est pas neutre? L'autre revendication est la constitution, à l'usage personnel du parlement, de systèmes informatiques capables de renseigner sur les effets directs et indirects des législations sur lesquelles les parlementaires sont consultés. C'est ainsi que le Congrès américain a constitué un gigantesque rassemblement documentaire qui emploie plus de 800 personnes parmi lesquelles des experts de la plus haute qualité; le budget 1977–78 ne sera pas éloigné des 20 millions de $. Doté d'un système informatique qui en fait l'une des plus importantes organisations contemporaines d'accès automatisé aux sources d'informations, il va recevoir, en outre, des moyens très importants adaptés à l'analyse prospective.

Ainsi l'informatique intervient comme facteur de changement dans l'exercice même de la démocratie. Il faudra encore beaucoup de temps pour voir comment s'établiront les nouvelles lignes de partage entre les différentes localisations du pouvoir, sans parler des échelons locaux ou décentralisés dans les régions ou les municipalités et des contre-pouvoirs que constituent les grandes entreprises multinationales, les syndicats de cadres, d'ouvriers et d'employés, les unions de consommateurs qui tous se dotent de leurs systèmes propres d'information.

(d) *Informatique, violence et terrorisme*

L'attitude de l'opinion publique à l'égard des fichiers de personnes peut évoluer sensiblement selon que seront contenues ou exacerbées les manifestations de violence et de terrorisme. Sur le sujet d'Informatique et Liberté, l'opinion publique est aujourd'hui indécise et les mass média l'ont surtout avertie des dangers qui menacent le secret de la vie privée; il suffirait que la violence diffuse se développe, que drogués, attentats, viols, hold-ups, prises d'otages s'accroissent pour que l'opinion bascule et que soit exigée la protection des biens et des personnes par toutes les mesures adéquates, en particulier de nature informatique, si telle devient la condition de la sécurité. Or, on doit considérer comme probable cette poursuite de la montée de la violence diffuse dans les pays industrialisés car elle est entretenue par les mass média, par l'urbanisation, par l'incompréhension de la complexité et par la protestation des pays pauvres en état de surpeuplement.

Eu égard à la violence, la vulnérabilité des systèmes informatiques n'a probablement pas fait l'objet de soins suffisants. La concentration des informations dans des grands systèmes localisés, les points de passage obligés des réseaux que sont les centraux de commutation, l'écoute clandestine des informations et l'insertion d'informations erronées, la fraude informatique constituent des points sensibles du système nerveux de la société post industrielle dont, dans la hâte des mises en service de moyens nouveaux, la protection n'a pas été assurée avec un coefficient de sécurité suffisant. En dehors des protections physiques des installations, bien des progrès techniques restent à faire pour garantir la confidentialité et l'inviolabilité des informations stockées ou en transit.

Si le glissement des mœurs conduisait à la surveillance des personnes par des moyens informatisés, alors la revendication pour que soit en même temps protégée la vie privée prendrait tout son sens. Mais on voit bien à quel point les droits de l'homme seraient menacés par la seule existence d'un instrument capable du dire à chaque instant tout sur chacun. Nous trouvons là une contradiction de plus dans le difficile problème qui lie l'Informatique avec la Liberté.

Indiquons au passage qu'en ce qui concerne la protection des biens, rien ne s'opposerait techniquement, pour les années à venir, à ce que disparaissent les signes monétaires et que toutes les transactions qui s'expriment en termes d'argent soient exécutées par cartes magnétiques accompagnées d'identifiants inviolables. L'opérateur pourrait, par exemple, se faire reconnaître d'une banque d'identification, à distance, par ses empreintes digitales ou par sa voix. L'extorsion de fonds ou les prises de rançon deviendraient alors impracticables. Mais presque rien de ce que ferait l'individu ne pourrait alors rester caché...

La pollution par l'excès d'information

Une boutade résume assez bien la menace de pollution intellectuelle que fait peser la disponibilité de quantités énormes d'informations. Parlant de la gestion comptable des entreprises, on a pu dire «qu'avant les ordinateurs on attendait deux mois pour obtenir un document qu'on lisait en trente minutes, alors qu'après l'ordinateur on obtient en trente minutes des documents que l'on met deux mois à déchiffrer.» Et il est bien vrai que l'homme peut être atteint dans sa capacité de jugement car la machine contribue généralement à rendre plus difficile une opération cérébrale qui est cependant essentielle : trier dans les informations que l'on reçoit celles que l'on doit s'empresser d'oublier, celles qu'il faut garder en mémoire pour un message ultérieur et celles qui sont immédiatement utiles. Mais il n'est pas évident que cette difficulté persistera pour nos enfants familiarisés dès l'adolescence à l'emploi des automates programmables; elle est peut-être attachée à ceux qui ont été éduqués dans la civilisation du livre et de l'écrit.

Il devient nécessaire, cependant, d'appeler une réflexion en profondeur, qui pourrait faire l'objet d'une recherche, sur la lutte contre la prolifération de l'information.

Le besoin d'une remise en ordre est particulièrement sensible dans le domaine des publications scientifiques et techniques. L'explosion de la documentation issue des laboratoires a fait l'objet d'un rapport très documenté du Prof. Anderla, édité en 1973 par l'OCDE, sous le titre *L'information en 1985*. Dans une approche globale nous pouvons considérer que les conclusions en restent, aujourd'hui, pour l'essentiel, valables. Elles conduisent à envisager une croissance annuelle moyenne voisine be 12,5% en volume qui, pour être traitée utilement, exigera une automatisation croissante, progressant à l'allure de 30% par an. Pour l'année 1985, la production de documents scientifiques atteindrait 6 à 7 fois le volume produit en 1970, ce qui reviendrait à consulter, pour se tenir au courant du travail de cette seule année, autant de documents scientifiques qu'il en avait été publié depuis la Renaissance jusqu'au milieu des années 1960. En conséquence, on voit se constituer des équipes de documentalistes qui tendent à travailler en vase clos afin de prendre en compte l'information reçue sans s'interroger sur leur utilité à l'égard des chercheurs qui s'en nourrissent. Et dans le même temps, de jeunes chercheurs noient leur imagination sous le flot des informations qui les inondent sans être capables de séparer le bon grain de l'ivraie.

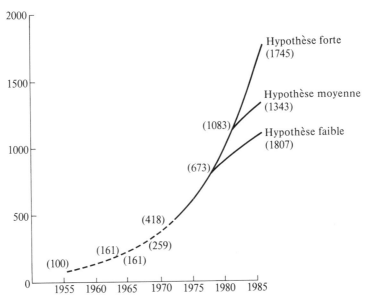

Fig. 14.9. Volume des informations scientifiques et techniques: tendance 1955–70 et projections 1979–85 par ajustement de la courbe exponentielle et de deux courbes logistiques, au taux 1955–75 de (l+r) = 110 (indice 1955 = 100). [*Cf.* rapport G. Anderla – OCDE – *L'information en 1985.*]

Le mirage du transfert du savoir-faire par l'information automatisée

L'idée qu'il suffirait d'être branché sur des canaux d'informations scientifiques et techniques automatisés pour progresser dans le développement et pour opérer des transferts de technologie est à la base du système UNISIST que l'UNESCO s'efforce de mettre en place. Il est difficile aux pays industrialisés d'expliquer aux pays en voie de développement qu'il s'agit en grande partie d'un mirage, car cette affirmation, en dépit de sa validité, serait interprêtée comme un refus de partager la connaissance. Mais il faut insister sur le fait que la capacité à communiquer de l'information dépend au moins autant du récepteur que de l'emetteur. Si le récepteur ne dispose pas des moyens de décoder le message, de juger ce qui est pour lui pertinent et de rejeter l'inadapté, l'information ne fait qu'ajouter une perturbation à une ignorance en donnant l'illusion, ce qui est très grave, qu'il y a eu enrichissement dans la connaissance. Cette remarque soulève le problème fondamental des relations entre les nouvelles technologies de l'Information et les valeurs culturelles que nous examinerons en guise de conclusion.

14.7. Technologies de l'information et valeurs culturelles

Si l'on résume en quelques lignes le constat qui vient d'être fait sur les conséquences du développement des technologies nouvelles de l'information, force est de conclure qu'il s'agit, non d'un problème technique ou économique, mais d'une question de civilisation.

Voici, en effet, un faisceau cohérent de technologies encore presque à l'état naissant qui, en moins de cinquante ans mais tout particulièrement au cours des 20 dernières années correspondant à l'anniversaire aujourd'hui fêté par l'OTAN, passe d'une contribution négligeable du Produit National Brut à plus de 5%, bouscule le vieil équilibre des monopoles d'État des Postes et Télécommunications, creuse dramatiquement l'écart entre pays développés et pays en voie de développement, cristallise la capacité industrielle autour d'oligopoles américains, pose un problème d'éthique quant à l'exercice de la concurrence à l'intérieur de l'économie libérale, apporte des progrès décisifs en matière de productivité des industries et annonce une potentialité, à la fois prometteuse et embarrassante, de réduction des emplois dans les activités tertiaires. En outre, cette mutation des technologies oblige à prendre en compte le problème de la répartition du pouvoir dans l'entreprise, retentit sur la déshumanisation du travail, tout en proposant des moyens de progrès pour l'enrichissement des tâches, modifie les attitudes mentales des chercheurs et des praticiens en médecine, en sciences humaines, en mathématiques et dans bien d'autres disciplines où la conception peut être assistée par les ordinateurs.

La matière traitée, l'information, ne peut pas être séparée de l'instrument et voici que, dans un climat empreint de passion et de mystère dû à l'inassimilation d'une technique par une culture, se mettent en mouvement les structures de pouvoirs et que sont mis en question les concepts traditionnels de liberté et de responsabilité. Chemin faisant, nous avons vu que l'information elle-même tendait à se concentrer dans de grandes bases de données et des modèles prévisionnels connectés à des réseaux de télécommunication et que, là encore, se formait une structure en oligopoles essentiellement basés aux États-Unis d'Amérique.

Dans la mesure où la puissance politique n'est plus contrôlée aujourd'hui par l'occupation militaire des territoires, mais par l'efficacité technique et économique, la division internationale des capacités en matière de technologies nouvelles de l'information et en matière de contrôle de la circulation de l'information, définit de nouveaux rapports de forces et retentit directement sur les situations d'indépendance, d'interdépendance ou de domination des États-Nations. Bien davantage encore on peut parler d'une mutation au sens où les biologistes l'entendent pour décrire l'évolution des êtres vivants. L'homme n'est pas muté génétiquement, mais il est muté dans son comportement de groupe et ses réactions sociales deviennent radicalement différentes selon qu'il appartient déjà à la «civilisation de l'information» ou qu'il reste attardé aux âges pré-industriel ou industriel. Ce phénomène doit être regardé comme beaucoup plus important que la supériorité militaire conférée par la disposition d'armes perfectionnées grâce à l'électronique et aux calculateurs, car il s'agit d'un clivage entre une adaptation culturelle à progresser dans les activités de la connaissance et une inaptitude du système de valeurs intellectuelles à faire un bon usage des moyens technologiques nouveaux.

Nous avons vu dans les pages qui précèdent, combien différaient les comportements de l'Amérique du Nord, du Japon, de l'Europe et des pays socialistes. Il faut s'attacher à comprendre ces différences car elles sont le signe d'une mobilisation de valeurs culturelles divergentes.

Aux États-Unis, en dépit de l'existence de structures ayant le caractère d'oligopole ou de monopole, c'est le grouillement de la vie créative par un processus qui reste, pour l'essentiel, de nature darwinienne. Il ne semble pas que l'on se pose de questions sur l'utilité sociale ou culturelle du mouvement de croissance explosive des nouvelles technologies. Elles naissent dans les laboratoires, dans une large mesure au hasard des essais scientifiques et techniques; elles sont transférées sur le marché pour y subir leurs essais de sélection grâce à la vie interstitielle intense qui existe dans les Universités et l'Industrie. Le processus de sélection est largement accéléré et amplifié par l'intervention des contrats de développement passés par les organismes chargés des grands objectifs nationaux comme l'Espace et l'Armement. Les succès sont exploités internationalement. Il s'agit, d'un développement, en quelque sorte, *sui generis*; la société américaine «engendre» la «civilisation de l'information» grâce à son esprit d'entreprise, au goût de ses populations pour l'innovation et à la sélectivité de son immense marché.

Le Japon s'est trouvé jusqu'ici placé devant le fait accompli américain, mais il a bandé sa volonté sur le désir culturel de participer à la création de l'avenir. Singulièrement abrité par le protectionnisme de fait que lui donne l'isolement de sa langue et de ses traditions, il a rapidement développé une consommation intérieure pour les produits de masse, ce qui lui a fourni la base nécessaire à la technologie des composants et des microprocesseurs. Il aborde maintenant la phase de consolidation par la conquête des marchés extérieurs.

La situation de l'Europe de l'Ouest est caractérisée par l'homogénéité dans les attitudes de retard; elle ne s'explique pas seulement par un phénomène d'échelle dû aux divisions de son marché, mais par des raisons culturelles. L'Europe comprend intellectuellement l'importance de l'évolution liée à l'apparition des nouvelles technologies de l'information mais, en quelque sorte, elle les refuse sentimentalement, viscéralement, car elle a peur de leurs conséquences sociales et politiques. L'Europe manifeste ainsi son attachement au passé; sa réaction devant les conséquences du nouvel outil technologique témoigne de ses difficultés à accepter le changement. Il serait erroné de chercher à localiser des responsabilités; c'est tout le corps social qui réagit dans un sens négatif. Les mass média amplifient la perception des dangers sans en montrer les aspects bénéfiques; les syndicats ouvriers organisent l'immobilité des structures en refusant de remettre en cause certains avantages acquis dans le cadre de la civilisation pré-informatique; les gouvernements sont constamment obligés de se défendre lorsqu'ils interviennent par des aides cependant notoirement insuffisantes eu égard à l'enjeu et comparées à celles consenties, sans obstacle, pour maintenir en survie des activités dépassées; les chefs d'entreprise préfèrent contrôler leurs risques en important procédés techniques et équipements; les hommes de laboratoires sont lents à se reconvertir vers les nouveaux domaines et refusent de sortir du cocon protecteur que leur fournissent les institutions publiques; il ne leur est d'ailleurs proposé presqu'aucune structure opérationnelle pour la vie interstitielle entre la recherche publique et l'industrie. Et cependant, contradictions éclatantes, la compétence scientifique et technique est presque partout présente et, le besoin d'utiliser les outils nouveaux s'exprime par le succès d'une technique presque exclusivement importée ou inspirée par les modèles étrangers.

Le retard des pays du bloc socialiste en techniques informatiques et électroniques mérite aussi qu'on y porte attention, car si pour Lenine le communisme c'était «le socialisme plus l'électricité», le désir de dominer la planification aurait dû conduire à rajeunir cette définition par le nouveau slogan «l'organisation plus l'informatique», ce dont témoigne d'ailleurs l'intérêt que portent les dirigeants soviétiques à des organismes internationaux tels que l'IIASA.[1] Or, la contribution des pays socialistes à la naissance des nouvelles technologies de l'information est quasi négligeable. L'explication réside probablement dans le fait que la planification s'oppose à l'éclosion de l'imprévisible et que l'on ne saurait concevoir, en l'absence de la sélection par le marché, l'établissement d'un courant d'innovations étrange et impétueux caractérisé par des sélections de nature darwinienne.

Si telle est l'explication, elle s'étend bien au delà de l'aptitude à engendrer de la technologie, car les nouveaux concepts issus de la théorie de l'information, de l'analyse de système et de la thermodynamique des fluctuations montrent que la croissance de la complexité ne peut s'absorber que par l'exfoliation de structures vieillies et par l'apparition de nouvelles structures. Ces nouvelles structures dans leur nature et dans leur date de naissance, ne font pas partie des événements prévisibles. De tels événements, parmi les plus fondamentaux pour notre vie sociale, échapperaient donc par nature au contrôle par la planification. L'ordre, en quelque sorte, ne pourrait se nourrir qu'à partir d'un certain désordre. Les phénomènes que nous avons examinés dans cette étude nous conduisent donc à reformuler la notion de liberté.

La liberté ne serait plus seulement réductible à une philosophie des droits de l'homme; elle apparaîtrait comme la condition nécessaire au déploiement de l'innovation technologique et aux adaptations structurelles qui en découlent.

Il n'est pas acceptable pour l'homme de se laisser emporter par le courant d'une technologie sans influer sur ses orientations et sans chercher à la canaliser. Dans la mesure où l'Europe tend à s'opposer au mouvement, l'apport de sa contestation peut constituer une force utile pour maîtriser en partie l'événement et pour optimiser ses conséquences sur la société.

Plus sensible que les États-Unis aux phénomènes sociaux en raison de l'attitude de ses syndicats ouvriers et plus préoccupée peut-être de la qualité de la vie au travail, l'Europe semble pourvue de dons particuliers pour proposer des formes

---

[1] IIASA – International Institute for Applied System Analysis – Laxembourg près Vienne (Autriche).

d'adaptation des nouvelles technologies de l'information aux impératifs culturels. C'est pourquoi l'on pourrait s'attendre que certains pays européens s'installent en leaders dans plusieurs domaines d'études tels que l'ergonomie administrative, l'enrichissement des tâches, la participation aux responsabilités, grâce à l'aide des systèmes informatiques, les relations entre les problèmes de gains de productivité et d'emploi.

L'Europe pourrait également apporter la puissance de sa tradition juridique. Le Colloque OCDE de Vienne de Septembre 1977 sur les flux d'informations transfrontières a bien montré le clivage entre les tendances des européens, par les législations appropriées, à contrôler le droit à la propriété et à la circulation de l'information et à interdire sa concentration et son traitement dans des sanctuaires informatiques qui joueraient un rôle analogue à celui des paradis fiscaux, par opposition aux tendances américaines plus proches, me semble-t-il, d'un esprit de laisser faire – laisser passer. L'Europe est-elle trop conservatrice, l'Amérique trop aventureuse? il est difficile aujourd'hui de trancher tant le sujet manque de maturité, mais l'antagonisme des positions peut être valablement considéré comme une contribution à une meilleure compréhension des situations et à la recherche de compromis utiles.

14.8. Vers une civilisation de la connaissance?

Les impacts que nous pouvons dès aujourd'hui recenser de la poussée de la complexité et de ses instruments de traitement que constituent les technologies de l'information entrainent-ils irrésistiblement les pays les plus développés vers une nouvelle forme de la civilisation, une «civilisation de l'information» ou, mieux encore vers une «civilisation de la connaissance»?

Beaucoup d'observateurs en parlent d'une manière suffisamment convaincante pour que l'utopie commence à prendre quelque consistance et devienne, peu à peu, un projet. D'importantes contributions ont été données sur ce sujet par le Prof. Parker de la Standford University et par le Prof. Ithiel de Sola Pool du M.I.T. lors du colloque organisé à Paris en Février 1975 sur Informatique et Télécommunications [6] rejoignant les thèses du Prof. Porat déjà cité. L'importance des moyens installés, qui constituent à la fois le squelette et le système nerveux de cette nouvelle civilisation en cours d'enfantement aux États-Unis, est bien décrite par une récente publication de la Documentation Française intitulée *La Révolution Documentaire aux États-Unis* [7]

Cette civilisation de la connaissance proposerait une sorte de limitation de la consommation des biens matériels et de l'énergie à des niveaux proches de leurs niveaux actuels – éventuellement à des niveaux inférieurs si des contraintes de rareté s'exprimaient avec une nouvelle force. Mais la croissance de l'activité humaine ne s'arrêterait pas pour autant; elle se situerait dans une consommation de substitution qui serait celle des biens culturels dont le support est l'information et dont le catalyseur est la communication entre les hommes. Alors, il n'y aurait plus de limitation pour insuffisance de ressources primaires, sinon la saturation des cerveaux humains. En somme, l'humanité découvrirait que la culture est un bien fondamental, qu'elle est la source de jouissances quasi illimitées à la condition de ne plus être le privilège d'une élite et le développement culturel s'accomplirait un peu à la manière dont s'est accompli le développement matériel de masse après que H. Ford ait découvert que ses ouvriers pouvaient devenir ses clients à condition qu'ils disposent d'une capacité économique suffisante.

Plusieurs conditions nécessaires à l'avènement de cette civilisation de la connaissance paraissent également sur le point d'être remplies, notamment les deux suivantes:

(a) Pour se cultiver et jouir des possibilités que donnent la pratique d'arts majeurs ou mineurs, l'exercice personnel d'un artisanat, l'acquisition de nouvelles connaissances ou le perfectionnement, il faut disposer de temps libre. . . Mais les gains de productivité, en provoquant du chômage, ne montrent-ils pas que le temps libre devient une denrée qu'il faut savoir distribuer et employer? (cf. Fig. 14.10 ci-après).

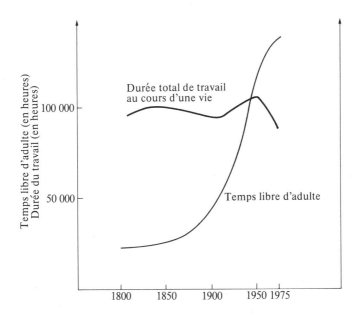

Fig. 14.10. Étude de la variation de la durée totale du travail et de la durée du temps libre pour un individu moyen en France entre 1800 et 1975. La croissance du temps libre pour la moyenne des francais au cours du dernier demi-siècle est un phénomène caractéristique de l'évolution de la société des pays développés (le temps libre est obtenu après déduction du temps physiologique, de travail, de transport, et d'enfance). Au cours de la même periode, l'espérance moyenne de vie est passee de 36 ans à 72 ans, mais en relatif, le temps libre est passe de 8% à 26%. [*Source*: INED–IRIA.]

(b) Pour que la culture devienne un produit de consommation de masse, il faut pouvoir la distribuer, provoquer une excitation des dons avant l'adolescence, poursuivre l'éducation tout au long de la vie. . . Mais les «technologies de l'information» ne fournissent-elles pas des possibilités énormes par des méthodes nouvelles d'enseignement assisté par moyens audio-visuels et informatiques, par la transmission de la connaissance à distance, par des jeux éducatifs et des outils d'aide à l'auto-contrôle des performances?

Ainsi l'évolution technologique dont nous avons signalé la violence et l'étrangete conduirait l'humanité vers l'avènement de l'esprit? Cette perspective contraste singulièrement avec le sentiment d'inquiétude et de pessimisme qui imprègne notre époque. Pouvons-nous l'expliquer? Un sociologue ou un philosophe, ce que je ne suis pas, nous dirait peut-être que l'événement arrive trop tôt, pour un homme impréparé, pour qui agressivité et jouissance sont encore essentiellement rattachées à la notion de biens matériels. Au reste, cette civilisation de la connaissance n'a même pas de support idéologique et les religions n'y ont pas délégué leurs phophètes, sauf pour parler de l'amour dans le désert de peuples sourds. . .

En revanche les obstacles concrets que peut prévoir l'ingénieur ou l'économiste ont été, pour beaucoup d'entre eux, identifiés au cours de cette étude: influences sur l'exercice du pouvoir et de la démocratie de la capacité de concentrer abusivement l'information, tendance naturelle à la constitution de monopoles ou d'oligopoles encombrants, évolution vers la rigidité des systèmes sociaux gérés par des systèmes informatiques complexes. On doit y ajouter la perturbation à prévoir dans la notion de propriété et dans la relation:

$$(argent) = (pouvoir \ de \ consommation)$$

Cette relation suppose, en effet, que l'on peut s'entendre avec une approximation acceptable sur la «vérité des prix». Or, quel est le prix d'une information? N'est-il pas davantage dépendant du contexte — secret, fraîcheur, effet sur les informés, capacité de divulgation. . . — que de l'objet? Et comment se règleront les désordres monétaires et l'inflation lorsque plus de 50% des actifs manipuleront des symboles auxquels on saura très mal attribuer un coût?

Quant à la notion de propriété et au contrôle des richesses nationales, l'absence de brevets d'invention en matière de logiciels et l'incapacité des douanes à. contrôler les flux de données transfrontières démontrent que nous pénétrons dans des terres inconnues.

La «civilisation de la connaissance» ne nous promet pas le bonheur paradisiaque et nous avons peur des difficultés qu'elle révèle. Peut-être avons-nous peur aussi de la période de transition, celle pendant laquelle il faudra abandonner la course à la croissance du PNB par augmentation des consommations matérielles pour commencer à faire croître la consommation culturelle. Car nous avons appris à gérer la phase que nous quittons et nous n'imaginons pas encore comment gérer la suivante?

Pour le bon fonctionnement d'une alliance comme celle qui est concrétisée par l'Organisation du Traité de l'Atlantique Nord, il me semble que deux sujets particuliers de préoccupation devraient retenir l'attention des responsables, le retard européen d'une part, l'écart qui s'est encore creusé avec les peuples en voie de développement d'autre part. Commençons par ce dernier problème car il agit comme une sorte de clé de voute.

Le Nord ne peut pas proposer au Sud, pour les 30 ans à venir, son modèle actuel de consommation. Le gaspillage mondial d'énergie et de matières premières, dans l'état actuel de nos connaissances techniques et de notre capacité de formation de capital, conduirait évidemment à un échec.[1] Cet échec, les peuples en voie de développement nous en rendraient responsables; il exciterait les ambitions de revanche, manifestées probablement par une violence diffuse incontrôlable. Mais si le Nord, freinant sa consommation de biens matériels, s'orientait vers une civilisation de l'information et de la connaissance, il proposerait alors un modèle généralisable aux 8 à 12 milliards d'hommes qui peupleront la terre dans 50 à 70 ans car ce modèle ne se heurterait pas à la rareté des ressources et à la finitude des richesses accessibles. Personne ne doit se faire d'illusion sur les difficultés à vaincre pour propager au sein des PVD les moyens d'une croissance culturelle, mais au moins un tel objectif ne dirigerait-il pas à priori l'humanité vers une voie sans issue.

Et c'est ici que le retard européen est préoccupant car, si le décrochement persiste, et tout montre qu'il s'accroît, jamais les États-Unis ne pourront, seuls, devenir les éducateurs du monde entier, ni résoudre pragmatiquement et théoriquement tous les obstacles qui s'annoncent sur la route. Il est des batailles qu'il faut savoir ne pas trop fortement gagner et si la victoire s'est offerte trop facilement, il faut accepter d'encourager ceux que l'on a distancés, à vous rejoindre. Ainsi se pose, me semble-t-il, les termes du dialogue entre les États-Unis et l'Europe en matière de compétition dans les technologies de l'information et dans leurs applications. En conséquence, il faudrait regarder avec faveur des mesures prises par les nations européennes et par la Communauté des 9 pour favoriser l'essor préférentiel de leurs industries et organiser la récupération de leur retard. Sous certains aspects, ce serait une sorte d'extension atlantique de la notion de loi anti-trust afin de donner toutes ses chances à l'émulation, à la sanction de l'essai par la sélection du marché dont le jeu est devenu trop inégal aujourd'hui.

Mais la première condition serait évidemment que sur ce domaine capital des technologies de l'information, le plus essentiel peut-être de tous par son impact sociétal, l'Europe veuille réellement prendre en main son destin. En toute honnêteté, je dois, sur ce point particulier, laisser le lecteur sur une interrogation?

14.9. Scénarios pour l'Europe

Il est, en effet, difficile aujourd'hui de prévoir comment l'Europe réagira. On peut, cependant, essayer de raisonner sur deux scénarios extrêmes qui constituent, en quelque sorte, les limites à l'intérieur desquelles se situeront les États Membres de la Communauté Européenne en présence du défi de la civilisation de la connaissance.

Le scénario du «retard contrôlé». L'Europe subit, elle ne participe pas à la conduite de l'Évolution.
— Les industriels choisissent la voie la plus sûre pour faire des profits immédiats et minimiser leurs risques. Ils maintiennent un potentiel de recherche juste suffisant

---

[1] Ceci est indépendant du fait que la priorité, pour les Peuples en Voie de Développement, continuera de se situer, évidemment, dans la satisfaction de certains besoins élémentaires – nutrition; vêtement; habitat; santé; éducation.

pour leur permettre de comprendre les innovations américaines et japonaises et un potentiel de développement apte à l'application d'une politique du «suiveur proche», c'est-à-dire du suiveur décalé dans le temps de 2 à 5 ans.

L'effort est alors orienté essentiellement vers l'assemblage de composants et de matériels pour une bonne partie importés et vers une bonne adaptation, aux cas particuliers du marché européen, des technologies inventées à l'extérieur, notamment par une forte activité en logiciels.

Afin de ne pas être excessivement débordés par les multinationales contrôlées par l'étranger, les industriels européens font effort, chacun pour leur compte, pour obtenir de leurs gouvernements nationaux le protectionisme d'une préférence d'achat au profit de leurs opérations d'assemblage, de distribution et de maintenance.

Il n'existe pas, en Europe, de source de conception liée à une production de composants et de microprocesseurs pour certaines applications de masse—calculatrices de poche, microcalculateurs programmables, électronique d'horlogerie et, en conséquence, les microprocesseurs employés en automatisation, en bureautique, en miniinformatique, en automobile sont importés. En revanche, des sources non compétitives, subventionnées par les gouvernements, subsistent pour l'électronique d'armement et les équipements nucléaires dont les performances et le degré de sophistication restent compétitifs.

— Chaque gouvernement prend isolément la charge des frais de la politique du «suiveur proche» de manière à préserver son indépendance politique (liberté de choix des solutions pour l'Armement, les Télécommunications, les programmes spatiaux, etc.) ou économique (exportation de biens d'équipement intégrant les technologies nouvelles).

Ces frais ne sont pas négligeables car ils supposent la préservation d'un potentiel de recherches suffisant pour constituer un pôle de compréhension et d'échanges, d'enseignements et de formation et l'entretien d'ateliers pilotes de circuits intégrés, de microprocesseurs et de mémoires largement déficitaire, sans parler des «préférences nationales» qui ne vont pas toujours sans pertes d'efficacité.

— Les différentes administrations des Télécommunications européennes jouent de leur monopole pour favoriser la politique de préférence d'achat et pour freiner certaines initiatives techniques qui ne leur conviendraient pas, en rattachant à leurs propres décisions le développement de réseaux d'ordinateurs et les systèmes de transmission de données et en pratiquant une politique autoritaire de tarification.

— Les États européens s'habituent à devenir clients des États-Unis pour l'accès aux grandes banques de données internationales concernant la prévision économique et les connaissances scientifiques et techniques. Ils essaient d'obtenir quelques contreparties de réciprocité en se constituant eux-mêmes en sources de données, mais ils importent les grands logiciels d'interrogation des bases de données dans les cas particuliers où ils ne les interrogent pas directement par télécommunications spatiales.

— Le retard sur les États-Unis et le Japon s'accroît pour tout ce qui concerne les applications Grand Public de l'Informatique — terminaux domestiques, jeux intellectuels, aide à l'enseignement.

— Les «technologies de l'information» sont ressenties comme le support d'une culture importée. Syndicats et mass média en stigmatisent les dangers et exercent une pression de retardement sur leur implantation de manière à repousser l'échéance des conséquences sur les réductions d'emplois et sur la modification des conditions de travail. Un clivage en deux secteurs apparaît; celui des industries qui veulent rester compétitives et automatisent tous leurs facteurs de production, y compris leur secteur tertiaire; celui des «activités protégées» par leurs appartenances directes ou indirectes à l'État ou par leurs attachements géographiques (Assurances, Sécurité Sociale, Santé, etc.) où l'on se hâte avec lenteur d'introduire des phénomènes socialement mal compris et mal tolérés.

– L'existence de la Communauté Européenne n'est pas mobilisée comme élément majeur d'une politique. Chaque État Membre procède à ses propres normalisations, établit indépendamment ses législations sur le droit de l'Information, par exemple lorsqu'il s'agit d'assigner des limites à l'emploi des solutions informatiques à l'égard des Libertés. On demande seulement à la Commission des Communautés Européennes d'assurer une coordination souple entre les politiques nationales et de prendre en compte quelques réalisations expérimentales.

Il ne faut pas avoir beaucoup d'imagination pour préssentir les dangers auxquels ce scénario expose l'Europe en matière de compétitivité globale et de retard d'intégration des technologies nouvelles dans sa civilisation.

Le scénario de l'offensive. L'Europe décide de participer activement à la création de la nouvelle société et à la nouvelle culture issue du développement des technologies de l'Information. Elle décide, en particulier, de s'appuyer sur sa dimension globale pour ouvrir réellement son vaste marché intérieur.

Cette offensive qui aurait pour objet de donner aux industries européennes le goût de revenir sur la vague la plus avancée de l'innovation, appellerait plusieurs décisions:

– un accroissement très notable des moyens accordés par les États à la R et D publique et industrielle. Pour sortir du retard, il faudrait passer pendant quelques années au delà du «seuil d'originalite», à définir par des études convenables, mais que je situerais à environ 3 fois le niveau moyen actuel dans la CEE. Il s'agit donc de mobiliser, en supplément, seulement quelques % du PNB et de déplacer en faveur de la Recherche-Développement dans les technologies de l'information une petite fraction de ce qui est consenti au profit de disciplines moins essentielles.

Cet effort financier devrait s'accompagner d'un effort sur les structures de manière à réinstaller une vie active dans le vide intersticiel qui s'est créé entre les instruments de la recherche, les industries, et les utilisateurs.

– une reconstitution du potentiel de conception et de production de masse pour les composants électroniques et les microprocesseurs, de manière à parvenir à l'existence d'au moins 2 producteurs européens de stature internationale. Cet objectif imposerait une relance du marché de consommation de masse par des productions intérieures d'horlogerie électronique, de microcalculatrices, de terminaux domestiques, de moyens informatiques d'enseignement pour le premier cycle et un large développement des applications des microprocesseurs et de l'électronique dans certains industries support, telles que l'automobile.

— la constitution d'un ou plusieurs centres européens pour la concentration et la diffusion, par moyens informatiques, de données de caractère économique, scientifique et technique, de manière à équilibrer les oligopoles américains par quelques oligopoles européens en matière de stockage et de distribution de l'information.

— une politique très active et diversifiée, non monolithique, des monopoles d'État des Télécommunications en vue de prendre des initiatives ou de favoriser des initiatives dans les secteurs privé ou semi-public pour la fourniture de moyens de transmissions, de réseaux et de services informatiques, ce qui paraît supposer également une intervention très énergique dans le domaine des télécommunications spatiales.

— un effort intense de formation des décideurs, des responsables des groupes de pression, des syndicalistes, des journalistes, afin de démystifier l'informatique et d'équilibrer les dangers que l'on imagine par les espérances que l'on peut attendre. Cet effort de formation appelle un vaste débat public sur «l'écologie de l'information», analogue au débat qui est nécessaire en matière d'énergie nucléaire. La familiarité avec les nouveaux outils devrait être acquise dès l'enfance par les jeux et l'enseignement.

— la préférence accordée aux sources de conception et de production européennes dans les politiques d'équipements, mais sans cloisonnement nationaux, de manière à relancer l'émulation intra-européenne.

— le financement préférentiel de l'étude et du lancement industriel d'applications nouvelles spécifiques des besoins européens : automatisation de l'industrie et des services, introduction dans les techniques de Santé, dans l'enseignment de post-formation ou de reconversion des adultes, dans la gestion des grandes administrations sociales, etc., secteurs pour lesquels les marchés européens correspondent à des caractéristiques structurelles ou culturelles différentes de celles que l'on observe aux États-Unis.

— la Commission des Communautés Européennes agissant par un organisme spécialisé approprié à la manière d'un catalyseur, pourrait jouer un grand rôle dans l'établissement de normes communes, la mise en place de réseaux, l'harmonisation des législations et le financement de certains opérations telles que celles liées à la constitution des grandes banques de données.

Ce scénario de l'offensive n'est évidemment pas hors de portée si on le compare, notamment, aux sacrifices consentis par chacun des pays membres pour sa Défense Nationale. Or, il s'agit bien, fondamentalement, d'une défense de la situation culturelle et économique des européens, c'est-à-dire des paramètres profonds de leur identité qui ne peuvent pas être dissociés de la Défense Nationale. Ce scénario ne se heurterait aujourd'hui ni à l'obstacle du manque de spécialistes, car l'humus intellectuel est constitué, ni à celui d'un marché artificiel : le marché européen existe et il serait encore plus demandeur s'il était décloisonné.

Il est impossible de dire aujourd'hui où se situera la ligne de partage réelle entre le «scénario du retard contrôlé» et le «scénario de l'offensive». Ce que j'ai seulement voulu souligner dans cette étude, c'est qu'il s'agit, pour l'opinion publique et pour les décideurs politiques, non d'un choix banal devant un événement sans lendemain, mais d'un choix de destin — glisser, insensiblement, vers l'inadaptation à la civilisation du XXIème siècle ou concourir à en définir les forces principales.

Quant à l'Alliance Atlantique, il s'agit pour elle d'une question d'équilibre ou de déséquilibre global. A cet égard, je ne peux, en guise de conclusion, que revenir sur la déclaration du Président Carter à Paris le 4 Janvier 1978 «L'Amérique» a-t-il dit, «a besoin d'une Europe forte». Parmi les facteurs prioritaires qui fixeront la force de l'Europe, figure la place qu'elle occupera dans les technologies de l'information et l'usage qu'elle saura en faire au service de la démocratie.

References

[1] Porat, M. *The Information economy,* These de Doctorat, Institute for Communication Research, Standford University, (Aout 1976).
[2] Les figures 3, 4, et 5 sont extraites d'une communication faite par M. Maurice Guernier au Club de Rome en Septembre 1977. La figure 5 résulte d'une étude confiée par l'IRIA à la société PAC (Paris).
[3] cf. Document OCDE *Impact des entreprises multinationales* Industrie des Ordinateurs et de l'Informatique (1977).
[4] JACUDI – Japan Computers Uses Development Institut (TOKYO Mai 1972) – Le JACUDI a été ultérieurement (1976) fusionné avec le JIPDEC (Japan Information Processing Development Center).
[5] Chiffres approximatifs – Sources: GAPSET/CESA Jouy en Josas et BIPE Neuilly sur Seine – estimation des erreurs dues aux imprécisions des nomenclatures ± 10%.
[6] cf. Document OCDE du 27 Janvier 1975 ref. DSTI/CUG/75. 1 et 75.4.
[7] La Documentation Française – Paris – No 321 – 14 Oct. 1977.

Commentary
Lewis M. Branscomb

Professor Danzin and I share the view that information technologies represent an extension of human capability of a sufficiently profound character such that it will, in time, radically influence the development of a given society. Information technology can be thought of as permitting a vast acceleration in human evolution, from the very slow process of gene mutation and biological selection to the much more rapid process of technology development and selection.

This is, of course, not a new phenomenon. The rise of modern science brought with it the development of machines that contained their own sources of energy, extending man's physical capability beyond the limitations of his bones and muscles. The extension of such capability to the human brain is likely to be of even greater human significance. Professor Carl Sagan of Cornell University in his book *The dragons of Eden* makes the observation that the human brain has grown physically about as much as it can without impairing the mobility of the females who must give birth to children with increased cranial capacity. He believes this limitation would ultimately have slowed the process of development of the human species to increasingly higher orders of intelligence, were it not for the capacity of the human mind to create extensions of human intelligence though information machines. While I am not inclined to exaggerate the potential of artificial intelligence,

I do believe there is some validity to Professor Sagan's view of man and his information technology as a sociobiological entity.

What are we to make, then, of this radical change in the rate of increase of human capability? I believe we can take comfort from an examination of experience in the past with social consequences of new powerful tools for doing man's work. The advent of self-powered human machines in the seventeenth and eighteenth centuries, for example, was accompanied by a rapid decrease in the prevalence of slavery. Slaves became uneconomic in competition with machines. Similarly, I believe that information technologies can bring an even more important dimension to human capability, permitting a sufficiently wide sharing of access to knowledge that knowledge will indeed become 'common heritage of mankind'.

If one wishes to be sceptical about the benefits of information technologies, one can find much to be concerned about. Nevertheless, I believe that many of the tensions and anxieties of today's world are only evidence of the readjustment process towards a world in which aspirations for a better life are shared on a truly global basis, by people in all forms of societies. Aspirations necessarily precede their fulfilment. The frustrations of the underprivileged are more eloquent testimony to the potential benefits of technology than expressions of concern about the disruption of technology-driven social change. Information technologies can, then, be a strong liberating force from ignorance, drudgery, and inequality.

Professor Danzin moves rather quickly from such general observations to a concern with computer technology specifically, and indeed a rather narrow focus on the economic and political issues associated with the development of the computer and related industries in different countries. This view, it seems to me, tempts one to underestimate seriously the importance of the communications side of information technology. Looking at the world of today, television has had a much more profound social impact than has the computer, which has until recently been relegated primarily to increasing the productivity of mathematicians and reducing the drudgery of accountants and stock clerks. Marshall McLuhan and many others have analysed the impact of television on the generations that have grown up with it. Television in the United States has had a profound role in the evolution and communication of new social values. It has played a constructive role in the development of race relations, for example, that could not possibly have been accomplished by churches, schools, or political institutions alone.

One must then recall that 1976 marked the hundredth anniversary of the invention of the telephone by Alexander Graham Bell. I suspect that the telephone, the most ubiquitous realization of information technology in the world today, has had, in its turn, an even greater social effect than television. The telephone has changed the force of laws of modern human interaction by adding a distance-independent communication capability between individuals. The effect on society, I believe, is quite as great as would be the effect of long-range forces in chemistry when an electrically neutral gas is rendered conducting by ionization, thus changing laws of interaction between molecules from electrical exclusively short-range forces to a condition of long-range electrical interactions. Indeed, in other writing I have been bold enough to draw the analogy between the effect of communications added to human interactions and the addition of long-range electrical forces to a gas when it is ionized into a plasma. The plasma, we know, has many important collective properties over and above the properties associated with the individual molecules. These collective properties give the plasma a great power, but among them are significant forms of instability. The properties of a plasma are also heavily determined by conditions in the boundaries of its environment. So too with modern societies.

What the computer brings to telecommunications technologies that is new and important is more than just the ability to make calculations, that is to manipulate information according to prescribed mathematical rules. Computer technology above all provides the capability to control the organization and flow of information, and does so with techniques for storing it that greatly increase the flexibility and reduce the cost of many types of information systems. Thus computer technology will add new dimensions to many information technologies—publishing, mail, television, telephone, telegraph, facsimile, etc. and will enhance science, literature, education, and news.

One profound consequence of the combination of telecommunications and computer technologies is their tendency to increase the interdependence of global societies and equalize the human aspiration of different societies, given the evident possibility of all societies benefiting from the heritage of human knowledge. Information, after all, is an unusual kind of commodity. When information is used, it is generally not consumed; it can often be used by an arbitrarily large number of people without reduction in value. Thus the natural tendency for information to flow is a universal equalizer.

One can be impressed, even disturbed, by the evident inequalities in the level of development of knowledge-oriented institutions in different societies. But one must be even more impressed with the often repeated experience that people from primitive societies can, through exposure to educational and cultural opportunities, make the jump in a single generation from their primitive heritage to the mastery of the latest discoveries of modern science and the powers of contemporary technology. Each society will choose in what way it wishes to take advantage of that modern knowledge, but its exercise of choice is itself a source of interdependence. However much one may worry about the loss of 'information sovereignty' and other threats to nationalistic aspirations, the rise of interdependence is a stabilizing force in the world today. It represents an evolution from an outdated reliance on national structures for organizing human affairs toward a more global community whose political institutions have yet to be fully developed. Compared to such effects, I believe, the social impact of information technologies in the fields of automation, management, productivity, employment, etc., are relatively modest. Nevertheless, a few comments on these topics may be worth while.

The computer is a tool for increasing control of information processes. Computer automation provides the machinery to reduce the very 'information pollution' that Professor Danzin is concerned about. The potential to deliver precisely the information that an individual needs, discarding the irrelevant is illustrated at the popular level in a Japanese experiment with facsimile news in the home. Information technologies will soon make this potential a realistic possibility, which may be either accepted or rejected by various sectors of society.

Regarding automation, it is important to appreciate that the proper use of computers for automation is to enhance human control over the production process, not to replace it. Computer automation can permit the personalization of every article in a production process to the specific requirements of its intended end user. The principal loss in human value associated with introduction of conventional mass production is the loss of this varied capability to specialize a product to its intended customer. All Model A Ford cars were not only mechanically the same, they were initially even painted the same colour. The power of computer automation to restore the capability of customization is illustrated within the computer industry itself.

Modern electronic circuits are designed by engineers using computer programmes that automate the design process. These programmes can be used to drive automated tools that produce the specific patterns required on each silicon wafer as it is processed through the production line. Thus, with the 'personalization' of the wafer produced by an electron beam device, driven by a computer whose instructions came from the original designer electronically, every item off the production line can be different. The computer also provides the high level of logistic control to ensure that the specific device manufactured reaches its intended destination. All this may sound very capital-intensive, but it should be appreciated that this process is, on the contrary, labour-intensive. In the IBM Corporation, for example, well over half of the total operating expense, including depreciation on capital equipment, is for salaries and benefits of the people employed. More importantly, on a typical product, nearly all of the labour content is supervisory, engineering, or technical, with very little labour content associated with traditional factory shift labour. The effect then of automation is not to displace jobs but is to upgrade them very substantially, increasing the interest content of the work, the financial reward to the worker, and the opportunity for exercising judgement and intelligence in the course of the work.

Obviously, the skills required impose an obligation on the employer to engage in very extensive internal education and training. IBM employees benefit from over a million student days of classroom instruction each year. Fortunately, information technology makes this learning much more productive than it has been in the past. The process of developing this kind of job enrichment involves the gradual upgrading of the demands of the job and the skills of the workers simultaneously. Thus the proper use of information technologies in the workplace enhances the dimensions of the job, increases the scope for human control of processes, and adds to the quality of the lives of those involved.

Regarding the effect of information technologies on individual liberty, the principal effect has been to draw attention to the need for the refinement and extension of the more limited concepts of information liberty in the past. In the United States, the First Amendment to the Constitution concerns the freedom of speech and press. The notion is now gaining wide acceptance that citizens also have a right to receive information, to have access to the media of communications, as well as to be protected regarding information about themselves. New technologies enhance the possibility of satisfying such legitimate desires; this in turn accelerates the effort to reach a political consensus to define them.

Thus the concepts of information rights and human rights are deeply intertwined. The attention being given to this subject in most of the nations of the world is evidence of a healthy response to the potentials of the new technology. For not only does the new technology in some cases accelerate the requirement for clarification of the rights of citizens and the obligations of government, it also makes possible the realization of rights that were out of our reach with more primitive information technologies.

Some of the social effects of new technology can be disruptive within individual communities. For example, Professor Danzin suggests that computer-aided medicine may tend to emphasize the distinction between the artistic and the scientific aspects of the practice of medicine. I suspect this is true, and the effect may be troublesome to the physician, but it is certainly beneficial for the patient. This may be typical for the effect of information technology on society as a whole, for changes in structure of knowledge-generating and managing institutions reflects itself in changes in influence of these institutions.

This brings me to Professor Danzin's concern with which he terminates his chapter. While on one hand the new technology is bringing advances in concepts of human rights of privacy and access, there are also new threats to the unimpeded movement of information across national borders. The Balkanization of knowledge for political reasons could frustrate the global benefits of enhanced diffusion of information and the development of world-wide co-operation and interdependance.

I would like to make one observation about the arguments advanced by Professor Danzin concerning strategies for developing information industry capabilities in Europe. These strategies include a mixture of concepts for encouraging (a) the industries that provide the tools and communications facilities and (b) the institutions that provide the information and programming content that will be carried or provided by these products and services. In the United States the first of these two problems has been addressed quite successfully through highly competitive industrial activity, up to this point largely driven by the requirements of the business community and of the government's directly operated services, such as military and space activity.

Relatively little attention has been given by the United States Government at federal or state and local levels to the provision of the new kinds of information-oriented social services that the new technology makes possible in principle. As a result, in some respects we in the U.S. are behind Europe in defining what those services should be and how they should be provided. It is my conviction that there is a very proper role for government in the sphere of information generation for social purposes. Just as it is inescapably the obligation of governments to foster educational and research activities, library services, and supplementary education beyond the school environment, other social and cultural activities of an information nature can and should be encouraged through government action. Some areas of information technology, such as interactive CATV, are retarded in development because of the lack of development of sources of programming content.

In the final analysis, it is the quality, appropriateness, and availability of the information itself that will determine the character of the contribution that information technologies make to society. The boxes, wires, and antennae are, after all, only means to an end and do not of themselves determine the character of the end to be served.

15

Science, technology, and man: population and education

Lord Vaizey of Greenwich

15.1. Introduction

In about 10 000 B.C. there may have been between 2 and 20 million human beings in the world. By A.D. 1750 the number had grown to 750 ± 100 million; in 1850 it was 1200 ± 100 million and by 2010 perhaps 6000 million. In 1750, Europe had one-sixth, and Asia two-thirds of the population; in 1950 Europe had about one-sixth, Asia one-half. The Russian population had doubled, but was still only 6 or 7 per cent of world population. These trends will continue.

It is estimated that world population may double by the earlier part of the twenty-first century from its present level. In this doubling, two problems arise. The first is the obvious pressure of one species, *Homo sapiens*, on available resources – a pressure which natural history would suggest is bound to end in some catastrophe, unless the species, by intelligent adaptation (that is largely by education), can adapt its behaviour to modify the pressure of numbers.

The second problem arises because the most highly developed part of the world, broadly the NATO countries, has come virtually to the end of the population explosion, the explosion which it initiated. As a result, its population is diminishing as a proportion of the world's total. This brings in its train military and economic consequences. Further, the transitional form of the switch from population growth towards stability brings with it important consequences which are dwelt on below. In particular, one aspect of the population movement has been a major shift into urbanization. This unprecedented degree of urbanism has results which are of sufficient significance to merit special consideration.

It is within the context of the study of demographic trends that attention is turned to education – a process which has been through massive changes in the past generation, and which is about to assume massive responsibilities in the light of changes which can clearly be envisaged and which are due in part to the population changes which are discussed here.

The movements of demographic change seem to have been above and beyond the power of government intervention, which has been the usual response to the various social and economic questions that are uncovered by research in the social sciences. The upward and downward movement of population has been rather like a climatic change; something that happens which seems to make governments helpless spectators. There has been an amazing fall in the birth rate of the Western countries and of the Communist countries in the past 20 years, even though the population explosion, as it is commonly called, has continued to take place in the Third World. It is therefore inescapably true that the bulk of the population which falls still within the Third World is likely to become even more dominant in that area that it is now. Furthermore, the shift of the world population, even in the developing countries, into urbanization makes a substantial impact on the prospects of world food and energy production, since towns use much more energy which is not home produced. This growth of urbanization also of course affects the balance of food production, and might well lead to the position where the world runs out of food even more rapidly than it seems to do at present.

On the other hand, it must of course be acknowledged that many of the most heavily urbanized countries, like the United States, are themselves very substantial net food exporters. Furthermore, the fall in the population of Europe will in itself lead to the possibility of agricultural self-sufficiency in that continent. The development of new sources of energy, such as North Sea oil, increases the possibility of increased self-sufficiency for energy as well, if only in the short term.

The next question which arises from the consideration of demographic change is the balance between age and youth, and the fact that the way in which the world has coped with the explosion of population – and, indeed, one of the major causes of the growth of population – is by the development of new and different forms of scientific knowledge and the wide development of formal education. This has led to a very substantial expansion of education which has spread through large parts of the Western world in the belief that there should be a significant shift towards social justice or even equality between different social groups; in particular, this has led to fears that excellence might well be penalized in the educational system. The question that then arises is whether or not the changes in the education system will in themselves cause the West to lose the scientific and technological lead which has for many years been among its principal advantages.

In this particular context, the chapter comes to conclusions which, while agnostic, are broadly speaking optimistic.

15.2. Population and society: values

Birth and death, awareness of them, and an understanding of their place in the pattern of life are essential concomitants of being human. A chapter, therefore, which is dedicated to the study of population in the NATO countries, together with a further study of education, which in essence is the handing on of the cultural and technological heritage and its development, is bound to be concerned ultimately with the deeper purposes of life. Indeed, it is a digging down into the reasons for which like-minded nations have come together for purposes of defence and, going beyond that, into a much more fundamental consideration of the ultimate purposes of human civilization. This, perhaps, is an unconventional way in which to open a scientific contribution. Nevertheless it is important that we should see the achievements of the modern physical sciences, which are almost entirely the achievements of a culture which has its roots in Western Europe and which spread across the Atlantic into North America, in a broad perspective. In particular, in recent years there have been fundamental changes in the attitudes to death and to birth which are already profoundly affecting society at all levels, from the productive process to the attitudes towards consumption, investment, and defence. These changes are as surely rooted in that culture as the physical sciences themselves have been.

In a community as large as NATO, embracing as it does nations which, though fundamentally united against a common series of enemies, are incredibly diverse in size, in levels of *per capita* income, in political attitudes and outlooks, and in religious and cultural affiliations, it is impossible, within a short space, to say things that cannot be contradicted, and rightly so, for the experience of one part of the area is bound to differ from that of another. Indeed, it is perhaps this diversity which is the most striking feature of the Western countries, especially when it is contrasted with the ideological homogeneity and the monolithic structures of the so-called 'socialist' countries. Indeed, the diversity of cultures is even greater in NATO than the diversity of economic and political forms.

This has been, however, the part of the world with the highest scientific and technological achievement in the history of mankind. Its culture also, however flawed it may be, has become in some respects the dominant culture. Its allegiance to constitutional, democratic processes of government is one key to its structure. Its belief patterns are clearly related in some respect to the scientific achievement, and both science and culture arise from the common experience of the Renaissance, the Reformation and the Counter-Reformation, and the Enlightenment, when the 'Western world', as such, emerged.

In recent times, however, major shifts have occurred in values. At the same time as organized Christian religion has been on the retreat, and constitutional democratic ideas on the defensive, massive social change has taken place, and this social change has been related to major shifts in attitude to birth, life, and death.

15.3. Life, birth, and death

The changed demographic picture, indeed, is partly related to a growing degree of social tolerance with respect to human sexual habits, even in countries like Roman Catholic Ireland, which still effectively prohibit divorce and which are stern about contraception, while at the same time there has been a growing prudishness about the discussion of death.

One aspect of this shift of values is a growing concern with individual human life — for example, the death penalty has been almost completely abolished in this community — and a greater value attached to each birth (reflecting declining infant mortality rates), and a changed attitude to death, now a taboo subject. This attitudinal change is discussed more broadly below, but the attitude to death highlights it. Death and violence are seen daily on TV, though death on an ordinary day-to-day scale has become much less common than it once was, as infant mortality rates have declined and as the rate of survival of people through middle years has increased.

Death has become both unfamiliar in ordinary domestic life for many parts of the NATO area, while at the same time in many individual lives it has become much more catastrophic. Death is now largely due to three causes: to cancer, which in turn is largely due to cigarette smoking; to heart disease, which is largely caused apparently by much of the consumption patterns of modern industrialized mankind; and to road accidents. The old epidemiological basis of infectious illness has been altered by public health practices and by the development of immunization techniques, so that to a very large extent death is now a product of our way of life. The consequence is that apart from those who are killed by that way of life, more people now survive into old age; and although in recent years in most countries the average length of life after the first year has not significantly increased, there is growing evidence in many societies that the number of people surviving beyond the age of 75 is increasing. The significance of death, then, has altered; and old age is more common than it was.

15.4. Economic consequences of demographic change

The changes in modern demographic patterns in our societies have thus caused most people to have a family in which the partners survive until they become eligible for the old age pension, in which one of the partners tends to die somewhat beyond the age at which the pension is received, while the other may well continue on into the eighties or even, increasingly, the nineties. In most countries the growing proportion of people beyond 65 (the usual male age of retirement) which has been a feature of the past 25 years, has ceased to be as powerful an influence in leading to growing social security budgets as it recently was. The proportion is now stabilizing. Within the total of the elderly, however, the number of the extremely old who require special attention seems to be increasing disproportionately. It is a society which has put a premium on the normality of a working life, which has created 'retirement' as a way of life, complete in itself, but which has lost many of the organic social links between birth, youth, working life, old age, and death. These formerly dominated societies united by a religion which 'explained', or made meaningful, all these events, and where people lived surrounded by relations and neighbours, which gave meaning to daily life.

Until recently, the growing proportion of those over 60 years of age imposed a burden on the working force. This proportion has ceased to grow, broadly speaking, in most NATO countries, and for that reason the adverse effects of demographic shifts have been borne in the years 1950–75. The reduction in the birth-rate leads to a diminution in the number of children; so that the two major dependent groups — the elderly and the children — will, for the next generation, be a stable, or even declining, 'burden' on society. The rise in female participation rates increases the potentiality of the labour force for work, though the tendency to extend education, to 18, 21, or later, and the tendency to reduce the retirement age, act in a contrary direction. The size of the labour force, within these demographic limits, is (from day to day) determined by the level of effective demand, and this — though in turn affected by the demographic pattern — will be considered mainly in Chapter 16.

It may be feared that as the birth-rate falls, the burden of the social security budget, falling on those who work, will rise. This is not at present correct, for the reasons given.

These remarks are chiefly true, of course, of the advanced industrialized countries of North America and north-western Europe, but they will become increasingly true of other countries as they achieve present north-west European standards of affluence and begin to emulate north-west European demographic trends.

Within the next 20 years, therefore, the 'burden' of the elderly is unlikely to increase. Indeed, it is likely in some sense to stabilize. This does not for one minute obviate the necessity of thinking carefully about the possible difficulties and problems of people beyond retirement age in the society which is coming into existence, but it does suggest that one of the handicaps under which public finances have suffered for two decades or more has ceased to grow at such a significant rate. The public finances may be relieved but what, indeed, should we ask of the mental and moral characteristics of the society which has pushed death into a corner? What price has been paid?

This is especially true of the elderly. As more and more people survive to 70, and to 75 and beyond, the problems of senility have become an increasing social issue. As nuclear families grow, and extended families diminish, as the population becomes more geographically mobile, as the world fills with gadgets that have to be mastered if daily life is to be lived – the telephone replacing letters for example – so the problems of the very old become more chronic.

But the difficulties of those aged 60 to 70 increase, too. Work, the centre of life for many people, is removed from them by retirement, and the whole balance of work and leisure is in urgent need of review. People aged 40 or so are subject to immense pressure, people aged 65 feel useless; the need to reallocate work and responsibility and to reassess leisure, is deeply caught up with the meaning of life, a subject already raised, because ordinary daily life has now become fragmented, and birth and death also have a changed meaning.

Within this broad picture, therefore, attitudes to death have gradually and dramatically changed, and it has become a taboo subject with a growing emphasis upon the immediate and the satisfaction which is to be derived from an affluent life. This, perhaps, may lie behind a profound change which is taking place not only in the industrialized countries but far wider, in the Third World, and also in the Communist countries of the East, towards birth.

15.5. The birth rate

In many countries there are still large numbers of families where the birth of children is regarded as a biological inevitability, a necessary by-product of human sexuality. There is growing evidence that in a number of north-west European countries, notably France and England, and also in the United States, this was no longer true in the eighteenth century or even perhaps in the seventeenth century. Studies have revealed early contraceptive practices which showed an unwillingness to regard unlimited childbirth as a necessary concomitant of adult sexual behaviour.

Nevertheless, it is only within the past century that the open practice of contraception with mechanical means and subsequently by pharmaceutical procedures has become a possibility. It was only, indeed, in the 1960s that the pharmaceutical revolution began to make a significant incursion into birth-control techniques, and it is broadly speaking within the past ten years that legalized abortion has become a major source of birth prevention. This in turn has stimulated (or perhaps been stimulated by as well) the introduction of more effective medical procedures for abortion. Rapid scientific development is taking place in this field at the time of writing.

Birth thus has ceased to become a by-product of sexuality, but has become for increasing numbers of women a deliberate act, and it is fairly clear that from the late 1950s in some cases, but from the mid-1960s in others, whole societies have increasingly taken the view that the size of the completed family should be significantly lower than in the past.

We have therefore witnessed not only the culmination of a great cultural change of attitude about death, but one even more dramatic about birth. Both changes, in their magnitude and immediacy, are by-products of the pharmaceutical revolution, which in turn is dependent upon the biochemical discoveries of this century. Both changes are profoundly important, though their repercussions cannot be even dimly perceived.

One major change has to do with sexuality. Let it be said at once that throughout the NATO countries there is no evidence to support the view that marriage or stable sexual partnership has become less popular than it once was. Despite the evidence of rising divorce and separation rates, there is also the evidence of rising marriage rates to suggest that the typical nuclear family of a man and a woman has never been so prevalent as it is now in the industrialized countries. What has happened, however, is that the great swing towards the popularity of a larger family, which tended to occur in France, the United States, and the United Kingdom, for example, towards the end of the Second World War, has been succeeded by a regression to a higher age for the first birth, followed by a smaller total number of births, reaching in some cases the situation where the population is no longer fully replacing itself. And do we not see here, once more (as in connection with death), a development where what is sought is personal fulfilment in adult life, a concern with the immediate, and not with the future or the past?

The rise of concern with an individual's life is bound up with broader social considerations. For example, people make careful provision, through social security and insurance, for old age. Parents are more and more concerned with the need to ensure a good education for their children, in order to give them 'a good start in life'.

The reasons for the changing attitude to birth may range from the proximate to the profound. Included in the proximate would be the development of the oral contraceptive together with the relatively easy availability of legal abortion, which has made the decision not to have a child more capable of being made effective than it used to be. The commonest cause which is usually given for the decision not to give birth is the increase in the proportion of young married women working.

The growing move towards opportunities for women, which are in some degree equal to those available to men, has meant that the premium upon remaining childless has grown. It is also said that a further reason is that the economic difficulties which have affected many countries since the late 1960s have made many couples dependent upon two incomes for their standard of living. The process of childbirth and child-rearing, together with the additional costs in housing, food, clothing, and travel, of having a family, has put that standard of living at risk – a process disguised by the growing affluence between 1948 and 1968 – and therefore has led to the decision to have fewer children. But underlying this is a much more profound shift in attitudes, surely, towards the question of human continuity and understanding of the place in which each generation stands in relation to those which have come before and those which go after. It is in this context that fundamental changes appear to be taking place in attitudes towards past and future, and in particular the revaluation of the present as an end in itself, in which immediate or virtually immediate satisfactions must be sought if life is to be regarded as having meaning. This is not to say that the satisfaction is shallow. Often it is the search for the profound, for the sublime, for the true, and the beautiful which (for example) causes people to abandon a bickering relationship in marriage (in previous years tolerated if not tolerable) in order to find a second, or third, or fourth union a marriage of true minds.

Underlying this basic shift in values is the growing economic independence of women, allied to their far higher educational and skill attainments than in the past, and the changes in the laws dealing with marriage and divorce, and towards sexual equality.

15.6. Migration

The movement of population which arises from changes in the death and birth rates is complicated enough, but from the perspective of an attempt to forecast the future so that social, economic, political, and military policies may be based on adequate surveys of the likely scenarios, it is necessary to move from the macro to the micro scale in considering population change. It is, of course, important to know the general long-term considerations that affect the relationship of society to demographic change, but the detailed year-by-year, region-by-region, city-by-city changes are of greater significance in the short term. This inevitably raises the central question of migration.

Individual migration has, of course, occurred since the beginning of recorded history, and it is well known that mass migration, that is to say, a movement of whole communities, has been one of the major agencies of social change since the earliest known movements of mankind: the incursions into the Roman Empire, the original Arab conquests, or the mass migrations across the North Atlantic provide examples. Astonishingly enough (at least to those who regard recent times as in this respect settled), the demographic history of the world (from the point of view of migration) since the end of the Second World War has been one of the major cultural shifts in the settlement patterns of mankind. The recent movement of whole populations, though now rare, has not been as rare as it was in the nineteenth century, when it seemed that individual movement had replaced mass migration as the major source of population change. Nevertheless, a drift of people across frontiers has taken place steadily and at a high rate, despite the fact that the relatively *laisser-faire* world of the nineteenth century has been replaced by formal bureaucratic processes involving changes of domicile and nationality, which have been more formidable for many countries than at any time since the Reformation. For example, the migration of Russian Jews, which in the period from 1880 to 1914 was one of the most important population shifts of the period, involving (especially) the buildup of a large Jewish community in the eastern United States, has been followed by formidable controls on emigration by the USSR. The USA, since the early 1920s, has replaced its open-door policy for immigrants by one that is far more restrictive, and in the 1970s both Australia and New Zealand have also discouraged immigration, relative to their earlier policies. The growth of state socialism has generally meant restriction on emigration and immigration and, certainly, on the transfer of assets.

There has been a major change in attitude to migration, however, with the development of 'long-stay', job-seeking movements of working people.

Some migration has, of course, been on a permanent basis, either voluntary, such as that from the Caribbean and Pakistan to the United Kingdom, or that from the United Kingdom to Canada and Australia, or involuntary, as that of the German populations of the occupied territories in Eastern Europe into the Federal Republic. Other migrations have, however, been more than nominally temporary; the so-called guest workers, who have moved from Portugal, Spain, Turkey, Greece, and Southern Italy to Germany, Switzerland, Northern Italy, and France, as well as to the Scandinavian countries, have been formally temporary migrants, but in many respects they have the characteristics of those whose intention it is to migrate permanently, whether *de facto* or *de jure*. This subject will be returned to.

The arrival of migrant workers has highlighted the difficulties of urban life. The guest workers, and the coloured communities in the United Kingdom and in the northern cities of the United States, are often blamed for urban problems of crime, slums, poor health, poor educational attainment, and racial tension. Yet many analyses have shown that those difficulties (save the last) exist in cities which have no such immigrant communities. That is not to deny, however, that these communities by themselves (but not of course by their own fault) are flashpoints of social disturbance.

15.7. Theories of migration

The causes of these individual and tidal movements in human settlement patterns have been much analysed. They are, of course, in many respects overwhelmingly political – the flight of the oppressed from oppression, or of the fearful from war and civil disturbance. But leaving that on one side, there is a clear tendency for countries where the rate of growth of GNP *per capita* is accelerating to encourage immigrants, while those countries whose rate of growth *per capita* is rising less fast tend to provide the emigrants. Professor Brinley Thomas[1] has indicated that there has tended to be a reciprocal relationship over many years between the countries of northern Europe and North America, which is best represented by changes in the building cycle. This cycle of economic activity measured by the level of domestic construction tends to be longer than the normal trade cycle, and tends to have a reciprocal relationship between the two areas. What Professor Thomas has discerned is a mass movement of population which has tended to stimulate the longer-term economic movements of the North Atlantic economy, and in turn to generate still further population movements. This synchronization is no longer between north-west Europe and North America; it clearly now embraces a larger geographical area, on a different basis, and it would be useful to apply the Thomas techniques and theoretical apparatus to the wider North Atlantic Community and its total trading area.

In many countries, Yugoslavia and Spain for example, the rate of growth of the economy in the period 1950-70 was high, yet the rate of emigration was also high. This was because those economies were becoming more and more adapted to modern industrial and agricultural technology, using more capital goods, thus creating a surplus of labour in circumstances where (as in all preindustrial societies) a surplus of labour almost always exists for much of the year. Thus, accelerated growth in a poor country tends, paradoxically perhaps, to accelerate emigration, as well as to speed up domestic labour absorption.

What is clear, however, on an impressionistic basis, is that the power of population to act as an exogenous variable of considerable strength in the economic growth model has not diminished.

15.8. The brain drain

One of the characteristics of migration which is singularly important is that it tends – in some cases overwhelmingly – to consist of the relatively better educated and trained, and certainly the more able in a general conversational sense of the word, even where they appear to lack formal educational qualifications. Migrants tend to be disproportionately male, from the younger and more vigorous age groups; they tend to have fewer dependents than the population which they are leaving or the population which is receiving them. The net addition to material output on the part of migrants to their host countries is therefore disproportionately great, since the surplus value which they generate tends to provide additional revenue for their host country, beyond what they themselves use. A case in point is the social security system, where they make fewer demands for old age pensions, for invalidity benefits, and for other social assistance than workers drawn from their host community. Thus, countries that are the source of emigrants tend to lose relatively, and countries that are on the receiving end of immigration tend to gain from the movement of population. This reinforces the relative disparity of rates of growth between emigrant and immigrant societies, which was itself part of the original process generating the migration.

The migrants tend to be those people who are 'success oriented', or at least not prepared to accept failure and hopeless poverty without a fight. This reduces the whole 'tone' of a society, which in turn makes it more defeated, a phenomenon well-attested in Ireland from 1845 on, in the United Kingdom from 1945 on, and perhaps in Canada in the 1980s.

Two categories of migrant may be particularly discerned. One is the highly skilled professional or technical worker – the so-called 'brain drain'; the other is the entrepreneur. Minority communities have for long been famed as sources of entrepreneurial activity in different societies, whether it be the Greeks in Turkey, the Jews in the United States, or the Asians in the United Kingdom.

A considerable amount of attention has been paid to the brain drain, especially the migration of scientists and engineers, over which a near panic took place in the 1960s; far less attention has been paid to the movement of entrepreneurs which probably has greater economic significance.

There is no general model of the causes and consequences of migration in modern society, though manifestly it plays a major role in the changes that have been occurring, not only at the economic, but at the social level, throughout the NATO area.

15.9. Internal population movements

This lack of a general model is particularly apparent when migration is not so much a question of movement across national boundaries as a migration within one economy within a national frontier, where the model is well established. There has been since the Middle Ages a significant shift of population from rural areas to the towns. This process, of course, was significantly accelerated during the process of industrialization, until in most advanced countries the proportion of the population devoted to agriculture. is well under 10 per cent and in most cases under 5 per cent. This is not the same as the proportion of people living in rural areas, but it has become rare in most advanced industrial countries for more than one-fifth of the population to be rural. This is a process which it may be assumed will be continued for a substantial period in those countries with a rural population substantially in excess of this proportion.

Not only has the movement been between the countryside and the towns, however; it has also been between regions in a country – for example the enormous shift of populations in the United States into California and other states of the south-west, and the massive shift of Italians from the south to the Po valley.

15.10. Urban issues

Movements of population within countries – that is internal migration – have been of significance in recent years because of the accelerated rise in the past century of urban centres as the dominant form of human settlement. In particular, this century has seen the growth of conurbations. In the past 25 years the conurbation has changed to a new form, of which Los Angeles is probably now the most significant example. It is tending to become a polycentric series of human habitations linked by rapid internal transit systems, like freeways, and (significantly) not dependent upon its regional hinterland for much economic succour. The regional hinterland is often, indeed, turned into a sort of park-like landscape which is used partly for recreation and partly for commuter settlement. Originally a conurbation was centred on one or two big cities, like London, or the New York–New Jersey complex, and the lines of communication radiated from the centre into a hinterland which formed an essential part of the economy of the region. But now the old historic centres of cities that have become parts of conurbations have tended to decline, turning first into purely business centres; now business itself has tended to migrate from the centre of cities to other parts of the conurbation.

The mobility of the population within urban complexes depends not only on transport but on opportunity for jobs, for homes, for leisure, and for education. In other words, the creation of an urban complex itself creates the opportunities that make people more mobile. And, by making them more mobile, it enables the more affluent, politically powerful, and socially adept to choose a life-style that incorporates a desirable home in a good area, satisfactory schools for the children, suitable jobs for the parents, and access to leisure – golf courses, yachting marinas, theatres, and second homes – all largely by car, or perhaps express train. For these people, the advantages of the conurbation immensely outweigh the disadvantages (the sense of isolation, the lack of a sense of community). But, for the failures of the system, the conurbation tends to maximize disadvantages. The failures – the poor, who tend to be the elderly, the unskilled, the social isolates, the handicapped, the social minorities – tend to congregate in what are or become slums, where social problems are high and incomes and employment are low.

The problems of urban deprivation and desolation which it might have been thought the rise in real incomes would solve (since housing and other social expenditure is highly income-elastic), have in many ways become more severe as a result of the way that urban society has developed. This is largely because of the inadequate planning of urban infrastructure, rather than because of inadequate spending. There is little evidence, for example, that deprived inner city areas are more than marginally worse off in spending *per capita* on public health and education, and total public outlays (including welfare payments and housing subsidies) are of course far higher in these areas than those elsewhere. It is, rather, the lack of an adequate response to the growing freedom of individuals and their families to move to desirable locations, relatively free of cost and other limitations, leaving behind what rapidly develop into ghettos. The consequences of this for social development are serious, largely because the speed of public response has not caught up with the rate of resettlement.

This rate of change of population settlement, this shifting of population to a series of suburban nuclei, has been accompanied by a growing division (a division which has been apparent for the past two centuries to a lesser extent but which has now become acute) between the areas where people live and the areas where they work. The most obvious consequences on transport, the use of energy, and upon the economy are much discussed. The demographic consequences, however, have been less often discussed. The pressure of the difference between workplace and home has put growing pressure upon the nuclear family, not only because the various components of the extended family tend to be ever more geographically dispersed, and therefore unable to offer support, but because members of the nuclear family itself tend to work in different places and the children to go to school or college in yet still other places. Thus the network of family relationships becomes to a very important extent a network of transport, and the competing demands of work on husband, wife, and children make the nuclear family subject to powerful centrifugal forces.

The tendency for internal migration to the new-style conurbation, of which the Ruhr is an example, has furthermore been accompanied by a very high degree of mobility during most people's own lifetimes over very large distances. This personal mobility is best typified perhaps by the employment practices of large national or even multinational corporations, which – in the United States for instance – may well move its employees from Cleveland to Los Angeles, and then to Chicago, and finally to Atlanta in the course of a working lifetime. It may also even send them on extended overseas tours. This modern peripatetic substitute for permanent migration corresponds to some extent to the pattern of behaviour of those people who served the former imperialist powers in executive or military capacities. It has become a potent source of the international culture which has become so much a dominant force in contemporary life, particularly through its constant interpretation through the television, radio, cinema, and the printed word, so that the modern élite is not indigenous to the country, but a group of corporation executives who are in a special sense 'rootless'. The typical folk heroine nowadays is perhaps an air hostess; frequently the typical successful creative artist tends to be a tax exile, living quite divorced from the culture from which he derives the substance of his work, and where those who enjoy it and pay for it live. The effects of this rootless culture on the family, extended and nuclear, are of course profound and, in particular, serial marriage, which is now the predominant mode for the international community, affects attitudes to child-rearing and childbearing.

15.11. Genetic and social questions in demographic change

Underlying these shifts and changes of the population geographically, socially, and culturally must be discerned two fundamental considerations which have been the subject of much scientific study in the past 20 years, and which obviously offer a major realm for further detailed scientific examination. The first of these is the question of the genetic endowments of the population, and the other is the relationship of this genetic endowment and geographical mobility to the process of social change. It is in this context that the recent and prospective changes in education must be seen.

Genetic inheritance is a subject of considerable complexity. Oversimplifying, purely in order to put the matter into perspective for present purposes, it is fairly clear that physical characteristics are inherited and that they are to a very large extent manifested – leading to the mistaken view that they are subject to adaptation as a result of environmental influences – according to well-understood interactions between genetic processes and the environment in which they find themselves. The difficulty that has arisen is whether or not mental characteristics are inheritable. This again is not a subject of dispute where certain major characteristics are concerned. There is no doubt that all processes subject to genetic laws are indeed inherited. The question arises as to what is meant by a 'mental characteristic'. The predisposition to agree with (say) a new technology, or social relationship, unknown before this generation, is not in any meaningful sense 'inherited'.

15.12. Intelligence

The genetic endowment of the individual is random. That of a population is not, of course. The issue is whether societal characteristics affecting the population make their way to the genes so that the environment can be said, in that sense, to pass on to future generations. Socio-biologists and biologists differ on this issue.

The point of agreement surely is the conditions in which certain human characteristics are manifested in a social environment. What is at issue is that whole subtle, complex social process which is generally understood by the term 'intelligence', that is to say, the relationship between people and each other and their environment, which has to do with the general ability of the person to cope intelligently with circumstances in which he finds himself; intelligence is, in fact, a mass of genetically determined characteristics, which are uniquely combined by an individual, and manifested in an environment.

Geneticists refuse to accept a 'nature–nurture' conflict, since the point is that the manifestation of genetic processes is possible only in an environment, and nature and nurture are therefore indissolubly linked in principle.

Inheritance is not, of course, purely genetic, but includes social forms or environments. The way that these environments adapt to a new technology can lead either to success or disaster socially, and thus, by affecting the environment, alter the physical characteristics of the population – as when the native races died out in many areas of European settlement.

Nature and nurture are indissolubly linked, above all, in what is an imprecise notion like intelligence, as just defined. To some extent, it must be obvious that this capacity is inheritable according to the complex processes of genetic change which are now being interpreted, but it leaves open to a very great degree the extent to which this hereditary process manifests itself, or put more vulgarly 'is affected by' the socialization process that broadly goes under the heading of 'learning'. 'Intelligence' is imprecise partly because it is measured within a society, and there is a lack of comparative norms to deal with differences of intelligence between societies. In what sense, for example, are people of different ages, or different cultures, or different social groups equally 'intelligent'?

This is not the place, nor is this the author to give an authoritative account of this complex question, which has been the subject of dispute not only between biologists and sociologists, but particularly between those who would themselves claim to be either biologists or sociologists. The changes in the environment that have already been discussed will certainly change the human characteristics that present themselves.

The differential birth and survival rate of the various regional and social groups within national societies must also be considered here. It was this, indeed, that led to the popularity of eugenics, a study with which many great scientists were associated earlier in the century. It became fashionable, up to some 50 years ago, for a great deal of anxiety to be expressed in the scientific community that the relative decline in the birth rate among the more highly professionally successful sections of the community would ultimately lead to diminution in the average level of intelligence, as the less able groups had a higher birth rate and, it was argued, would tend to overwhelm the more able groups. This crude formulation has been abandoned. The spectrum of ability goes across a wider range in each social group, than the gap between the average intelligence of the groups. Thus the more radical fears have been abandoned as not based upon scientifically valid premises.

15.13. Social change and genetic change

Thus, to sum up the argument so far, mankind has been changing in directions that have been largely uncharted during the process of physical and social migration since the end of the Second World War. This substantial process of shift in direction must be assumed to accelerate over the next quarter century, which must surely see an acceleration of many of those trends which have led to the experience of the past 30 years. In these circumstances, the consequences in the long run for humanity and its complex social structure must be substantial, though unknown except in the very broadest of broad outlines.

15.14. Social structure

The whole relationship of genetic change and social change, just discussed, has of course been closely related to a substantial process of social change independent of its genetic basis. This process of change of social structures throughout the industrialized world – and above all the non-industrialized world – has been accompanied by an acceleration of the shifting economic base of modern society, which of course, in Marxist reasoning, would be sufficient in itself to change the social and political superstructure. The process by which genetic change becomes incorporated into the human community is obviously through the medium of the social structure. The independent study of social structures – independent, that is, of biological and genetic study – has a series of conclusions of its own which are not necessarily related directly to genetics.

The NATO world consists of a series of more or less closely related social structures incorporated into nations, these nations again forming the political components which make up NATO; each nation, in turn, has its own political, cultural, social, and biological links going beyond the Alliance through a network of formal alliance arrangements (like the European Community or the Commonwealth), or informal relationships (like those between Canada and the United States, or between the Scandinavian partners). Within each of these political entities, sovereign states, or nations, the social structure represents the equivalent of a biological ecosystem which arranges the human community of that nation in a structure that has a relative degree of permanence, but which is linked to other systems in a total human ecology, within which major changes occur with major or minor, direct or indirect, effects on all other parts of the system. Social study over the past 30 years has revealed both the depth of the social structure (in several senses, which will be discussed later) and its extraordinary degree of persistence, stability, and relative permanence over long periods of time. The depth and stability takes on, and may be seen, in several dimensions. Not only is it something that appears to be at the heart of the very nature of the culture – and of the processes of change of the culture – of each community, but also the social structure itself appears to be integral to the understanding of each individual human personality. A personality in isolation is inconceivable. That is not to say, of course, that all people are socially conditioned all of the time in all of their actions. The notion of the uniqueness of the human personality and of self-will has a deep scientific and religious basis. Nevertheless, the extent of this uniqueness, and of freedom of choice, has in recent years had its theoretical philosophical, and scientific basis much modified. The notion of a unique desert island sort of personality is an abstraction, and is now seen to be a misleading chimera.

The question that arises is whether or not the study of social structures has in itself enabled a sufficient series of advances to be made in interpreting the relationship of different societies with each other, and understanding of the extent to which each society might be thought to be stable. This is necessary in order to interpret individual thought and action in such a way that causal connections can be suggested with sufficient firmness to allow social engineering of a subtle kind to be undertaken. It is this question which lies at the basis of the interpretation of modern educational practice and reform. For if it is so, then much pedagogic change that has occurred, or might occur in the future, can be tested empirically. But if it cannot, then the issues are more open.

15.15. The persistence of social structure

In a period in which the technology available for human beings to use has changed more rapidly and more profoundly than ever before, and in which the range of ideas on social and ideological issues has stimulated debate and even conflict between different social systems, without parallel since the sixteenth and seventeenth centuries, it is perhaps somewhat bold to state that there is a persistence of social structures which is of great importance and significance beyond the level that is affected either by technological or ideological changes. Nevertheless, it must be said that this seems to be the case.

The past 20 years have seen clashes of ideology, as in the Soviet Union and in China, and physical clashes which, while not as great as those of the Second World War, have repeated on a smaller scale its more devastating effects, such as those in Korea and Vietnam. There have also been internal conflicts, like the French student rising of 1968, or the conflicts in the United States, which have appeared at some time or other to give an indication that a radical uprooting of social structures was about to take place.

Even so, after the period of stress, accompanied often by the most violent physical change, the persistence of social forms has been amazing.

This persistence (which is certainly apparent to those who study education) is a strong argument for genetic conservatives. It used to be a commonplace for historians to assign characteristics to whole populations – the Protestant Ethic of Max Weber, for example, or the high motivation to success of the Jews. In recent years, for obvious political reasons, ideas of this sort have fallen out of favour. But they do indicate a problem about the extent to which a society is manipulable. Do the packed churches of Poland and Russia, for example, show that once the Communist regimes collapse, the nineteenth-century and early twentieth-century norms of these two societies will reassert themselves? Or, on the other hand (since this is a realistic example and a highly relevant possibility), will the collapse of Communism lead, first, to the sort of alienation that has been seen elsewhere following political and social change, and will the changing nature of the techno-logy (especially urbanization) lead to a complete break with the past? It is hard to imagine, for example, Cambodia reverting to its pre-1970 condition even if it were to be freed from its present tyranny.

15.16. Education

This persistence of social forms is of particular significance for the study of changes in educational structures, changes which are likely to have been at their height in the past 20 years, and their consequences. Though they have yet to be discerned in their full amplitude (and cannot be so, of course, necessarily until the children and students affected by the changes have completed their working lives), they have nevertheless led to the view that education by itself is less likely to affect the social structure than was at one time thought. The effectiveness of education in terms of its more narrowly pedagogic consequences has, of course, been the subject of in-tensive though largely inconclusive study.

In respect of the relationship between education and social change, it has of course been regarded by ideological conservatives, and particularly those with a deep interest in the alleged strength, simplicity, and directness of hereditary influ-ences on measured intelligence, that since, human characteristics are genetically determined, all ideological and social change is bound to be froth on the surface, so far as the fundamental qualities of a population are concerned. It has already been shown that this apparently 'scientific', though in fact extreme, position is not tenable, since genetic characteristics can only be discussed in a social environment. Nevertheless, even those who do not hold this extreme position, or may even per-haps tend towards an opposite one, would concede that the durability of social structures has been such as to lead to a growing doubt whether educational practice is by its very nature likely to lead in one or two generations to the social and other consequences at one time thought to be the prime mover and main justification for the educational changes that were occurring. The question for education, thus, is whether or not its position in the social structure is such that it can play an effective role as an agent of change. And indeed, with the relative inconclusiveness of pedagogic investigation of teaching techniques and the curriculum, it is hard to establish a firm basis for educational development, other than argument *a priori* or on ideological grounds.

15.17. Education and social change – the problem stated

Thus, while at one level deep and profound changes have been taking place in attitudes to life and death, and people must necessarily have been changing their attitudes to the nature and purpose of life itself, and this has been mirrored by radical changes in the population, arising from changed family structure and consequently changed birth-rates, it is equally true to say that these profound changes have not so far been shown in major changes in social structures throughout the world. The changes in social structure that have occurred, as in China, have been imposed politically, and how profound, in fact, are these changes? If Chinese communism collapsed would Chinese traditions reassert themselves? Is Chinese communism perhaps extremely traditional in certain crucial ways?

15.18. Demographic and educational change

At first sight this might seem a surprising statement. But even in countries that have passed through major technological transformation – and this has happened to all but a few countries in the past two centuries, the social structure appears to have been remarkably permanent and enduring. England, for example, changed from a rural society with links that could be seen to lead directly back to the sixteenth century, and with less certitude but still plenty of folk memory to the eleventh century, or even earlier, to a society that can barely remember yesterday. The same is true of societies that have been through a remarkably rapid process of technological change, like Spain and Portugal, where within a generation the physical and social environment has been in many places completely altered.

That being said, however, the demographic change through which this generation has passed is of great significance. The achievement, however recent, of a population structure that is relatively stable after a period of radical instability (for a century or more in most cases), together with a radical shift in the balance of population between the member states of NATO on the one hand and the Third World on the other, is of exceeding importance, although difficult to assess.

15.19. Is Marxism a solvent?

To add a more speculative note, it might well be that the period of stability of social structures, which appear to have been so strong that they will survive even radical technological and ideological changes, may come to an end because of pressures from outside. The incursion of European invaders led to the collapse not only of pastoral societies in North America but of apparently firmly based ancient cultures in Central and South America, and above all in Asia, where the apparently hermetic societies of Japan, Mandarin China, and Hindu India virtually dissolved after comparatively insubstantial periods of European contact. Thus the impact of several generations of established Marxist ideology and the rise of new militant forms of Third World nationalism may be of so radical a character that they will cause the cultural crisis in Western countries to take such a form that the social structures themselves will begin to dissolve and change, and to change in a way and with a profundity that the ideological disputes within the Western world and the technological changes of the past two centuries have not in fact yet succeeded in effecting.

15.20. Education — success or failure?

To raise these deep issues is profoundly relevant to the issues that have led to the changes in education of the past generation, and to the deep sense of disillusion and even anguish with which parents, teachers, and students look at their daily lives. The failure or success of education is not only to be assessed by success or failure in the acquisition of one or other skill or item of knowledge, important though that is, but by other matters of the kind now alluded to.

Since education is concerned not only with the transmission of a culture from one generation to another, and the persistence and permanence of the culture, but also with the handing on of skills and the development of new skills, with the development of processes by which ideological conflict is both made interpretable and acceptable within the culture, and still further with the development of ideological struggles, it is necessarily in education that much of the understanding of the process of social and technological change must be sought.

The past 25 years have seen the most extraordinary rate of growth of education in terms of numbers of students and of teachers that has ever been seen in the history of the world. Education has now become one of the major sectors of the economy, employing as it does in most countries a high proportion of the qualified manpower stock.

15.21. Causes of the educational explosion

The pressures that were behind this explosion of education were manifold, but they may be conveniently summarized under several headings.

(a) *Knowledge accumulates.* In the first place there has been the sheer rise in knowledge. It has become a commonplace to say that of all the scientists in the history of the world, 90 per cent are now alive, but this extraordinary fact, and it is almost certainly still a fact, as it was from 1850 to 1970, sometimes leads to the conclusion that the rate of accumulation of knowledge has been such that the requirements of maintaining and furthering the stock of knowledge themselves impose a substantial burden upon the education system. Knowledge does not survive of and by itself. It requires tending.

(b) *Demographic boom.* Another factor of considerable importance has been the demographic boom discussed earlier in this Chapter, which has now come to an end in the advanced industrial countries. Nevertheless, until the late 1950s, the birth rate in many industrialized countries was still rising, and in consequence the education system was expanding from demographic causes. Thus expansion was rapid in relation to the period of stability or slow growth which it had experienced in the earlier part of the century, following an earlier expansion at the end of the nineteenth century.

(c) *Economic change.* A further (and perhaps the principal) factor in the expansion of education had been the rapid growth in the levels of national income *per capita* which was experienced, especially in Western Europe, during the period 1948-69. This rise in national income *per capita* had been accompanied historically by a very substantial increase in personal consumption, not only of material goods, especially consumer durables, and cars, but principally (in the more advanced industrial countries) of services ranging from transport, holidays, and health to education.

(d) *Revaluation of the child.* Within each family, too, there had been a significant shift of expenditure towards the more dependent parts of the family, and in particular towards the children. The steady revaluation of childhood has been a phenomenon of industrialized countries since the eighteenth century. Up to the seventeenth century, generally speaking, children were treated as imperfect adults and few, if any, special arrangements were made for them. They wore scaled-down versions of adult clothes and they ate small portions of adult food. But with the coming of the industrial revolution and the irruption of romanticism as a major factor in cultural life in the Western World, the cult of the child became a phenomenon of some significance, culminating in the writings of many sages, typified perhaps by Dr. Spock, in which the interests and welfare of the child came to be seen as equal and often dominant in any family, even above those of the principal income earner, and certainly above those of adults in general.

(e) *More education per child.* This radical rise in the standard of living of children compared with that of the standards of living of older people has been accompanied by an intense increase in expenditure per child on education. There may be an end of this trend in sight, since in recent years the standard of living of the retired has risen very fast relative to that of the employed. Nevertheless, at the critical period for expansion of education, it was childhood and youth that formed the object of a concern that became a cult.

(f) *Complexity of life.* Underlying the growth of education, there has further been the growing complexity of modern industrial, social, political, and military life that has required a growing body of skills, which in turn have been rooted upon the acquisition of a body of qualifications, and these qualifications have in turn been built upon a substantial level of educational achievement. It has been common knowledge that the level of attainment in mathematics, for example, at any given age level has had to rise in order that the level of attainment at first-degree level might be sufficient to enable specialists to proceeed from there to the manifold and difficult tasks of the modern scientist and engineer.

(g) *'Rationality'*. Underlying all this pressure for an expansion of education has, of course, been the rise to prominence in world culture, both Communist and non-Communist, of the view that a rational explanation of the physical and social world was the proper way in which mankind might understand its place and destiny in the scheme of things. This has implied that, since the task of education was knowledge, and since knowledge was the key to life, it must be through education that modern societies would come to realize themselves, to achieve what it was that would in some sense fulfil them.

15.22. Summary of the consequences of educational expansion

Since the causes of the expansion have been broadly speaking similar (though of course differing in detail), it is perhaps not as surprising as it might be that the form of the expansion of education has been strikingly similar in different countries, considering the extent to which its original structures varied. It has first of all been the acceptance of the universalization of compulsory education between the ages of roughly five to the ages of roughly sixteen. Even in countries where the formal age of entry to school is six or seven, the *de facto* provision of pre-school education from the age of four or so has become such that (in the towns at least) substantial proportions of the population go to school from the age of four onwards. Thus universal compulsory education, supplemented by high voluntary attendance before and after the compulsory period, has meant a rise of school life *de facto* to 12 years or so, though of course in the poorer countries four to six years are still the norm. In the rural areas, there has always been a major relative handicap in this respect; but, however, throughout the NATO states, in rural areas there has been a significant move to achieve levels of provision hitherto only reached in urban areas.

15.23. Education and equality

At the same time, there has been an attempt to incorporate all social groups into the educational structure, not only in the compulsory period, but at the higher age levels. Evidence of this is that even where the compulsory school-leaving age is still less than 16, there has been a substantial increase in the rate of voluntary staying, followed by a hesitation several years ago, and a slower resumption of the trend. Thus the length of school life throughout the NATO countries has now increased substantially, for reasons already given, and to this increase in length has been added a very substantial increase in the rate of survival into higher and further education beyond 16. Again, this process of growth has recently shown signs of hesitation, but it is one of the major features of the period 1948–70.

Though a principal reason for the expansion of educational provision has been the attempt to provide more equal educational opportunity, it has to be said that, while the extension of education to girls, to rural areas, and to the less affluent social groups has been dramatic, the relative rates of achievement by these groups, with the exception of girls, has not significantly changed in many respects. That is to say, though more ethnic minority children, more manual workers' children, and more children from rural areas are progressing to higher education, those from the more socially affluent groups are doing so as well, so that there has been no marked narrowing of the gap. This will be a matter which will be discussed later on in the chapter, but it is important to note it at this juncture, because it plays a significant and important part in the interpretation of what is actually happening, and what might come to pass in the next generation.

15.24. How has education grown?

The first point, then, is that the system has very substantially expanded. How has it expanded?

(a) *Buildings and teachers.* Within the education system, there has been a very substantial rise in the quality of building and of school equipment of all kinds. Most children in most NATO countries go to modern, well-equipped schools. And at the same time, very substantial improvements in the pupil–teacher ratio have occurred at all levels. Thus the number of teachers per 100 children has increased very significantly over the period, indeed it has improved by between 50 and 100 per cent in the NATO area as a whole since before the Second World War. The training of the teachers has also been very substantially lengthened, and in many countries the possession of a university degree is now a prerequisite for admission to the teaching profession. Up until comparatively recently, the preparation of teachers for the non-academic schools was in a segregated sector of higher education. This is no longer the case in most countries. It is, of course, a matter for debate whether the lengthening of training has increased the effectiveness of pedagogy.

(b) *Educational expenditure.* The second point is that not only has the system expanded, but that it has become far more lavishly endowed. This is revealed by the doubling, and in some cases the tripling, of the proportion of the GNP devoted to formal education in the several countries of NATO; the magnitude is demonstrated by indicating that the GNP has doubled in some countries, and in others risen by six times, over the same period.

(c) *Changes in education structure.* The achievements of this education system must be interpreted against the structure that has been developed for education as a whole. Broadly speaking, the age undifferentiated primary school based upon 'standards', to which pupils were promoted, has been replaced by a primary school that is based upon the notion of a common curriculum for each age group, but not differentiated either by attainment, ability, or for sex, for social origin, or for ultimate educational or occupational grading. This process of structuring the primary schools has extended well into secondary education, where the arguments for and against differentiation have raged in many countries. The solution appears to have been accepted that, broadly speaking, differentiation on any of these grounds would not be generally acceptable before adolescence, though exceptions are occasionally made for artistic 'giftedness' and for gross handicap. In some countries, this process of non-differentiation has been continued until the undergraduate period in higher education. The future of this specialist secondary education, and even of specialist higher education, is still much in dispute. There are several major strands to the argument which need to be sorted out before this process can be interpreted in a broader sense.

15.25. New pedagogic theories

It must be understood that a major philosophical and pedagogic theory has lain behind this newly adopted structure, which owes much to the English primary school, the American high school, and Swedish research and practice. It is, first, that differences between children are not as important as their similarities; next, that shared experience is socially valid; third, that children learn different things in different ways and at their own pace; fourth, that each child needs, in fact, a highly individualized 'package' of education, to bring out his latent talents. The obvious requirement for such a theory to be effective in practice (even assuming it to be valid in whole or in part) is that the schools should be lavishly staffed and equipped and the teachers extremely highly skilled. While the first condition has been met, it is doubtful whether the second has. Furthermore, the teaching profession itself incorporates those who hold older theories, and those who have more revolutionary political and pedagogic views than the great majority of the profession.

15.26. The 'certification' process

In order to understand the process by which the 'modern movement', if it may so be termed, has succeeded in getting its views so widely accepted, it has to be appreciated that the accumulation of careers in which a formal qualification at a high level is required before entry, and where continuous processes of updating of knowledge are then required, has ensured that the processes of individual educational and social advancement and of economic and social change have become very closely intertwined. Modern society is dominated by highly educated, professionally qualified people. The extension of the public sector, and admission to its senior- and middle-level posts by competitive examination, has been a potent force in creating a society dominated by formal qualifications outside the bureaucracy itself; the formalization not only of the medical profession but also of the para-medical professions, and the development of a whole range of scientific and quasi-scientific specializations has been notable.

Together with specializations in the business world, e.g. accountancy, and in the legal profession such as corporate law, they have cumulatively led to the view that the beginning of specialization for a career should be based upon earlier educational qualification. This has been grafted on to an older tradition that beneath the high intellectual level of preparation necessary for high university attainment, characteristic for example of imperial Germany before 1914, there should be a very highly articulated structure of craft training of all kinds. And this, in turn, has led to a consideration that, within the secondary system, there should be the beginnings of the process of preparation which would enable high standards to be achieved, and attained within a measurable span of pre-career preparation over a whole spectrum of occupations, all requiring formal qualifications.

15.27. Society requires educated people

At the same time, however, there has been a growing conviction that the development of modern society depends more and more on a very wide basis of general education and culture. Modern society requires not only well-trained producers, but well-trained consumers. Thus career preparation for a rapidly growing range of professions, trades, and crafts, requires high secondary education attainment, but modern society requires high educational achievement so that its members can function — not necessarily well, or effectively, but just 'function, period' as the Americans say — just as much as citizens as producers.

15.28. Selection and social differentiation

This powerful and incontestable view of the facts has been uneasily allied also to the conviction held on social grounds by a wide range of people that the early selection of children for educational courses is *de facto* a selection for them of specific occupational grades.

This process of educational selection is in itself a confirmation and probably an accentuation of the social differentiation which in our society is regarded (following the arguments advanced by such bodies of opinion) as an unacceptable manifestation of the hierarchical structures, and the prevalence of former cultural modes, which have been called into question in the whole of our present society. (Attention has already been drawn to the persistence of those social structures). It has therefore seemed that the necessary, and indeed inevitable, course which should be adopted by those who wish to promote social mobility on the one hand, and on the other the necessary degree of occupational flexibility that is suitable for a society which is highly responsive to cultural, social, technological, and scientific change, should be the broad preparation of children and of young people in modern knowledge. Thus many students in high quality United States universities take degrees in a wide range of subjects, which to some extent mirror the wide background which is expected of cultivated continental Europeans who take the *baccalauréat* or the *Abitur*. In this process of reasoning, it must be noted that a prior assumption is that the differences in society are not genetic and unalterable and, furthermore, that social mobility and occupational flexibility is compatible with a society which practises social and economic equality to a far greater degree than any actual society does at present.

15.29. Secondary schools

It will be seen, therefore, that the difference to which attention is often drawn by those who are dubious of the effects of the recent educational changes is a matter more complex and profound than that between specialization and generalization at the secondary school – and it is more an apparent than a real conflict. There is an old continental tradition – extended to the USSR and the USA – of a broad-based secondary curriculum. There is a narrower English tradition. Attempts to modernize both traditions have been accompanied by the already discussed and growing desire to equalize opportunities in eudcation, and to go further towards achieving an equality of the condition of education. In order that this might be done, there has been both a far wider provision of education and also a growing uniformity of curriculum, at least up to the age of 16, together with a growing tendency towards an undifferentiated structure of secondary education and, to some extent, of tertiary education. These tendencies have been carried furthest in Sweden and in some of the Canadian Provinces, but similar characteristics of reforms of educational structures can be shown to be evident throughout the whole of the NATO area.

In many respects, these tendencies have also operated in other countries, particularly in the Third World and in the Communist countries, though with some significant differences in the latter which will be mentioned later.

It is important to realize, however, that the arguments about specialization and generalization are different from, though to some extent linked with, the arguments about growing equality of access and growing equality of condition within the educational system. These can only be considered in a broader social, economic, and technological perspective. It has already been suggested that the evidence seems to indicate that education has not been, as it was once thought that it would be, the key to social change. It is equally important to realize, however, that a knowledge-based society and economy is nevertheless dependent upon the attainment that comes through the education system.

15.30. Education and social mobility

The formal education system in all countries has thus passed through a process of adaptation and change that has accompanied the massive increase in educational outlays and the very substantial increase in the number of people involved in education, both as pupils and students on the one hand, and as teachers on the other. This process has raised two major questions which will now be considered in the light of the arguments so far presented. The first concerns the relationship of the provision of education, and its organization, both with respect to structure and the content and nature of the curriculum, and social mobility. This matter has now been extensively investigated, and the results of a substantial amount of economic, sociological, biological, and genetic research are to hand, which enable an informed judgement on both the causal relationships between educational and social change, and also (to a considerably lesser extent) the processes which are at work.

15.31. The effectiveness of education

Secondly, however, there is a vast area of knowledge that is accumulating which must be brought to bear upon the central question of the effectiveness of education, both in its narrow pedagogic sense of whether or not a particular form of teaching a particular discipline is more or less effective than any other, and the broader question of the effectiveness of education in relation to the attainment of society's objectives through the development of research and learning, in so far as they affect the cultural, political, and social stability of different societies. It will at once be apparent that in the USSR and the Soviet-occupied countries, many aspects of education that are taken for granted in the NATO countries cannot conceivably be allowed to exist. It is no part of education's task in those countries either to change society (which has ossified), or to call its values and its achievements (positive or negative) into question. The task of education in Soviet-occupied territories is to train people in skills and indoctrinate them politically.

Of course, even accepting the far wider terms of reference that education has been set in NATO countries, there is difficulty in interpreting the achievements of education in relation to the experience of social mobility and shifts towards greater equality in the relatively short period over which these relationships have been studied. In biological work, the generations of rodents or insects which are observed follow each other with rapidity, but human generations are so widely spaced in time that it is impossible to make definitive judgements which would stand up to biological criteria on the basis of clear-cut evidence.

15.32. Social and educational mobility – conclusions from the evidence

The most that may be done with what evidence is available, is merely to allow a certain number of possible inferences to be drawn. There is, in addition, the extreme complexity of the relationships that are being posited, since it is known that there are close positive correlations between genetic endowment, including the complex of characteristics broadly known as 'intelligence', family background, including the marriage of parents from similar circumstances, parental occupation, parental income, geographical location, racial and religious group, the intelligence and motivation of the children, their choice of school, the style of home-life and family relationships, and their school achievement, together with their subsequent careers. To trace a causal relationship in this complexity is exceedingly difficult, which is not of course to say that it cannot in principle be done. But the longest chain of evidence so far does not exceed three generations, and the models must be judged less by reliable scientific statistical tests than by their inner logic and apparent reasonableness. And since reasonableness alters with intellectual fashion, and with striking rapidity, it cannot be a very firm guide to scientific validity. On the one side there are those simple-minded models which would argue that specific groups, for example Black Americans, are less genetically well-endowed than other groups, say Jewish Americans, and that in consequence the difference in average income – and everything else that is correlated with income, from crime to Ph.D.s – between Black Americans and Jewish Americans is genetically determined. At the other extreme there are those who would argue that the differences that are apparent between the general family and social environment of the two groups, together with the quality of the schools that they attend and the neighbourhoods they live in, is almost exclusively responsible for the observed differences in their relative educational attainment and economic and social performance.

It has already been argued that the agreed scientific evidence for the correlation of intelligence with social and racial origin is strong, but the dispute that inevitably arises as to whether or not the origins of these observed differences are primarily genetic or social in part hinges upon whether in fact the intelligence tests, despite all attempts to make them culture-free, are to a considerable extent bound up with the dominant culture in the society in which they are formulated. This latter hypothesis as to their inherent bias seems less strong than it once was, in view of the sophistication of the work that has proceeded on intelligence tests.

15.33. Some possible developments

That being so, and while it suggests a limitation, possibly permanent, but certainly effective in the short term, on the degree to which education (in the academic sense) can be widely dispersed, it is nevertheless still true that the proportion of more able children among relatively disadvantaged groups who do not proceed to make the fullest use of educational opportunities is, in all countries, even the most affluent, relatively high; furthermore, the relative rates of drop-out of those who obtain entry into advanced education coming from the relatively deprived social groups is still significantly higher than those coming from the more advantaged groups.

The evidence tends to support the view that the positive relationship between educational attainment and family background, while it may well be in whole or in part genetic in origin, must be due in individual concrete circumstances to a very substantial extent to the acculturation process within the family and its immediate social environment, including the school. Interestingly enough, differences in this respect between countries, though they mirror the total availability of education to all social groups, seem to point nevertheless to substantially similar differences between social groups, which seems to suggest that these social characteristics are relatively unaffected by the particular social and educational institutions which prevail in different countries.

It follows, therefore, that the process of educational reform and enlargement has not to any significant extent diminished the relative disadvantage of substantial numbers of the population coming from the less affluent sectors of society, even though it is true, of course, that in absolute terms, the level of achievement and attainment has increased throughout all NATO countries.

The evidence for increases in social mobility as a result of educational reform is not particularly strong, though there is historical evidence that rates of social mobility have always been higher than has been commonly supposed, and largely determined by the rate of growth of occupations in the semi-skilled, skilled, professional, and managerial groups. At times when the economy has been changing rapidly, so that these occupations have been growing and becoming more freely available, the rate of upward social mobility has increased.

15.34. The reduction of inequality

It is of course perfectly correct to argue that the influence of education cannot be assessed by itself, but must be seen in the context of other social and economic processes. For instance, a generation that is rehoused in better houses, gets better jobs and higher incomes, has more access to adequate medical care, and is given political and social responsibility is likely to do better at school even if there is no improvement in the education system *per se*. All that can be said is that where such simultaneous change has occurred, the evidence tends not to support the argument, which is a little paradoxical, since at first sight it seems so reasonable.

There is, of course, relatively little evidence that can be brought to bear across the different societies in NATO to indicate whether or not the degree of social distance between one sector and another has to any extent diminished, though there are strong and substantial reasons for believing that this must be so. Evidence from the distribution of income over time and between societies suggests that the distribution of income — and of wealth — has become more equal in many significant senses as societies have become more affluent, and the relative income differential between the upper quartile and the lower quartile is less the more affluent a society is in international comparisons. There is thus evidence over time of a substantial reduction in inequality in that sense.

There is, further, the circumstantial evidence that patterns of expenditure, and the kinds of goods and services in common use by different social groups, have tended to become more similar as technological progress has taken place. There is not nearly so much difference between a good specialist car and a popular car now as there was forty years ago. Differences in dress and eating habits are much less marked between social classes than they used to be. All this indicates the development of a culture which spreads around a norm, or a mean income pattern, from which only the very poor and the very rich are in any real sense excluded. This in itself is a revolutionary change in the social habits of NATO countries, and is in striking contrast to the patterns in many Third World countries.

Inevitably, education has played its part, both in creating this kind of society and in reflecting it and its values. It would be difficult to say, however, that education at this stage played a crucial or critical role in the creation of this type of society.

The earlier objective of equality of opportunity, which came to dominate educational thought was a powerful one; it is based, of course, upon ideas of social mobility. If every private soldier has a marshal's baton in his knapsack, a good many are in the event going to be disappointed. But it has been superseded, in radical circles, by the doctrine of equality of status, which (to the more conservative) looks like a process for handicapping the talented. Certainly education has not been particularly successful in attaining either objective.

15.35. Education and skill

What could be more adequately and satisfactorily said, is that the education system fulfils other roles in society than those connected with social mobility and social change and the creation of equality, and it is to this particular aspect of education' social and economic role that the discussion will now turn. It is perhaps safest to assert that the pursuit of equality in education has been a major task of the past century, and in particular of the past two decades, but that the success in attaining this equality has not been in any degree marked. In other respects, however, the more traditional role of education, in the transmission and furtherance of knowledge, has perhaps been more successful, though somewhat eclipsed, than the manifest social role it has assumed during the period of the great reforms.

Indeed, this relative failure of education in achieving equality, or some of its other social objectives, and its frequent reception of criticisms that it has not been pedagogically successful in giving skills, suggests that both the education system and the social system as a whole should be changed in such a way that the acquisition of skills, both of those which are relevant to occupations and of those which are relevant to social values and activities, should be acquired outside the school and during the working lifetime – in other words, that 'deschooling' should be taken seriously. There is striking evidence that large numbers of pupils and students 'vote with their feet' against education, by absenting themselves. In many senses, the experience of the French education system, a country which, second only to the United States, has placed its education system at the centre of its set of manifest social values, in consequence of the student revolution of 1968, led to the accelerated development of a series of new ideas that have been enshrined in the notion of *éducation permanente*. It is this development which is perhaps of some significance, and will require extended discussion.

15.36. Permanent education

The massive technological shifts that are taking place are making many skills redundant, ranging from those of agricultural workers to bank clerks. As the communications revolution speeds up, much of the existing clerical paper-based technology will begin to disappear. There is much experience of re-education, continuous education, adult education, and retraining. The problems are, broadly speaking, the following. First, while the motivation of the students is high, they find it difficult to prepare themselves, often lacking the correct educational background or diplomas, the logistics are difficult, and there are difficulties over finance. Consequently, even when arrangements are adequate, drop-out rates tend to be high. Furthermore, many education authorities and professional and employing bodies are unwilling to switch their main activity from initial training to post-experience courses, which require special and complex arrangements. There has been growing emphasis on this work, but it requires an administrative effort well beyond the capacity of many organizations, as well as a rearrangement of the working life of adults and their remuneration, which is not easy to organize.

15.37. Is education getting worse?

In fact, one basic premise of those who have advocated *éducation permanente*, namely that the existing education system is unsuccessful, is not necessarily soundly based.

The level of attainment of different occupational and specialist groups within society has perceptibly increased in recent years, as is evidenced by the much higher number of people qualifying at each level of the education system, in almost all fields. It thus would not be correct to say that the expansion of the education system has been in general accompanied by declining levels of attainment in any sector of the NATO economy, unless independent evidence can be adduced that the quality of the certification process has itself been devalued over the course of the expansion. Again, the cumulative range of evidence that might be adduced for this alleged deterioration of the process of certification is not powerful. On the contrary, there is evidence of rising standards measured by the content of the material which has to be presented at the relevant examinations for qualification. Thus judged by the major formal criteria of 'success' in education, namely examinations for qualifications, the education systems have been markedly successful.

15.38. Education and science

There is, too, a further point of major significance. It remains true that the great bulk of scientific and technological achievement throughout the world is generated in the NATO countries, though within those countries the weight of scientific and technological achievement has shifted from Western Europe to North America, notably over the period of the past quarter century. This to some extent represents the shift in relative population weight, and it also of course has its origin (above all in certain sciences) in the migration of highly qualified scientists, first as a result of the Hitler persecution, and more recently, as a result of the relative salary differentials between the two areas and the vast expansion of the NASA space programme and its associated scientific programmes in the United States.

Recently, the opinion has been expressed that the United States is about to lose its scientific and technological lead. The rise of Japan, itself largely stimulated, of course, by the United States, is the most clear example. It would be simple common sense to expect that societies where rapid growth and change take place, and which heavily endow science and education will make a contribution. It is clear, however, that the new countries will be Japan and Western Europe, with perhaps a few other centres; they will not be in the Soviet Union and its occupied territories. Why not?

It remains true that, for all the vast expansion of the education systems in the Soviet Union and the Soviet-occupied countries, the rate of scientific innovation in those countries has remained relatively low, and the initiation and development of technological procedures has been small. Two major relevant examples may be given. Virtually no progress has been made in agriculture and associated industries in the USSR and Soviet-occupied territories, while it has been revolutionized in

the West; in the more spectacular industrial development based on chemical research, in particular pharmaceuticals, there is virtually no evidence of Soviet
technological advance independent of that which has been copied from the NATO
countries. No single major new drug has emerged from Soviet countries, during
a period when world medicine has been revolutionized by pharmaceuticals. The
sole exception, indeed, to this lack of scientific and technological progress might
well be the space programme of the Soviet Union, where it is evident that German
engineers have played a significant role.

It is not necessary to attribute all this difference to the different performance of
the education systems of the two power blocs. The connection between scientific
achievement of all kinds, technological advance, and economic progress and military strength has been much examined. Broadly speaking, the conclusion is that
while the underlying rate of economic development of civilization as a whole is
clearly ultimately due to scientific achievement, there was a lengthy period during
which technological advance outstripped the scientific knowledge which underlay
the development of the technology; it is only in the past century that the rate of
scientific advance has outstripped the technological frontier. This is a matter of
supreme importance. This civilization is the first to be based on scientific knowledge; in that sense the present civilization has a history of only a century.

15.39. The developing world

The developing world comprises an extraordinary group of societies, some very
old, some new, some extremely poor, others adequately fed and housed. They are
united, however, by a thirst for education. In almost all of them there is a traditional form of education, usually religious in basis, and there is also a subtle and
complex structure of training in traditional techniques. But the thirst for education is not for this; it is rather for the science, technology, know-how, and culture
of the West. Consequently the growth of enrolments has been phenomenal. The
growth of education is part of the great explosion of expectations in the Third World

The most important implications of this are incalculable, for the obvious reason that nobody can foretell the impact of new Western technology and ideas on
so great a variety of old cultures: the impact of the past five centuries, and especially the last two, has been probably the most dramatic and profound the
world has ever seen. If, therefore, we are at the beginning of the impact of a yet
more revolutionary technological era, the consequences will presumably be pro-
portionally the more striking. In the specific field of education, however, Western-
style education, when transported to the developing nations, has two manifest
defects. In the first place, it is extremely wasteful, largely because drop-out rates
are so high, mainly owing to the seriously adverse conditions under which the
students live, and the inadequacy of the teachers. In the second place, much of it
is strictly irrelevant to the short- and long-term needs of the developing nations.

For any significant impact to be made, therefore, radical reorientation of the
education systems in the developing countries is necessary. The demands that this
makes upon the NATO countries are potentially many; partly in teachers, partly

in goods, including materials for buildings and books, but above all in ideas. Educational technology (television, computer teaching, and other devices) enjoyed a vogue in the 1950s and 1960s; it has shown no cost savings and little, if any, increased effectiveness over existing teaching techniques. But the analysis of pedagogic effectiveness does lead to improvement, as the Open University in the United Kingdom has shown, and there is reason to suppose that technological developments in pedagogy might be successful in the future, especially in developing nations, if the managerial capacity of educational systems were improved. In this way, the developing world might fully enter the scientific civilization.

15.40. Science, technology, and education

In this new civilization, certain interesting fundamental matters have been discerned. which suggest that at times when developments have been particularly strong in a science — when fundamental advances are made, like Rutherford's work on the atom — there is a growth in the gap between a scientific achievement and the technological progress that is associated with it. It is this which in principle makes the possibility of a technological leap in that sector more substantial than elsewhere. At the moment the gap between scientific advance and subsequent technological development seems to be most obvious, and is most obviously being filled, in biology and in some associated specializations. The achievements of the chemical revolution have already been stupendous, and the achievements of physics, particularly in the nuclear and electrical fields, have been equally striking. The full achievements of biology, though already apparent, for example in pharmaceuticals, have yet to begin to be seen in their fullest extent.

Scientific advance has to a very important extent become an exogenous variable in the economic and social structure, with a momentum virtually of its own. The reasons for this are complex, but may boil down to the theory of the 'critical mass'; once a group of scientists, which has an inner coherence of its own, is at work, it seems to develop autonomously. It is true, of course, that much applied research is based on manifest need, but pure research — and the two concepts, as Rothschild has pointed out, are ambiguous — seems to develop autonomously. It follows from this reasoning that the main effort which is necessary at any point in time for the incorporation of scientific achievement into the economy lies largely in the technological field, and it is the relative differences in this area that particularly affect the rate of growth of different sectors and different economies. The education and training of technologists is thus of critical importance. There is at the moment no strong evidence of any serious shortage of scientists and technologists in relation either to demand, or some hypothetical 'need' based upon a hypothetical maximum or optimum growth rate. Indeed, there is some reason to suppose that there is, at the moment, a serious over-supply of scientists and technologists in the West in relation to actual demand, even if the economy were to resume the growth rate of 1948–68. This arises partly because the rate of expenditure of the US NASA and defence programmes has been severely cut back, and further, because the economic recession has particularly affected those more heavily technologically dominated industries such as aerospace and heavy engineering, which have been particularly heavy employers of more highly qualified scientific and technologically trained personnel, while at the same time the output of scientists and technologists has substantially increased.

Nevertheless, it might well be argued that over the longer period these comparatively temporary factors, of NASA and the recession, may well decline in significance, and certain longer-term trends towards a higher demand may appear. It appears fairly clear from work which has been done by economists and others that the rate of incorporation of technological innovation into society has become mainly a function of the managerial styles of different firms or industries, and it is therefore on the quality and character of management that attention must chiefly be focused if this argument is to be acceptable. Again, management which is biased towards the greater use of technologists must be seen most significantly in the broader social context of the acceptability of innovation and change. It is in this particular context, therefore, that the role of the educational system in the future must most seriously be considered.

All projections of the 'need' for highly qualified scientists and engineers are highly speculative. It seems almost certain, however, that the most gifted are – as it were – gifts from God, and cannot be planned for. At the next level, the highly competent, the educational system in the NATO area is capable of producing far more than can reasonably be required, though there may be specific shortages of specific skills at particular times. (Thus, to take a parallel area, though there are more potential musicians than jobs for them in Europe and North America, there is a shortage of highly skilled violinists for major orchestras.) Indeed, like the Swiss and British navies, science now has more admirals than able seamen.

In the development of a culture that accepts technology it is – to pursue the nautical metaphor – the able seamen who are needed. Some countries, like Germany and the United States, and, to a lesser extent, France, seem more successful than others in creating a body of competently educated citizens who can cope with technology and are not frightened by it.

15.41. Education, management, and technology

If, as might be supposed, the development of science and of technology has depended chiefly upon the genius of certain gifted individuals and the experience of these gifted individuals has owed less to formal education than might be supposed and more to experience of life as a whole, any changes in the education structure (from this point of view) are to some extent irrelevant. On the other hand, if the educational system plays its part in altering the attitudes of society at large to the rates of change of the social acceptability of technological innovation, and in particular to the development of managerial styles that lead to the growing acceptability of and enthusiasm for technological innovation, then it might well be supposed that the educational system plays an important part in generating a thirst for change, not so much with what it does for the more highly intellectually gifted, but with its teaching of the average citizens of the country.

The revolution in agriculture is a very striking example of this process, which deserves — and to some extent has received — intensive study. This has been due to the application of science and technology to agriculture, but it has only been possible because of the high calibre of farmers who use the new techniques. Despite the growing urbanization of the NATO countries, the agricultural potential has grown and it has become possible for the whole of NATO to become self-sufficient. Indeed, the Soviet Union is a very large net importer of United States and Canadian grain, for example. The absence of a market economy in the Soviet empire is the major cause of its agricultural backwardness.

Agriculture, to be successful, requires highly motivated and well-trained small businessmen. Large-scale employment of large numbers of people, requiring bureaucratic styles of management, is curiously unsuited to the varied needs of agriculture. No large-scale employment structure in agriculture has ever been successful.

Those parts of the Third World that have been most directly influenced by the technological innovations which have had their origins in the NATO countries have themselves broken out of the cycle of agricultural insufficiency. This is vital. But the growth of mouths and stomachs exceeds the growth of food. So it is highly pleasing that it is likely that those countries which break out of the agricultural bind by modernization will also be those which seem most ready rapidly to adopt contraceptive techniques which will limit the rate of growth of their population.

The technological gap between the NATO countries and those Third World countries which are not within the NATO countries' orbit, and also between the West and the Soviet empire, is growing, not lessening. To this extent, therefore, it is highly likely that the relative process of technological change will itself accelerate in the NATO area, yielding a growing surplus which can be utilized, whether for the NATO countries' own purposes or for overseas aid, or whatever. (Of course, in absolute terms, the rate of growth may slow down.)

15.42. Resistance to technology

At the same time, it must be noted that there has been a significant counter-movement to the acceptance, enthusiastic or unthinking, of the process of technological change, especially in respect to its impact on cultural and social values, and this particularly among the more sophisticated and affluent sections of the populations of the NATO countries. The very affluence which technology has yielded has, as it were, bred as one of its children the counter-culture that has become a potent factor in the higher education institutions and intellectual circles of the whole NATO area. It is of course true that some aspects of this counter-culture have proved just as potent in disrupting the acquiescent attitudes of Soviet youth, and of youth in the Soviet-occupied countries, to the political structure there as the more conventional and traditional causes for dissent against Soviet practices, which arise from the love of liberty or the desire to practise a religion untroubled by persecution.

There has been a serious — some would say even catastrophic failure of the social sciences both intellectually, and practically; that is, both in understanding the world, and in devising acceptable ways of changing it. This somewhat forceful statement may need some spelling out (which has been done elsewhere), but it is perhaps best demonstrated that there are no agreed theories of society when it is recalled that the ideological division between the Soviet and Chinese states, on the one hand, and other societies, is precisely about the social sciences. There are two sciences of economics, sociology, psychology, and politics; there is one physics, chemistry, and biology. Further, in the area of applied studies, there is no agreed policy based on theory to explain economic growth, income distribution, inflation, crime, nationalism, or any of the more obvious problems of our time.

In this context, where the lack of an agreed ideology is of deep importance, and the existing creeds have diminished in their hold, revolutionary doctrines have spread their influence — though Marxism is probably of diminishing importance now that its political consequences are so well known — the general disillusionment with rationality as a guiding principle of life has some understandable basis. 'Ought' cannot be derived from 'is'. When to this philosophical issue is added the problem of the evident disadvantages that flow from technology (which is not to deny for one moment the evident advantages it gives to society), the basis for disillusionment is easily understood.

A counter-ideology to disillusionment, a revival of enthusiasm, cannot be artificially generated; it can only be hoped for, or expected.

15.43. Some speculative matters

The new interpretation of the meaning of civilization itself has been referred to earlier, both in connection with the change in attitude to the family, and to life, death, and birth, and also in the growing counter-revolution, which may be discerned within the higher education system itself.

A further factor, of course, leading to the concept of life-long education, has been the fact that the process of continual social and technological change has meant that substantial numbers of people have had to radically adapt themselves to a new environment. Perhaps the most surprising thing is not the failure of people to adapt themselves in the absence of formal procedures to enable them to do so, but the speed with which the most unlikely social groups seem to accept revolutionary technological and social change; for their social structures to persist despite this rate of change, is something which has already been discussed.

In this sense, of course, the profound ideological movements which have occurred in the acceptance of totalitarian Marxism in the USSR and the Soviet-occupied countries may prove in the long run to be more ephemeral than they at present seem. It may well be that the ideological battle which the NATO countries have to some extent seemed to have lost in the past years, as democracy, capitalism, and imperialism have taken some hard knocks, may in fact have been won at a more profound level.

Reference

[1] Thomas, Brinley. *Migration and economic growth: a study of Great Britain and the Atlantic community.* Cambridge University Press (1973).

Commentary
Sylvia Ostry

The contribution by Lord Vaizey is remarkable for the enormous scope of human problems and the sweep of global history which it encompasses. It raises questions that are among the most critical facing the world today but which, because they are so fundamental to the survival and development of humankind, are essentially timeless. They include the problems relating to world population growth and food supply, the relationships between rich and poor nations and the redistribution of wealth and income within and among nations, the interrelationships among individual human, social, and technological advancement, and the crucial role of education, which is at once the facilitating factor in the process of such advancement and a constraint upon its speed and magnitude.

Yet the very scope of the chapter, which is the source of its fascination, makes the process of summary and review extraordinarily difficult. The canvas is immense and the technique is one of broad brush-strokes. The reviewer is forced to select portions of the canvas, but the process of selection itself is frustrating because the detail and texture are implied rather than depicted. My selection has been dictated by my own preferences and knowledge, but I am also mindful of the responsibility of the rapporteur to stimulate discussion.

I must confess at the outset that I am unable to discern in Lord Vaizey's contribution a dominant theme which links the various subject matters he raises. There are, in fact, several themes, but they are introduced, drop out of sight, sometimes to reappear in an oblique and evocative fashion, leaving the reader intrigued but, like Oliver Twist, wishing for 'more'.

Let me cite one example. Lord Vaizey, at several points stresses his concern with the 'fundamental consideration of the ultimate purposes of human civilization', including 'fundamental changes in the attitudes to death and to birth which are already profoundly affecting society at all levels from the productive process to the attitudes towards consumption, investment, and defence' (p. 153). He writes of 'a growing degree of social tolerance with respect to human sexual habits . . . a growing prudishness about the discussion of death' (p.153). And, more significantly, he stresses the 'growing emphasis upon the immediate and the satisfaction which is to be derived from an affluent life' (p. 155), and sees as a significant development affecting demographic phenomena a view 'where what is sought is personal fulfilment in adult life; a concern with the immediate, and not with the future or the past' (p.155).

The idea of what the American journalist Tom Wolfe has called the 'Me Generation'[1] has been introduced, in various guises, by many contemporary social scientists. With typical panache, Wolfe cites the 1960s slogan of an advertising copywriter — 'If I've only one life, let me live it as a blonde!' — not as 'another example of a superficial and irritating rhetorical trope' but as a notion which 'challenges one of those assumptions of society that are so deep-rooted and ancient, they have no name . . . man's age-old belief in serial immortality.'

Richard Easterlin's gloomy question 'Are we locked on a hedonic treadmill?' and the surveys on the relationship between affluence and happiness that provoked it [2] have formed a backdrop to much of the current growth–no-growth debate. There is not really very much separating Tom Wolfe's cheekiness from Fred Hirsch's [3] grave moral misgivings that the pursuit and encouragement of unalloyed self-interest is the motive power that both fuels and ultimately may destroy the modern capitalist economy. The argument that hedonism is seriously *dysfunctional* is also explored by Daniel Bell [4] and Heilbroner [5] with different, though perhaps equally pessimistic conclusions.

This is not the occasion to review this complex and important literature, but only to suggest that one wishes Lord Vaizey, in touching on the idea of the permissive society in the context of a study of population and education, had explored more fully the linkages between that idea and socio-demographic phenomena — fertility, migration, women's liberation, female labour market participation, etc., — or educational problems and policy. For example, increasingly today (at least in North America) employers are finding it difficult to transfer young men or women in executive or professional positions unless they can provide jobs for their spouses. Impatience with authority and insistence on 'sensitivity and openness' have created tensions in the work environment and a pressure for changes in management style. The desire for self-gratification has stimulated more inter-enterprise or even inter-occupational mobility. The view is summed up in a quote from a young and up-coming executive: 'If I don't get present-day satisfaction, I won't stay. I won't work for the future. Nobody's going to guarantee its arrival' [6].

To cite another example of an important idea that is introduced but then left unexplored: the economic consequences of demographic change. In this instance, Lord Vaizey suggests that in developed countries 'the growing proportion of people beyond 65, the usual male age of retirement, which has been a feature of the past 25 years, has ceased to be as powerful an influence in leading to growing social security budgets as it recently was. The proportion is now stabilizing'. (p.154). He concludes that 'the adverse effects of demographic shifts have been borne in the years 1950-75' (p.154) and 'within the next 20 years — the burden of the elderly is unlikely to increase'. (p. 154).

Apart from the fact that in Canada and, to a lesser extent, in the United States, population projections do suggest some increase in the proportion of the population over 65 over the next twenty years — although, of course, the really large increases show up after the turn of the century — one may also question the assumption that the financial implications of a shift in the age composition of the 'dependent' population are inconsequential. One would want to examine *per capita* government expenditures by age category rather carefully before reaching this conclusion. One would also want to consider the effects of these demographic changes on capital stock — can elementary schools be turned into 'golden age homes' (to use the dreadful North American euphemism)? Is an ageing labour force less mobile, and does this represent a cost in social terms if major industrial and geographic adaptation is to be required over the coming decades? Will savings behaviour be affected by these demographic changes? As the tendency continues for the elderly to live apart from their children, what does that imply for housing expenditures?

In North America now, and perhaps in other countries soon, will the phenomenon of 'Grey Power' – the growth of political pressure groups based on age – become a factor in government decision-making?

These are examples of the kinds of questions that relate to the topic of 'the economic consequences of demographic change'. It is obviously unreasonable and unrealistic to expect the author to explore these complex questions in any detail. Yet the contribution does, in fact, suggest a quite firm policy position by concluding that there is no major problem. That may well turn out to be the case, but the evidence presented is, in my view, inadequate to sustain this conclusion.

Another intriguing idea which appears and reappears in Lord Vaizey's chapter is that of the stability or persistence of 'social structures'. The term 'social structures' is not defined. Clearly, as Vaizey himself has stressed, fundamental moral attitudes have changed profoundly and have induced changes in behaviour – marriage, child-bearing, and rearing, labour market participation and mobility, etc. Surveys in Canada and the United States over the past several years repeatedly show changes in the attitude to work among young people. Technological change has affected methods of production, spawned new industries, caused sectoral and geographic redeployment of labour, and changed consumption patterns. But Vaizey must mean something more pervasive and profound than these changes connote when he speaks of the persistence of social structure. Indeed, he dismisses the changes in England over the past two centuries as having had little effect on social structure and the changes in China, being 'politically imposed', as possibly ephemeral.

Further, the term social structure is associated by Vaizey with the idea of 'social engineering' – also not defined. Since he asserts that social structures are 'surprisingly' stable, social engineering is difficult – there is 'a problem about the extent to which a society is manipulable.' The major instrumentality of social engineering is (or was thought to be) education.

The idea of the persistence of social structure is extremely puzzling. Does it mean a continuity of status based on property rights? If so, then there have been marked changes in England, for example, over the past two hundred years. Does it mean a set of social arrangements by which a collectivity lives, including the values underlying those arrangements? If so, there have been marked changes in most countries of the world even over the last several decades. The introduction of free public education in the NATO countries was a radical change. It may not have produced the major results many expected – either with respect to the elevation of public 'taste' (so desired by the Utopian socialists of the nineteenth century) or the greater equality of opportunity and result hoped for by the educational reformers of the twentieth century. But certainly it had some effect on social mobility (I shall cite some North American data below) and on income equality. Many other examples of change – in values and behaviour – have already been cited. So what is meant by stability? Monogamous marriage has persisted for centuries. Would anyone argue that it is the same institution today as it was a hundred – or even fifty – years ago? All social arrangements exhibit qualities of both change and stability. Otherwise we would be living in a situation of permanent chaos (a contradiction in terms and hence impossible). The only possible meaning I can fathom in Lord Vaizey's assertion about the 'persistence of social structures' is something about 'national character', which borders on the mythic, and which I am particularly ill-equipped to discuss. No doubt the present Soviet bureaucracy may resemble the bureaucracy of the Czars; its processes and purposes do not.

Lord Vaizey provides a masterful overview of the causes and consequences of the educational explosion over the past twenty-five years, which I need not summarize here. He raises three important questions concerning this educational explosion: its impact on social mobility, its effectiveness as a production process, and – the broadest question of all – the extent to which education has contributed to 'the attainment of society's objectives through the development of research and learning' (p.175). As an aside, it is interesting – and possibly significant – to note that Lord Vaizey does not deal with several other issues which have concerned economists dealing with the role of education – for example, the impact of education on economic growth, or the burgeoning literature on the 'New Home Economics'. [7]

One must agree with Lord Vaizey that the question of the causal link between education and social mobility is extremely difficult to determine, both because of the complexity of the requisite model and the need for more data, especially more longitudinal data. I am, however, struck by the recent evidence in Canada of a significant degree of intergenerational mobility – both upward and downward. These data are, of course, not conclusive but merely suggestive. Thus, in Canada in 1973 about one-quarter of Canadian males aged between 20 to 64 years were of the same socio-economic class as their fathers, while 15 per cent had moved down relative to their fathers and 58 per cent had moved up [8]. The Canadian survey replicates the famous Duncan–Blau survey in the United States, which also showed substantial movement up and down the social ladder [9]. The causal links between mobility and education are, of course, still the subject of considerable debate. And, as Lord Vaizey points out, the rate of social mobility has been strongly influenced by the rate of growth of particular occupations, and of the economy as a whole. As we move to a slower-growth economy, upward social mobility will be impeded and downward social mobility will become more widespread. The implications of these developments for manpower policy and educational policy are likely to be extremely important, but one sees few signs in most countries of governmental readiness and planning in this regard.

Because of disappointment with the results of education in promoting greater equality and in skill-training, suggestions of 'deschooling' and/or permanent education abound in the recent literature. It's difficult, at least for me, to take Illich seriously and I hope Lord Vaizey's use of the term doesn't imply that *he* takes Illich seriously either. The controversy among human capital analysts as to whether, or to what extent, education *produces* market skills or simply *identifies* attributes which make people valuable to employers (especially trainability), is far from settled. As Blaug has said, if we can't answer the question – how efficient is the educational system in assigning people to jobs – we would be wise not to get rid of it [10]. The people who would suffer most from deschooling are Illich's alienated slum-dwellers. Middle-class kids would no doubt continue to be 'schooled'.

A more acceptable alternative – compensatory education for the disadvantaged and planned peer group mixing – has scarcely been attempted on a sufficient scale to evaluate the results. The reasons for this deficiency probably lie in the realm of politics, not education, i.e. the reluctance of middle-class voters, especially in periods of economic constraint, to redeploy resources to favour the disadvantaged. Thus 'policies which favour the richer fifty per cent of voters are apt to be approved over those which favour the poorer fifty per cent, and the poor will have to settle for the "trickle down" effects that are by-products of economic growth'. [11]

The arguments for 'permanent education' are more compelling than are those for 'deschooling'. Lord Vaizey has stressed the importance of skill redundancy as a consequence of technological change, and others have emphasized the need to provide 'second chance' opportunities for the drop-out or inadequately trained. Another view emerges from the concern with the equitable distribution of job opportunities – or opportunities for 'good' jobs – as the demographic bulge moves into the labour market of the 1980s. A recent American study suggests that 'as demographic and economic forces reduce the possibilities for career mobility and more equitable distribution of work, other social forces heighten the desire for greater opportunity. Significant current trends are greater affluence, higher levels of educational attainment, declining family obligations, and, among older people, better health and increased longevity' [12]. In this study, a variant of 'permanent education' is proposed as a form of life-cycle work sharing. The notion of cyclic life patterns – a redistribution of some of the time currently spent in school and retirement to the middle years of life, in the form of extended periods away from work for leisure or education – is both attractive and innovative, but would require major changes in public policy and institutional arrangements in industry.

The third question about the role of education which Lord Vaizey raises is both the broadest and, in my view, the most important – the extent to which it has contributed to the attainment of social goals through research and learning. Here his condemnation is virtually unqualified, and that condemnation is reserved for the social sciences. One can scarcely argue with the statement that 'there is no agreed policy based on theory to explain economic growth, income distribution, inflation, crime, nationalism, or any of the more obvious problems of our time' (p.183). Yet I am reluctant to lay the blame entirely on the shoulders of the social scientist. The policy-making process itself must be examined. Major social-policy issues involve dilemmas and trade-offs. Trivial problems have technical solutions and technocrats solve them daily. Social scientists can and do articulate and define problems and even explain the practical consequences of choice. The politician deals in values; facts are subordinate and supportive. The search for consensus involves a high level of generality and ambiguity. The politician's time horizon is dictated by the next election. Professor Eckaus brings this dilemma sharply to the fore in his discussion of the reasons for the neglect of long-term problems (Chapter 17).

There are so many other issues raised by Lord Vaizey's stimulating contribution that I should like to discuss, but my role, as I saw it, was to select and not to summarize. There is, however, one general comment I must add. This chapter leaves one with a feeling of optimism – a feeling so rare today that it comes as a kind of shock. I doubt we shall manage to sustain that mood during the discussion – but it is a challenge worth trying to meet!

References

[1] Wolfe, Tom. The Me Decade. *New York Magazine* **9**, 26–40 (23 Aug 1976).

[2] Easterlin, R. Does money buy happiness? *The Public Interest* Winter, 3–10 (1973).

[3] Hirsch, F. *Social limits to growth.* Harvard University Press, Cambridge, Mass. (1977).

[4] Bell, D. *The cultural contradictions of capitalism.* Basic Books, New York (1976).

[5] Heilbroner, Robert L. *An inquiry into the human prospect.* Norton & Co., New York (1974).

[6] The 1960s' kids as managers. *Time* 34–5 (6 Mar 1978).

[7] For useful bibliographic references see Juster, F. Thomas (ed.). *Education, income and human behaviour.* A report prepared for the Carnegie Commission on Higher Education and the National Bureau of Economic Research, New York (1975).

[8] McRoberts, H. Social stratification in Canada: a preliminary analysis. Ph.D. thesis, Carleton University, Ottawa (1976).

[9] Blau, P. and Duncan, O.D. *The American occupational structure.* Wiley, New York (1967). See also Brittain, J.A. *The inheritance of economic status.* The Brookings Institution, Washington, D.C. (1977).

[10] Blaug, Mark. The empirical status of human capital. *Journal of Economic Literature* **14**, 827–55 (1976).

[11] Buttrick, John A. *Who goes to university from Ontario?* Ontario Economic Council, Working Paper No. 1/77, p.7. Toronto (1977).

[12] Best, Fred and Stern, Barry. Education, work, and leisure: must they come in that order? *Monthly Labour Review* **100**, 3–10 (1977).

16

Technology, standards of living, employment, and labour relations

Willem Albeda

16.1. Introduction

There is an interesting flow of optimism and pessimism in the views of economists regarding the employment effects of technology. As is to be expected during a depression, technology is seen as an independent source of unemployment. During a boom period, optimism takes over and more stress is laid on the indirect employment effects of new products and of the income generated by the new technology. We live at this moment in a period of recession, and so pessimism prevails. It is typical that during the long period of almost uninterrupted economic growth in Western Europe between 1945 and 1972, the technology debate did not draw much attention. As a student in Rotterdam, in the 1940s, I read Schumpeter's *Business cycles* [1], and was immensely impressed by his command of facts and trends explaining recessions and boom periods by 'outbursts' of technological innovation. At that time, the development and the introduction of new technological possibilities seemed to leave the economy with a full – or even overfull – employment; was it now possible, as we thought modestly, that the new macroeconomics had enabled us to absorb new technology without cyclical unemployment? Since then, experience has taught us differently, and the technology debate in which there had not been much interest since the Second World War, in Western Europe at least, flared up again. With this debate the old issues have again come alive. Does technology, by a labour-saving bias, lead to a future of permanent unemployment? Must we look forward to unemployment as the one most fundamental problem during the last quarter of the century? Is there a 'long wave', in the sense that 25 years of relative prosperity, weak recessions and strong upswings are followed by 25 years of relative depression with weak upswings and strong depressions? The 'long wave' theory was introduced by Kondratiev in the 1920s [2], and is therefore named the Kondratiev cycle. However, it has never been possible to determine with any certainty the extent to which there are causes at work that are responsible for such a 50-year cycle, although the figures seem to point in that direction.

In Schumpeter's view, new technology comes in strong outbursts. In the first phase of a new technological development, there are important problems in applying the new methods and problems incurred by the design of new products and new machinery to make the products. There is opposition from existing technology; people oppose the new technology from the viewpoints of safety, employment, etc., management has to be convinced of new possibilities. The more revolutionary the new technology, the more important the obstacles that have to be overcome.

Once the resistance has been overcome, there is, in the first instance, a labour-generating effect. New branches of industry develop; new products are being developed and marketed. But of course there will be a time when the market becomes satiated for some of these new products. Economies of scale are working, standardization leads to labour saving, competition leads to price decline. In short, the work-generating effect of the new technology comes to an end.

However, the stagnation of employment opportunities which set in about 1971, both at home and abroad, cannot be explained by technology alone. Factors of supply and demand are interwoven; as a result the interrelationships are highly complex. In the following section, I shall point to some important factors that have influenced the advent of structural unemployment, viz. 'demand structuralism', by which is meant the long-term effect of technology; 'search structuralism', which means problems of bringing supply and demand in the labour market more into line with each other; and 'cost structuralism'. In my opinion, the policy for the combating of structural unemployment will have to be based on these three causes, and I shall try to indicate how this can be achieved. Apart from the structural causes, however, part of the unemployment is still attributable to cyclical causes. When dealing with the cyclical component, anti-cyclical measures should be taken, e.g. supplementary projects and demand management. However, this does not affect the structural component.

There appear to be many theoretical possibilities of explaining the present situation, but it remains extremely difficult to entirely unravel the interwoven cause and effect. However, I do not believe that this is necessary to be able to arrive at solutions, just as a child need not understand the process of walking in order to be able to walk. I am convinced that the economic–technological world of today still offers sufficient challenge to ensure that there will be a new wave of expansion, although it may be more difficult to take up these challenges than it was during the past two or three decades. Such a favourable combination of such technological factors as modern transport, modern data-processing and communication, and modern chemistry are not likely to present themselves again in the near future. It is therefore necessary, particularly at this time, that employers and employees create the conditions required for coping with the forthcoming technology.

An additional problem in several countries is the fact that the political situation is at a turning-point where the left and right have become about equally strong, with a consequent reduction in political stability. This causes an uncertainty which adversely influences the investment climate and the employment situation. Consequently, employers and employees have a marked tendency to entrench themselves in defensive positions, a process which, if continued, may postpone any solution to the unemployment problem until the distant future. The urge is felt on both sides to safeguard established income positions. I am therefore of the opinion that without a combined employment labour relations and incomes policy it will be impossible to arrive at a solution. I am firmly convinced that employers and employees will have to provide the key that will open the way to a new entrepreneurship.

16.2. Possible explanations of employment and unemployment

As previously indicated, a complex of causes underly the long wave in employment and the attendant alternation of structural excess demands and easing of the tensions in the labour market. The development of demand plays an important part in this respect, as does the development of labour costs, which in their turn are connected to the availability of a sufficient supply. The influence of technological development is an additional factor, and a more-or-less independent cause of unemployment may be the development of the labour market as such.

First I shall classify the causes as demand structuralism, cost structuralism, and search structuralism. This terminology is taken from Driehuis, who used it in a paper for the OECD conference on Structural determinants of employment and unemployment [3]. However, I shall not follow Driehuis closely, and in my policy conclusions in §16.6 I will consider the question of the conclusions we may draw from a distinction between demand structuralism and cost structuralism.

Demand structuralism
Supporters of demand structuralism relate the current widespread unemployment to the interaction of new technology and changing demand patterns. The debate on the effects of technological change is already old, and in principle two main theories are brought forward: the compensation theory and the underconsumption theory.

In its simplest form the compensation theory implies that a lower price for the product leads to an increase in demand, thereby increasing the labour force and compensating for the fall in employment originating in technological change. The more complicated reasoning says that a lower price for the products leads to lower spending by the public on this product, thereby creating a scope for expenditures on other goods and services, especially the latter, which may have a higher labour component.

By means of the compensation theory, it is possible to explain the upward phase of the Kondratiev cycle, particularly in the period after the Second World War. The post-war industrialization, including all the consequent improvement in labour productivity, which is partly the result of technological progress, has made possible an enormous increase in the standard of living and consequently a wave of expenditure which has strengthened the expansion process. Relative price reductions for industrial products in particular were, therefore, not found to be impediments to further expansion.

The underconsumption theory is more applicable to the years after 1973. This theory states that the saturation of demand in one industry leads to a fall in demand in another industry, to unemployment, with its effects, working in the same direction. Unless government takes measures to keep aggregate demand on a sufficiently high level, structural unemployment will develop. The underconsumption theory can only explain a new upsurge via the renewed occurrence of a situation of undersaturation. A new upsurge in which technology plays a part cannot be explained on the basis of the underconsumption theory.

In a recent paper for the OECD conference on Structural Determinants of Employment and Unemployment, Christopher Freeman re-explores the theory of the long wave [4]. He describes in detail the phases that may occur in the advance of new technology, and how they, through cumulation, may lead to the advance of a long wave. This line of reasoning may contribute to explain the boom years after the Second World War and the easing of the tension that set in in the 1970s. Freeman distinguishes (but I am using my own terminology here):

(1) A phase of scientific preparation of new technology: to be able to apply a technology, a combination of findings is often required and, only if this combination of findings is available, can the second phase be started.

(2) An introduction phase: the technology is applied in the production process. At first this is done in a rather cumbersome way, but fairly soon improvements are made. Goods and services must be supplied, process operators must be trained, and for all this, a great deal of investment and manpower is required.

(3) A phase of fully perfected technology: goods and services in this phase are produced with the greatest possible efficiency. The market becomes glutted, past investments appear to be overinvestment, and redundancy of staff develops.

If the flow of invention and applications could be spread out evenly over time, the long-term effects of technology would not need to result in an upward and downward secular movement. In actual practice, however, technological development comes to us in waves, in outbursts of technical renewal.

A timing of technical renewal could be ascribed to chance. (A random series is never regular. A random flow of travellers, for instance, will never arrive at the point of departure in a regular file, but always in bunches.) If the flow of inventions was determined by chance, there would still be periods with many inventions and periods with few inventions. But in addition, inventions are clearly stimulated in certain periods, which *a fortiori* applies to the practical processing and application of inventions. The tide may be moving in the right or wrong direction, and it was moving in the right direction in the post-war period. Many applications had been stopped earlier, in the years of the great depression, and then stimulated with the Second World War. For the same reasons, there was a large demand for new products, which explains the upsurge that occurred after the war, contrary to expectations. But the technological upsurge was to last more than just a few years; the multitude of products which after the war were applied to civilian purposes brought in their wake a train of secondary activities or operations. Mention can be made of the electronic communication industry (radio, TV), data processing and the automation of administration, plastics, chemistry, and many others. The influence of technical progress is shown in Table 16.1; it appears to have been of considerable importance in a number of sectors of industry, such as agriculture and textiles. The influence of technical progress is probably also reflected in the effect of upscaling

Table 16.1 Determinants of labour productivity—basic period 1951–1967

Industry	Scale		Costs		Working time		Annual influence of technical change as percentage
	Elasticity of production volume	Contribution to annual growth	Elasticity of the labour–capital cost ratio	Contribution to annual growth	Elasticity of working time	Contribution to annual growth 1960–62	
Agriculture	–	–	0.18	0.9	–	–	5.7
Animal food	0.30	1.1	0.33	1.2	-0.70	-3.2	1.2
Other food	0.25	0.8	0.29	1.1	–	–	2.2
Beverages and tobacco	0.30	1.5	0.41	1.6	-0.30	-1.4	1.2
Textiles	0.15	0.4	0.43	1.6	-0.40	-2.5	3.6
Clothing and shoes	0.30	0.6	0.16	0.6	-0.32	-1.9	1.7
Paper	0.35	2.6	0.43	1.6	-0.57	-2.6	0.5
Chemicals	0.30	3.2	0.42	1.6	–	–	1.6
Oil refining	0.35	3.6	0.78	3.0	–	–	0.0
Basic metals	0.10	1.0	0.49	1.9	-0.79	-4.9	4.7
Metal construction	0.25	1.5	0.24	0.9	-0.78	-4.9	1.6
Electrical equipment	0.25	2.7	0.42	1.6	–	–	2.5
Transport equipment	0.30	0.9	0.13	0.5	-0.95	-5.9	1.3
Other manufacturing	0.25	1.7	0.28	1.8	-0.92	-4.1	2.7
Public utility	0.10	1.0	0.44	2.2	–	–	4.8
Construction	0.30	1.5	0.18	0.7	-1.00	-6.2	1.1
Wholesale and retail trade	0.35	2.0	0.36	1.8	-0.80	-3.2	0.5
Sea and air transport	0.35	0.9	0.30	1.1	-0.86	-3.4	0.2
Other transport	0.15	1.0	0.11	0.6	–	–	3.5
Other services	0.15	0.6	0.20	1.0	-1.38	-5.5	0.0

Source: NCPB

and the improvement of efficiency, determining factors in the demand for manpower.

The crux of Freeman's line of reasoning is that such phases may be cumulative. This can be ascribed on the one hand to the fact that the length of the phases may be quite different, depending on the technology of the product concerned, while on the other hand certain causes or incidents may coincide, leading to the so-called long or Kondratiev wave. Freeman's explanation is interesting because it makes it possible to understand the employment situation in the post-war period. According to his theory, the excess demand is the result of phase 2 in a considerable number of sectors of industry. However, as more production proceeds to phase 3, excess demand disappears and gradually makes way for a decrease in demand for labour. At the same time, more and more symptoms of excess production appear, and the investment level leaves much to be desired, certainly from a point of view of demand creation.

For completeness, I repeat the description of the consecutive phases of the long cycle, which originates with synchronization in the progression of technology in a large number of areas of production:

(1) The 'preliminary phase': for the long cycle to begin, it is first necessary that there be available sufficient scientific and technical knowledge which can be used commercially. If a 'cluster' of knowledge is available, the up-swing can begin.

(2) The 'upswing': the new technology brings investment in its wake and, consequently, employment; new sectors of industry appear. Continued research leads to further applications and new techniques. Increasing competition leads to pressure on prices, and in due course pressure also appears in the labour market because of a lack of (adequate) manpower. Wages increase; labour-saving techniques become attractive.

(3) The 'downswing' is characterized by strong competition and increasing capital intensity, whereby labour-saving techniques gain in importance. In addition, the new technology makes itself strongly felt in other fields and negatively affects employment in these fields.

The alternation of periods of preponderantly boom and slump tendencies results from the accumulation of activities characteristic of the introduction phase and the fully developed production phase. While steampower, railways, and electricity, together with the motorcar, caused the first, second, and third Kondratiev cycles, so the radio industry gave the impulse to the fourth long wave. This occurred immediately before the Second World War. The subsequent 'electronic revolution' has not only led to a number of durable consumer goods, but more importantly to the birth of the electronic capital goods and components industries. At present the results of the growth in electronics are becoming perceptible in other industries: in process control, in telecommunication, in banking, etc. The labour-saving effects of this are evident.

Cost structuralism

When explaining current employment, cost structuralism emphasizes the supply side of the economy. In consultation with employers and employees, it was possible to keep wage increases reasonably within the margin of the increase in labour productivity for many years after the war, which made expansion possible. Through the prolonged overstrain in the 1960s, labour found itself in an extremely strong position. At the same time, attention shifted from the target of employment — jobs became easily available — to the target of a larger share of national income for employees. In the course of the 1960s and into the early 1970s this led to real wage increases which by far exceeded the increase in productivity. In this respect two factors are especially important.

First, technological progress in a number of sectors of industry made possible labour productivity increases that by far exceeded the national average; for example in seaports and the manufacture of electrical and electronic equipment. Consequently, these sectors could act as wage leaders. In industry sectors which could not expect such productivity increases, firms were forced to pay increased wages which had no valid basis in productivity.

Second, over the same period (from about the mid-1960s), public outlays and consequent collective charges also increased considerably. As a result, a large percentage of the wage increase was no longer available for private spending. This gave rise to a process of burden shifting, which made labour costs rise even higher. When the easing of the pressure on the labour market was felt in the course of the 1970s, it was impossible to reduce the increase in costs fast enough. The relatively higher wage costs may have contributed to a large extent to substitution of labour by capital and to the process of technological renewal. The Central Planning Bureau has investigated the influence of the relatively higher wages as far as the Netherlands is concerned, with the aid of a vintage model. One difficulty is that the processes of substitution of labour and capital, on the one hand, and technical progress, on the other, are difficult to isolate. In both cases, however, equipment may become obsolete faster because the rise in wage costs is steeper than the rise in labour productivity.

The results of the investigation actually draw attention to the importance of replacing old capital goods by new ones with higher labour productivity. As the labour productivity of the existing means of production becomes insufficient to pay the higher wages, the profitability of industry is impaired. Sufficiently favourable prospects are not available to justify replacement, and as a result, external financing becomes unattractive or impossible, just when internal financing is falling short.

The employment consequences are evident. Apart from the labour-saving which results from demand structuralism, companies are increasingly unable to maintain themselves because of cost structuralism. As a result, a great deal of activity is lost which cannot be quickly regained. For after all, with the closure of companies, an amount of know-how, organization, and market goodwill is lost, which can be rebuilt only with great difficulty. I consider this one of the most serious consequences of the present situation and am saddened by the employment consequences.

Search structuralism
The supporters of 'search structuralism' seek an explanation of unemployment in

the malfunctioning of the labour market. In my opinion, a major problem is also the fact that supply and demand in the labour market are not in line with each other. Under the influence of technological renewal and changes in product demand, there may be considerable changes in the structure of the demand for labour. Other qualities than those to which the current supply of labour is attuned are required. Normally speaking, people are trained for an occupational life of 40 to 50 years; consequently the nature of the labour supply changes only gradually. If shifts in demand take place at a much faster rate, they can only be counterbalanced by a forced rate in retraining and adjustment. In actual practice, such a quick process of retraining and adjustment cannot be organized, nor can everyone be expected to adjust himself automatically to changed conditions. All this results in frictional unemployment, in which changes follow each other rapidly.

On the basis of a comparison of, *inter alia*, unfilled vacancies and unemployment, Driehuis concludes that since 1970 frictions in the labour market have increased, particularly in France, Germany, and the Netherlands, but he believes that the proportion of frictional unemployment in the total unemployment is not more than 0.5 per cent. It is difficult to assess what value would be ascribed to such a figure. As the situation becomes less favourable, the dividing lines between the various types of structuralism become more vague. I am inclined to classify the more stringent selection requirements applied by employers and employees as frictional unemployment, resulting in part from relative overtraining. In that case, the proportion mentioned by Driehuis is certainly low, and in this case I am not considering that frictional unemployment may have remained partly hidden as a result of disabled insurance schemes and other forms of withdrawn supply.

Demand structuralism in addition to cost structuralism as a major cause of unemployment

Present unemployment and its component elements have been investigated in detail by Driehuis, according to whom unemployment has been structural since 1970 and is composed of a combination of the three aforementioned types. He believes that the most important causes are demand structuralism and cost structuralism. Demand unemployment is explained on the basis of four demand sectors:

(a) Private consumption. Compared with the situation prior to 1970, it has hardly increased;
(b) Export of goods and services: again little increase has been recorded;
(c) Autonomous expenditure (i.e. public expenditure and residential construction). There has been a marked decrease in France, Italy, the Netherlands, and Belgium;
(d) Investments. The decrease in investments with its strong effects on employment is an especially important source of 'demand structuralism'.

A complicated question is whether the shortage of employment may be a result of technological renewal and the substitution between capital and labour resulting from rising wage costs (demand structuralism), or a shortening of the economic life span of the means of production and the forced closure of companies (cost structuralism). Driehuis sees indications that the 1970s are characterized by less technological renewal, and as a result, the rise in costs can be coped with less successfully. According to him, cost structuralism, particularly in the Netherlands is, in addition to demand structuralism, a major cause of unemployment. This is true to a lesser degree in Germany and Belgium, and it is a negligible factor in France and Britain.

16.3. Analysis by sector of industry

It will have become clear from the foregoing that both demand structuralism and cost structuralism may have different effects on the different sectors of industry. After all, market conditions are different for different sectors of industry and this also applies to the development of labour productivity. In this section, more particular attention will be given to the problem for each sector of industry.

Both demand structuralism and cost structuralism can be regarded as elements of a neoclassical picture of the development of sales, production, employment, and fixed capital formation. Of particular importance for development in the sectors of industry are:

(a) The development of *overall demand*. If upswing tendencies prevail, it will seem that things are going well everywhere, although output may not be increasing equally everywhere. If downswing tendencies prevail, difficulties are bound to arise in certain sectors of industry. The employment effects may be regarded as a form of demand structuralism. An important factor in this respect is the elasticity of income or exports, in short the direct connection between the growth of total sales and the sales of specific products. This relationship has been studied widely. The best-known studies are, of course, those made by Engel resulting in the so-called 'Engel-curves'. In the following I shall refer to recent research carried out by the Central Planning Bureau with regard to the Netherlands published in the *Netherlands economy in 1970* [5] (see Table 16.2).

(b) A second factor is the *development of labour productivity*. There may be wide variations by sector of industry, depending on the long-term effects of determinant factors such as increase of the scale on which production is carried out, efficiency improvement caused by changes in the price ratio of production factors, the influence of technical change. All these are influences which co-determine the demand for work, and which together with the overall demand, have effects which should be seen as demand structuralism. The determinants of labour productivity (which are as a matter of fact determinants of the demand for labour) have been investigated by the Central Planning Bureau for the Netherlands [5], (see Table 16.1).

(c) As a third factor, *wage costs*, together with labour productivity determine the price level of value added in the various stages of production. (The price level of raw materials and semi-manufactures is also of importance, but has been left out of consideration for the sake of simplicity.) It goes without saying that there may be differences in wage level and wage development for each product as they are also

Table 16.2 Elasticity coefficients—Basic period 1951–67

	Private consumption		Exports	
	Sales volume	Relative price	Sales volume	Relative price
Agriculture	0.48	−0.67	0.2	−1.0
Mining	0.50	−0.20	–	−1.5
Animal food	0.37	−0.54	0.6	−0.6
Other food	0.37	−0.54	0.2	−1.2
Beverages, tobacco	0.93	−1.26	0.8	−1.0
Textiles	0.84	−1.15	0.6	−1.6
Clothing and shoes	0.88	−1.20	1.0	−1.1
Paper	1.36	−1.84	0.8	−1.5
Chemicals	1.29	−1.32	1.5	−1.2
Oil refinery	1.00	−0.70	1.1	−1.2
Basic metals	–	–	1.4	−1.2
Metal construction	1.53	−2.16	0.9	−1.3
Electrical equipment	1.36	−1.74	1.8	−1.1
Transport equipment	1.26	−1.67	1.0	−1.2
Other manufacturing	1.22	−1.63	1.0	−2.6
Public utility	1.50	−0.65		
Construction	0.66	−0.89		
Housing	0.66	−0.91		
Sea and air transport	2.47	−3.32		
Other transport	1.84	−2.43		
Other services	1.79	−2.11		

Source: NCPB

influenced by the training level and the scarcity of the occupational qualifications required, but on the whole, they follow the general trend in the wage level. When the wage level rises structurally in excess of the labour productivity, unemployment (cost structuralism) may occur.

(d) Finally, the elasticity of relative price is of fundamental importance for the cost structuralism, as seen from the industry point of view. Without an insight into the price elasticity, it is impossible to form a clear picture of the complicated compensation theory, and the same applies also to the underconsumption theory. The relative price elasticity has likewise been studied extensively such as the study recently done for the Netherlands by the Central Planning Bureau (see Table 16.2).

With the aid of such data it is theoretically possible to verify how demand structuralism and cost structuralism determine production and employment in various sectors of industry.

A major problem in this respect is the degree of accuracy which may be assigned to the coefficients. The general econometric assumption underlying this type of investigation is that these coefficients do not change, or show little change over longer periods of time. This assumption, however, only holds good to a limited extent. There are many reasons for fluctuations in these types of coefficients. The aggregation level of a sector of industry comprises products and production processes, each having a destiny of its own, so that if weighed together the co-efficients by sector of industry may fluctuate. It is also likely that for specific products and production processes there may be changes in the behavioural parameters which are determined by circumstances. This does not alter the fact that coefficients of this kind can be used as examples to clarify trends in production, employment, and the wage level.

On the basis of the data derived from Tables 16.1 and 16.2, Van der Werf [6] examined in further detail whether systematic tendencies can be recognized in the manufacturing and mining industries. The selection criteria are based upon differences in behaviour when formulating a representative grouping. In view of the existing variety of companies, it is quite possible that some might appear in the wrong grouping, from a behavioural standpoint. As starting points for regrouping of manufacturing and mining industries, consideration was given to:

(1) Determinants of cost structuralism: derived productivity functions for labour, based on Cobb-Douglas production functions and price elasticities for 14 branches of industries of manufacturing and mining, on which Table 16.3 is based;

(2) On the demand side: sales, elasticities for 14 branches of industry, of manufacturing, and mining (consumption and exports): compiled from the data derived from Tables 16.1 and 16.2).

Table 16.3 Selection criteria for industry grouping—basic period 1951–67

Sales elasticity	Elasticity of the labour–capital user cost ratio	
	< 0.4	> 0.4
< 1	Animal food Other food Clothing Non-coal mining	Textiles Beverages and tobacco Paper
> 1	Metal construction Transport equipment Other manufacturing	Chemicals and oil Basic metals Electrical equipment Public utilities

After some research, it was possible to distinguish a significant criterion in the influence of the ratio of wages and capital user costs on the demand for labour, higher relative labour costs calling for less employment of labour and more employment of capital. The elasticity of demand for labour per industry with respect to the labour–capital user cost ratio appeared to be clearly clustered around unweighed means of -0.23 and -0.48 respectively. From a demand point of view, a comparable division is possible into sectors with sales elasticity of over and under unity. The sales elasticities of consumer expenditure in each sector are defined with respect to total consumer expenditure. They are related to income elasticities, but the latter coefficients are not directly available, as the latest consumer surveys do not give income data. The sales elasticities for exports from sectors are defined in the same way with respect to total exports. Data have been investigated to find out whether other criteria could be used as well, but without any significant result.

In Table 16.3, manufacturing and mining industries are grouped into their respective classes. This grouping touches the relationship between recent industrial development and demand and cost structuralism. The reasons industries decline and companies fail appear to be:

they are stuck in markets that do not show expansion (low sales elasticity); the concomitant unemployment can be attributed to demand structuralism, and/or

they are trapped in a relatively bad competitive performance (slow reaction to increased labour–capital cost ratio); the concomitant unemployment can be attributed to cost structuralism.

In the best cases, the management of the company fails to react to new circumstances, but is still in a position to improve the product mix or to introduce new technology. In the worst cases, activity should be stopped in the present locations and transplanted to places where high relative labour costs are not such a pressing problem, at least not for the time being. On the other hand, industries in a relatively strong position (high sales elasticities, more possibilities to react to the production factors price ratio) could stay and expand in their present locations.

However, in both cases other developments have their impact on the labour market. Of course transfer of activities to other regions directly involves employment, but a reduction of employment also occurs in cases of maintained or expanded industries as they maintain their position by substituting capital for labour. The non-industrial sectors will absorb surplus labour, but whether total absorption of surplus labour is possible depends on the production function and their sales and price elasticities.

However, before any conclusions can be drawn with regard to the situation in the various sectors of industry, attention must be drawn to two aspects. It goes without saying that each sector of industry comprises activities which greatly differ in nature. The sectors of industry are based on averages, and thus not every activity of a sector of industry will fit into this picture. In the second place, a number of years have passed since the period over which the observations are based. This period only contains years from the upswing of the long wave, and it is likely that all kinds of activities which were then still in the introduction phase have meanwhile proceeded to the fully developed production phase. As a result of this, the coefficients of the production functions may have shifted; furthermore the product composition of a sector of industry may have changed, a factor which particularly influences values of sales elasticities.

These may cause the picture at the present moment to be significantly different, and I have taken into account the fact that the weight of the lower right-hand quadrant of Table 16.3 may have diminished. In this category are a number of sectors of industry in which, in the past, the labour productivity increases and the growth potentials were the largest, and probably larger than at the present moment. If so, this would have increased the other quadrants, from which we could derive fewer growth impulses and fewer possibilities for increasing labour productivity. In other words, the influence of demand structuralism is here reflected in a lowering of the average sales elasticity, and thus of cost structuralism, in the lowering of the average elasticity of the labour–capital user cost ratio.

The policy conclusion is obvious: it should be directed towards development whereby these negative tendencies are broken, thus making way for industrial development whereby the weight is shifted again to the lower right-hand quadrant. I shall revert to the possibilities of such policy later on.

16.4. The non-market sector

The Netherlands Scientific Council for Government Policy recently submitted a recommendation on long-term policy [7]. In this recommendation the absorption of redundant manpower was also considered. In the post-war period, the tertiary sector as well as its market sector absorbed an enormous amount of manpower. In the future too, this process will proceed, seeing that the sales elasticity should clearly be estimated as having a value of more than one, and the possibilities of buying the products of the service sector from abroad are limited, so that the problem of competition from countries with a lower wage level exerts a smaller influence. It is true that the Central Planning Bureau has estimated the elasticity of the labour–capital user cost ratio at 0.25, which is comparatively low, but possibly the influence of administrative automation has been underestimated. Should this actually be the case, the possibility of absorbing manpower in the tertiary market sector would be less favourable than was the case in the past.

The Scientific Council now advises an industrial policy directed towards strengthening non-labour intensive companies to maintain the national income *per capita* at the present high level or to raise this level, as a result of which possible problems in the balance of payments may be prevented. Consequently, the macroeconomic incomes problem can be solved, but this still leaves the macroeconomic problem of distribution, and the employment problem.

The Scientific Council sees sufficient possibilities for the creation of jobs through the existence of large unsatisfied needs in the non-market sector. This is not referred to as public service in an administrative sense, but to the large area of social activities that in a modern state are completely or in part financed by collective funds.

This sector of social activity is actually tending towards considerable impoverishment. Through the businesslike approach to life, including family life, many services that used to be unpaid have become professionalized. On account of high wage levels and the limited scope for increasing efficiency in this sector, these social needs have become far too expensive for large sections of the population. The suggestion of the Scientific Council is now to expand employment opportunities in this sector, financed from taxes on business profits. In my opinion, however, these proposals should not underestimate the difficulties which the problem of distribution already poses and which are practically insoluble. In this respect, two effects must be considered.

At present the Netherlands Government is trying to curb the growth of collective expenditure, which is a move in exactly the opposite direction. This is necessitated by the increase of taxation and social security contributions over a number of years. As the balance of government spending cannot be increased *ad infinitum*, a limit to the burden of taxation and contributions also means a limit to collective expenditure. The fact that the expenditure on income transfers tends to increase with the employment situation aggravates the problem. The entire curbing operation is an exceedingly difficult question of distribution, as it affects vested interests and established structures.

The second question which plays a part in this respect is the right of the social partners to establish wage levels in free negotiations between employers and employees. It is not quite clear how industrial labour productivity can be raised without the wages being allowed effectively to follow. This is irrespective of the problem of how the necessary laying-off of workers should be solved, now that the trade unions have become job-conscious. There can be no doubt that the supporters of the solution of expanding the non-market sector at the expense of industry are strongly underestimating the distribution problem, and with it the problem of labour relations as well. No doubt there are extensive possibilities for expanding the non-market sector and for meeting the needs not backed by the necessary purchasing power. In my opinion this should also happen. However, I only see solutions if in the market sectors (both in the manufacturing industry and in the tertiary sector) sufficient growth of production and employment opportunities are created to be able to solve the distribution problem without having to make too great an appeal to interhuman solidarity. Such major problems as the questions of distribution and employment cannot be solved over the heads of the parties concerned. Employers and employees together will have to come to grips with the demand and cost structuralism hand-in-hand. The solution must be found in the relationship between employers and employees, both individually and collectively, for without efforts on both sides, the machine cannot be restarted. It is obvious that governments can and must play stimulating roles in this respect.

16.5. Developments in labour relations

Questions of income distribution and employment are nothing new, and the correlation with employment is not being seen for the first time either. If I may confine myself to the Netherlands and the post-Second World War period, I find that the distribution question has continuously been the focus of interest. Immediately after the war there was a debate on levelling, and after an initial levelling process during the years 1945–6, wage increases were deliberately kept low to make room for the recovery. There is no doubt that the great majority of the Dutch people then backed the policy pursued, or at any rate endured it. People agreed that, above all, employment opportunities had to be created for our rapidly growing population. The means to achieve this was industrialization, and to make this industrialization policy successful, the international competitive position had to be strengthened as much as possible. Initial recovery, and later industrialization, demanded investments for which financial latitude had to be allowed.

All this required moderation in the field of consumer spending. The means to achieve this was an incomes policy, and in particular a wages policy. At first the government tried to control the average development of wages by means of a guided wages policy. After that, an evolution in wages policy took place, whereby a greater proportion of supervision of the development of wages was left to organizations of employers and employees. Finally a system of free wage bargaining was adopted, but the possibility of intervention was left to the government. Especially at the beginning of this development, it was ensured that wages would not be too much in excess of average labour productivity. Moreover, the degree to which wage increases were granted was also dependent on the development of productivity within the sector. This did not fail to exert a favourable influence on recovery. Gradually, and particularly during the 1960s, full employment was achieved, and even developed into overfull employment. In this situation, it became more and more difficult to keep paid wages in line with collective agreements.

We might describe this as a classical example of the distinction between a wage-market and a job-market. At first wages and incomes were mainly controlled by the government, but later on there was more and more participation on the part of the social partners. To this must be added that the supply and demand aspects of the labour market were no longer the decisive factors for wage determination; the industrialization policy pursued led to employment of more than the expected labour supply. Again and again the actual numbers of those who took part in economic activity were found to be in excess of the estimate. In other words: labour supply was found to be much more elastic than had ever been realized.

To put it somewhat differently, the job-market and the wage-market were found to have a certain autonomy with respect to each other. This also explains why the gradually developing sociology of the labour market in the Netherlands was mainly preoccupied with the study of the job-market and particularly when employment opportunities became scarcer, with the study of labour supply. Under the circumstances this development was more or less obvious. It was possible to study the job-market without bothering too much with the wage-market.

It took a long time to become evident that this was bound to have consequences for wage policy and wage levels. In 1964, the registered manpower reserve had already fallen below 1 per cent of the working population, but meanwhile there were still lengthy discussions in the Social–Economic Council about minimum differences of what had to be regarded as admissible wage increases, which of course were to be small. However, tensions appeared to have run too high without people realizing it, and wage explosion took place. There were wage increases not only for just one year, but for a number of years, that could not be compared with the figures for wage increases which had been hitherto agreed. Substantial consequences were immediately noticeable in the labour market.

The labour market in the preceding period had shown great elasticity, coupled with a controlled system of wage determination, but at a certain moment it was found that the tolerance limit of the system had been exceeded. This shows that there is a correlation between the wage-market and the job-market, a correlation which cannot be left out of consideration without consequences; I now deal with the period after the wage explosion. Although the great wage increases that took place originally could be explained as adjustments to what might be called the underlying demand and supply relationships, the entire period after the wage explosion can by no means be characterized in this way. This process of adjustment was completed in about 1970, but this does not mean that the high lack of balance in the labour market had come to an end.

Again, there was the phenomenon of a relative autonomy in the wage market on the one hand, and the allocation or job-market on the other, but the position was different now. In the period prior to the wage explosion, wages did not reflect a shortage of jobs because there was an increasing demand for workers, but subsequent wage increases were of such a nature that they encouraged a saving in labour. On account of the elasticity of the supply, this was not so conspicuous at first, although seen in retrospect, the phenomenon was there clearly enough. While during the first period we had mainly sought and created jobs in industry for our growing population, it now appeared that there was no longer growth potential in industry, but moreover in the long run it was superseded by a saving in labour. There was rationalization; investment in depth had to take care of contracting profit margins. Reorganizations and mergers occurred as well as closure and bankruptcy of certain companies; in the course of the 1970s, these phenomena became more and more apparent.

It is obvious that a distinction should be made between the increase in wages paid as such and the increase in labour costs, and furthermore, that a distinction should be made between an increase in money wages and an increase in real wages. This means that there may be a considerable margin between an increase in money wages and an increase in real wages, in other words, that there may be a considerable margin between an increase in labour costs and the real wage increase.

In the post-war period, social security schemes were expanded considerably. This development has had its repercussions on labour costs, although not for the full 100 per cent because of direct and indirect taxes, as well as national and employees' social insurance. In other words, the difference between labour costs and real wages is not unimportant. The process of increases in taxation and social charges has not come to an end in the 1970s; in addition to the high increases in money wages, the increase in the burden of social security services has continued. Although these social services in themselves should be seen as a form of real income, this only happens to a limited extent for those concerned. Perhaps we might speak here of a discrepancy in approach to the question of this collective burden between trade union leaders and members. One would get the impression, at least in the case of the average worker in the 1970s, that the increase of collective expenditure did not lead to a reduction of the increase of their real disposable income.

The increases in labour costs were considerable. They led to a reduction of profit margins, to losses, and to the symptoms that were discussed in detail earlier on: rationalization, investments in depth, reorganization, closure, etc. While in the first post-war period the wages and labour costs remained too low and led to strains on the wages front, the development of the wage explosion led to strains in the field of profitability. In the first period an increasing demand was made on manpower;

in the second period, labour-saving was implemented to a large extent, especially in industry.

As a result of the marked increase in unemployment in the last few years, employers and employees have become much more job-conscious than used to be the case. Recently, in collective bargaining, the trade unions even raised the question of unemployment; through the conclusion of job maintenance agreements, they hope to gain more influence on the company's policy with regard to employment opportunities.

The employers opposed these job maintenance agreements as they feel unable to commit themselves in respect to employment opportunities within the company. This may be too much of a burden for a company as it is already bound by agreements with suppliers, customers, and capital supplies. Only if the employees also make commitments, may we speak of a real agreement whereby a contribution comes from both sides. Such an agreement might contain cost-reducing elements to guarantee jobs on the part of a company, but this is understandably a rather difficult point for trade unions.

I believe that there is reason for dwelling on the sharpening of labour relations that has taken place precisely at this moment when the need for creating co-operation is most urgently felt. It is realized that the down-swing tendencies in the economy, retrenchment, closure of companies, and unemployment are imminent threats, both for the employer and for the employee, individually and collectively. Part of the tension that has arisen in labour relations can be explained by this.

Not only is increased unemployment having a demoralizing effect on those directly concerned, but it is also an imminent threat for large groups of the working population, who can only maintain themselves in the labour process marginally, both as employees and self-employed persons. This problem of persons having difficulty in maintaining themselves in employment has recently been occurring to an ever increasing extent. Advancing mechanization and automation in addition to other factors such as the relatively marked increase in the services sector have had substantial consequences for job content and working conditions in companies; the demand for humanization of labour conditions is increasing steadily.

In the first place, it may be stated that the job content of various occupations has changed drastically, and that certain occupations have disappeared, whereas new occupations have emerged. In the second place, there has been an increasing specialization of work, for which a relatively high degree of training is required. We can imagine how affected employees or self-employed persons feel about this; younger workers who have had a specialized training present an imminent threat to the older worker who has worked satisfactorily for many years. In other words, the safe feeling connected with having a secure job has been shaken. This situation applies to many more occupational groups in the Netherlands than we had ever known before and it strengthens many workers in their opinion that they are merely regarded as a means of production.

These tensions are reflected in the attitude of the trade unions. For after all, the trade unions are organizations of workers, and such organizations cannot afford too great discrepancies between their attitudes and those of their members. But these interests are often conflicting; they are both directed towards wages and towards the maintenance of jobs, and as a result, it is becoming increasingly difficult for the trade unions to be reasonable.

There is also the viewpoint of the employer, who is highly cost-conscious. Particularly in those sectors of industry where it is difficult to include all costs in the price of the product – and that applies to most sectors nowadays – the employer is led to regard his sales problem as a cost problem. To him, wage costs are a very important problem, and any increase in wage costs is an imminent threat to the employer who sees that trade unions are playing an important part in the determination of wages. In his opinion, the resulting high wages are impediments to the solution of the employment problem; they force him to economize on labour, although he regrets having to lay off workers.

The cost-consciousness of the individual employer is a clear example of a micro point of view. He is not in a position to realize that what applies to him, also applies to his competitor, and that if everyone is applying labour-saving methods at the same time, employment opportunities might be endangered. His company policy is directed towards maintaining employment opportunity to the greatest possible extent, so that an employer regards rationalization mainly as a means of strengthening his competitive position and of improving the financial position of the company.

Both employers and employees feel threatened in the current economic situation, so they tend to present their negotiating positions on working conditions in louder terms. Consequently, viewpoints in the negotiations on working conditions are formulated in harder terms, and differences in fact and in principle between employers and employees emerge more clearly. In this way, however, it is becoming more difficult to pave the way for the kind of co-operation required to set the economy in motion again. Employers and employees are thus inclined to assume a defensive attitude and to opt for solutions which may provide relief in the short run, whether or not they are solutions which in the more distant future may have a stultifying effect and, consequently, be impediments to a new economic take-off.

In the following sections I shall examine the possibilities that are open for developing policy to prevent these defensive attitudes and solutions, and hence to clear the way for more constructive co-operation.

16.6. Policy conclusions

It is now clear that the formulation of a policy directed towards the combating of structural unemployment will require pursuing a number of courses of action, each in itself inadequate to solve the problem, but which combined may provide a substantial contribution to a solution. When outlining these possibilities, a link should be established with the causes of search structuralism, demand structuralism, and cost structuralism. In addition, cyclical demand stimulation deserves attention, at the same time, since a further deterioration of the present situation must be prevented.

Policies for combating search structuralism

One of the greatest problems on the labour market is how to get the right man in the right place. In periods of fast change in production structure, an adequate functioning of the labour market requires adjustments in labour supply in a way that, humanly speaking, is not always possible. I have dwelled on this before. It should therefore not be expected that the policy directed towards counterbalancing the qualitative and regional discrepancies in the labour market (which means bringing supply and demand in the labour market more into line with each other) will necessarily lead to reducing unemployment to an acceptable level, yet it is not without effect. Recently, a new impulse was given to a placement policy within the framework of the 'New Style' employment office, which aims at more active methods of work placement. In this respect, the scope of action on placement is widened, regionally as well as with respect to vocation or social environment. On the other hand, placement will be assisted by integrating the problems of retraining, updating, and further training. Certain categories of workers are disproportionately hit by unemployment, particularly young persons, female workers, handicapped persons, persons over 45, and persons who have been unemployed for a long time, which emphasizes the necessity of creating equal job opportunities. These efforts are reflected in regulations under which jobs are created or retained by means of a state subsidy, either temporarily or permanently, for these categories. In addition, training measures are undertaken to strengthen the labour market position of the groups concerned to assist them to become more competitive.

In the Netherlands, policy is directed in this way and through other specific measures towards improving work placement and reducing frictional unemployment. The problem is that the number of jobs does not increase; in other words, the time and trouble taken to place a person in a job may result in the fact that another person does not get the same job. Seen in a more distant future, in which it is likely that a shortage of jobs will remain, consideration should be given to the ways in which a more acceptable distribution of available jobs can be determined. This is partly a question of better opportunities for those who are out of work and who want to work, and partly a question of how this will ensure better link-up with the needs of many people who have a job, but would like to work shorter hours with a proportional reduction in income.

When distributing work we may think of:

shortening the average annual hours of work (reduction of the daily or weekly hours of work, increasing the number of holidays, stimulation of short-time working);
shortening the active period (lowering of pensionable age, raising of school-leaving age, educational leave).

If other conditions remain the same, redistribution of existing jobs has significant consequences on the labour cost per unit of product. In a policy directed towards lowering search structuralism, the problem of costs indirectly comes up for discussion. In this connection I would point to the danger of defensive, rigid solutions, since they cannot be easily undone. In addition, they would inhibit a necessary future economic revival through an inheritance of costly regulations.

When seeking a new approach to employment policy, the following observations can be made. The employment problem gains a special character by the increasing participation of (mostly married) women. The social context in which this development occurs, the emancipation of women, also creates special problems for employment policy. There is still a wide scope of measures required to secure equal opportunities for men and women, partly due to meeting the challenge of creating working facilities that make it possible to combine work with family responsibilities.

During the last few years, more and more initiatives have been directed towards creating a more meaningful existence for those who are excluded from the working community for a considerable length of time. This includes hundreds of initiatives aiming in various ways at providing facilities to help the unemployed to spend their lives in useful ways. These projects often come under the category of 'socio-cultural activities for the unemployed'. In a broad sense a number of these activities could develop into social 'supplementary projects'.

Although it should be recognized that these projects are quite divergent, not only as to nature and content, but also according to their degree of success and viability, we may nevertheless speak of them as an important development in a general sense. With respect to such activities, governments have not remained inactive either. On the one hand, consideration is given as to in which cases the existing regulations with regard to performing productive work can be expanded to cover the period during which unemployed persons are in receipt of unemployment insurance benefits. On the other hand, care must be taken to avoid providing public spending for activities which normally fall within the sphere of usual employment in the market sector.

In all these cases, the organizations of employees and employers should be aware of the possible consequences in a more distant future. It would be an ideal situation if they could work together to find more effective operational solutions.

Policies combating cost structuralism
As the underlying trend in the increase in unemployment is of a structural nature, and the high level of labour cost is one of the determining factors, each package of measures, whatever the care bestowed on them, will demand sacrifices from various groups in our community. Therefore, a successful approach to all these problems can only be developed through consultation with the social partners involved. It requires the support of the parties concerned, based on their own responsibilities and their own views of the social development desired. The consultations should also deal with such questions as the burden of social security expenditure, the real growth of private incomes, worker participation in management, and employment policy. Whether it is possible to reach agreement between the social partners about longer-term cost-reducing solutions, an important condition for restoring production and employment opportunities, will have to be investigated. In this respect, whether an impetus can be given to a systematic and coherent approach to the policy concerning the determination and distribution of incomes becomes an extremely important question.

In broad areas, there appears to be a consensus of opinion about the seriousness of the situation. It is realized that restrictions in the development of incomes and in the growth of social security expenditure are unavoidable, but this consensus does not apply to the way in which it should be accomplished. The trade unions do not want an unconditional moderation. In their opinion, there should be a real improvement of the incomes of the lowest paid, and a direct relationship between moderation of income and improvement of employment opportunities.

This 'more profit more work' idea is not primarily concerned with the question as to whether there is a relationship between profit and employment opportunity. Of course, there is such a relationship. Profit is an important source of financing investments and necessary for creating jobs. But the question at issue is whether this profit is appropriated for this purpose, and whether the employees can derive certain rights from the 'profit' which can be made as a result of 'their' moderation. The incomes policy should therefore be incorporated in a more comprehensive policy, in which both material and non-material claims are given weight. With regard to the foundations of incomes policy, I expect to be able to continue the policy of my predecessor, Boersma, as explained in his inaugural speech for the IEA Conference on Incomes Distribution in 1977. [8]

One of the guiding principles in line with modern social views, but not at all new, is that all men are equal and that there is no reason for assuming *a priori* that people should have disparities in income. Differences in work performance should be recognized as such, if they are to lead to acceptable differences in income.

These differences should be based on the effort or sacrifices involved in earning an income. In this connection, we have in mind hours of work, the nature of the work (attractive or repulsive), responsibility, etc., in such a way that these factors can be interpreted qualitatively with the aid of job evaluation systems. The income differentials compensate for differences in sacrifices and efforts involved in the earning of income, and increase the supply of manpower available for this type of work.

There is a second category of income differentials, namely those arising from factors such as aptitude and training. As a matter of fact, these are not based on valid grounds, as they cannot always be regarded as a result of merit and, consequently, are more difficult to justify from an ethical point of view. Nevertheless, they cannot automatically be equated with the non-compensatory income differentials. Income differentials arising from aptitude and training call for a special approach, as they fulfil a function in attaining an equilibrium between supply and demand for workers and the element of scarcity clearly plays a part in this respect. I endorse the view of Professor Tinbergen and others who advocate increasing the supply through training, thus decreasing the quasi-rent.

By non-compensatory differences are meant factors such as social environment, knowing the right people, etc. In my opinion, differences arising from these factors are not acceptable. Incomes policy should primarily be directed towards levelling these non-compensatory differences.

Finally, the power factor plays an important part in income determination, particularly when the income differentials cannot be based on differences in effort, nature of work or training.

In the last few years income differentials have been reduced quite substantially, partly under the influence of government policy. Important developments are the increase of the minimum wage, the minimum wage of young workers, and the raising of national unemployment insurance benefits up to the level of the net minimum wage. The increase of the minimum wage may have been a factor in reducing the gap between the previous minimum wage and the higher salaries in industry. A too far-reaching policy of reducing income differentials might lead to a shortage of supply for jobs which are felt to be difficult or unpleasant, and surpluses in the supply available for the easier and more attractive jobs. This leads to disturbances in bringing into line the supply of and the demand for workers, and reinforces search structuralism. This means that the problem of scarcity is an integral part of the incomes problem.

In addition, there is the burning question of unemployment as a result of an insufficient overall demand for labour. Incomes policy should be prevented from leading to wage increases which in the longer run exceed productivity, as this would result in a further loss of employment opportunities. In this connection, I am referring to the unbalanced international competitive position and the substitution of capital for manpower, phenomena that are not restricted to the Netherlands, and which have led to unemployment.

Moreover, any increases in wage costs in excess of increases in productivity will inevitably raise prices. This leads us to ask whether incomes policy should be enlisted in support of the fight against inflation. As long as no progress has been made in the field of incomes policy, we are faced with the dilemma that it is impossible or very difficult indeed to combine the fight against inflation with full employment and free wage determination.

An essential factor is the reaction of workers and employers. The point of social acceptance is very important and it is, therefore, counter-productive to interfere in wage determination above the heads of the parties concerned.

Other very important groups are the self-employed and the liberal professions. A great problem is that the income of these categories is determined by market circumstances, although in the Netherlands the Prices Act makes provision for the possibility of enforcing certain tariffs and prices. The effect of such intervention is only relative, and it does not solve the problem of how to arrive at an acceptable distribution of incomes. It may be possible in the future to come to agreement in this field by co-operation.

In the development of the incomes policy, workers and employers' organizations and other social groups will be able and willing to perform an important role. I am convinced that by means of a workable consultation structure, the fight can be waged against the non-compensatory differences in income.

Special legislation directed toward the publication of the remuneration of classified jobs is in an advanced stage of preparation. In my opinion, this legislation will make for a better understanding of the relationship between jobs and incomes. In this connection, there is a need for a separate legislative framework concerning the determination of incomes directed towards extending and deepening consultations between, and with, the 'social partners' and other special interest groups and towards improving the unity of government policy in the matter of incomes.

A statutory framework is being considered for legislation on the determination of incomes. The consultative bodies will have to be shaped as the social need becomes apparent. Gradually, more concrete form will have to be given to the way in which the parties concerned can be represented in the consultative bodies in a representative way.

The scope of the legislative framework will have to be comprehensive. Specifically, consultations on all subjects directly affected by the incomes aspect will have to take place with those concerned. In this connection, mention may be made of:

(1) The field of labour relations (work in paid employment);
(2) Other acts and decrees having consequences for primary incomes;
(3) The social insurance acts;
(4) The taxation system;
(5) Government measures influencing incomes.

The desire to gain goods and earn income forms one of man's primary motivations. It is derived on the one hand from the need to feel secure and on the other hand from the urge to assert oneself, which to a greater or lesser degree is characteristic of all of us. Any incomes policy should allow for this; otherwise there is a great risk of opposition, either active or passive, frustrating the battle against cost structuralism. In my opinion, in this field the representative organizations of employers and workers, representatives of those practising the liberal professions, consumers' organizations, and other social interest groups bear a great responsibility.

Policies combating demand structuralism
The causes of demand structuralism (see §16.2, p.193) may be ascribed to a phase of the Kondratiev cycle in which insufficient technological innovation passes from the preparatory to the introductory phase. It is not unlikely that, in the period of stagnation in the 1930s, a large amount of innovative work was postponed because no means could be envisaged for carrying it out, and recourse was had to a defensive company policy. Such a thing may happen again. We should avoid a policy which does not use opportunities for production innovation, or leaves this initiative to other countries.

In my view, there are still possibilities for introducing new technology in the introductory phase, although this will be more difficult than was the case some 25 years ago. I have in mind the possibilities offered by nuclear energy to tackle the energy problem and of the challenges posed by the environment and the scarcity of raw materials. New technology in this field has been developed and means of production are available, although partly in experimental phases. The question arises, however, whether the impulses in these fields will be sufficiently vigorous, and whether not too great a risk is posed by the employer. As in so many other cases, however, it is important that great delays should be prevented, since, once behind, it is difficult to catch up. Possibly international co-operation within the EEC may help in setting up projects which can break the technological deadlock.

Meanwhile, new possibilities can only be followed up if there is no fear of introducing this new technology. Employers should see a possibility of bearing the investment risks incurred by the new means of production. This will occur only if they are not confronted with a cost problem such that there is a great possibility that the industrial results will be zero or negative. In that case, they will not enter these new fields.

The policy in incomes affairs is directed towards opening the incomes problem to discussion among all interested parties, among whom employers and employees are still the most important. At the same time, non-material affairs, about which no agreement has yet been reached between the parties concerned, should be brought up for discussion, for the solution of these is also a condition for successful bargaining. By this I mean questions of worker participation in management and co-responsibility in running the business, humanization of working conditions, problems of new methods of production, and all related problems, from safety to employment.

I would like to make the following observations in this respect. It would seem that the last 25 years of the twentieth century will be characterized by the strong desire for the realization of democracy in areas in which it has not been implemented before. Although it would be unfair to pretend that nothing has happened in this area, the industrial sector has thus far remained one of these. Some factors have played a part in this respect:

(a) Production on a company basis is particularly directed towards bringing production into line with demand backed by purchasing power. This has been seen as a form of economic democracy. A far-reaching democratization of the company would frustrate the participation of the consumers;

(b) The company operating for the market finds itself in a kind of permanent crisis situation. The environment is turbulent and many companies have to fight daily for their existence. Under such conditions, it is often said that participation does not fit into this pattern. This does not imply that leadership cannot be made dependent on democratic procedures;

(c) The mere fact that the company dates from a period in which autocratic leadership was the rule, and consequently has no tradition of democracy, plays a part. For the management it must be a risky business to let power out of their hands and to be dependent on the views and wishes of their employees.

If we take a broad view of industrial democracy, the consultation on conditions of employment may already be seen as containing an element of democracy. In that case, employers and employees should take up equally strong positions, which is not always the case. This greatly depends on the economic situation and on the subjects that are dealt with.

Thus far, employers have tried to fend off industrial democracy as much as possible; this has been possible by virtue of their position. Seen in the light of the need to arrive at a new form of co-operation between employers and employees in order to take full advantage of the technological possibilities, amongst other things, the question arises whether this attitude on the part of employers is so wise. In one way or another, it should be possible to find ways and means to discuss industrial problems without the trade unions usurping the position of the employer, and without the employer having to be afraid of being obstructed in his abilities to take action.

The attitude of the trade union is very important in this respect, and the way in which the trade unions themselves view this matter is crucial. There are trade unions that reject any form of co-operation with the company as long as 'the capitalistic enterprise' exists. In their view, working for democracy can only mean working for a change in the economic system. In some cases, this implies that the system is changed from within step by step, in others they have refused to co-operate towards measures for democratization, as these would tend to weaken the will eventually to achieve a transformation of the economic system. It goes without saying that with this view it is difficult to win over the employer to introduce a system of worker participation in management.

But, then, in the Netherlands, this view has never had many adherents among the trade unions. For instance, the 'Industriebond NVV' (Industrial Workers Union) in its leaflet called *Nice is different* (*Fijn is anders*) expresses its disbelief in change within the system; in practice, this organization has often been found willing to co-operate, although on the condition that employees should not bear co-responsibility for decisions which they have been unable to co-determine.

To achieve a positive attitude on the part of employees towards the introduction of new technology, employers and employees together should look for constructive solutions both for the employment problem, the cost problem, and other problems related to technological development. This presupposes a much greater willingness for employee participation in management on both sides than has been achieved to date. Decisions of importance for the development of the company in the more distant future, including those bearing on employment opportunities, should be open for discussion. As far as the costs problem is concerned, the ultimate result may be that employees accept part of the financial risks through profit-sharing schemes, capital-growth-sharing schemes, etc., instead of a fixed wage. On the other hand, a measure of employee participation can be organized as to the way in which a given innovation will be introduced.

Thus, the employees could recognize that a part of the entrepreneurial risk is born jointly. Of course this is already the case in the form of the unemployment risk. It is a fact that risks are an intrinsic part of life, which applies equally to employers and employees. Perhaps it would be best to recognize this, and instead of blaming each other for that risk, to make a joint effort to look for means to reduce and spead the risk.

I do not want to dwell on this any further, as these are highly speculative reflections on matters with which employers and employees have to come to grips. On the other hand, I am convinced that a certain amount of mutual trust is indispensable, and I am prepared to do all I can to contribute towards creating the climate in labour relations which is required for such a 'renewed entrepreneurship'.

16.7. Final considerations

Economic life has periods with predominantly upswing and predominantly downswing trends. These may be correlated with the advance of technology. If the advance of technology is impeded for some reason, downswing tendencies may prevail. The effect on employment may be seen in terms of demand structuralism. In addition, there is cost structuralism, the negative influence of wage increases on the demand for manpower, such as substitution of manpower by capital or accelerated obsolescence of means of production.

Policy must be directed towards breaking through the tendencies of demand structuralism, cost structuralism, and search structuralism, by substituting an approach to labour relations in which both the development of income and conditions of employment, and the development of job opportunities and production innovation are at stake.

In this chapter I have emphasized the correlation between the two, and pointed to the dangers which arise if insufficient account is taken of that correlation. It is important to keep the development of costs under the greatest possible control. This is, in fact, the object of the corrective measures which are taken with regard to collective spending and social provisions. The union movement has also demonstrated its appreciation of the importance of restraint in wage and income rises. There lies a danger, however, in the fact that the favourable effect of wage cost restraint is not immediately reflected in job opportunities. This may lead to tensions, and cause these to become so high as to result in an explosion in the field of wages under the motto that 'restraint gets us nowhere anyway'.

In order to be able to exert a more direct influence on employment, the union movement has now begun to advocate what are known as 'job maintenance agreements'. In itself this is an understandable and praiseworthy endeavour. But as yet the term 'job maintenance agreement' has not been accurately defined, and it is open to a variety of interpretations. Inasmuch as it is taken as referring to consultation between employers and workers about wages, employment, investments, and their interrelationships, it is unquestionably useful. It is also conceivable that agreements should be made on these subjects under certain circumstances. And in certain circumstances I am prepared to make some financial support possible, but conditions must be set to such support. Since a government subsidy can only be of a temporary nature, a 'job maintenance agreement' must not involve any permanent passing-on of wage costs to the government.

As a result of the rapid changes in production methods and differences in development from industry to industry, it is possible that the problem of harmonization of supply and demand, always a thorny one, may be intensified. In addition to the possibility of improving the mediation work and strengthening the position of vulnerable groups, other solutions are also under discussion, for example, whether it may be possible to bring about a different, more equitable distribution of work.

Always providing that there is no impediment to the functioning of the labour market, or that it does not have a stultifying effect, I am open to suggestions designed to ensure a better distribution of the available labour in so far as this is done jointly by employers and employees. These suggestions will be evaluated against criteria such as their effect on the development of the structural production costs, the method of financing by the parties, and the financial situation of the firms concerned. Furthermore, suggestions will be evaluated against criteria such as the labour market situation in the region, effect on workers' mobility and on natural outflow of labour. Finally, the experimental nature of this 'new-style' 'job maintenance agreement' factor must be emphasized in order to prevent possible precedents. Quite apart from the constraints imposed by the need for economy, I do not consider it to be desirable to provide wide-scale support for schemes where it has not yet been clearly established that their positive effects outweigh their negative effects. As for the income aspect, it is understandable that the union movement should be dissatisfied that only wage earners covered by collective labour agreements would be involved in income restraint. Incomes policy will have to include all groups, but cannot be successfully conducted over the heads of the interested parties; the representative bodies of the latter will have to be consulted. This may also be of essential significance from the viewpoint of employment.

But despite all this, there is still a risk that, under high unemployment, tensions will run high and other remedies will be sought. One such remedy which will certainly be considered is 'redistribution'. In a sense this is already a topical issue, for example in an experiment such as early retirement, and it is also more or less involved in cases where employment in the non-market sector is discussed. I feel that a problem such as this requires very thorough study. On the one hand, it is to be expected that in the future working hours will be cut yet further, but on the other hand one has to be cautious with measures that would be difficult to reverse or might even be entirely irreversible should circumstances alter. Redistributing work among a greater number of people could very well prove impossible without consequences in the field of income policy, i.e. the leisure time that one gets would be considered as a form of income. Whether people would be willing to look at it this way remains an open question, and everybody is entitled to his own opinion on this matter!

Meanwhile, there is the short-term problem. This year, the Netherlands is faced with over 200 000 unemployed, and it has to be assumed that an even larger number of potential job-seekers have not reported as such because of the lack of prospects of getting a job, or may have retired early, for example under the Disability Act, on the ground of a disablement which, in a more favourable labour market, might have been disregarded. The effective level of under-employment is probably therefore even higher. And of that number, a substantial proportion is out of work for cyclical reasons.

It is not easy to distinguish the cyclical unemployment from the structural. People are frequently out of work for more than one reason: partly structurally, partly cyclically. This complicates the diagnosis. The Central Planning Bureau has made an estimate of the cyclical component of the present unemployment. They find that it represents about one-third of the total, i.e. about 70 000 persons. There are no calculations available on the number of additional jobs which would be taken up again if the cyclical problem were to disappear, for example by means of short-term demand creation, and which may be in excess of 70 000.

This figure would give an interesting picture of the possibilities of short-term demand creation. Reference has already been made (§ 16.4) to the fact that the current policy is one of curbing public expenditure. The Dutch government wishes to achieve a reduction in the burden of taxation and other contributions, in the hope that it may thereby counteract cost structuralism. In addition, measures could be taken to maintain demand and directly promote employment. This is envisaged, *inter alia*, through job creation programmes which are financed outside the scope of the normal budgets.

Reducing the burden of the collective sector may, of course, theoretically be accomplished by means of reducing taxes and social security contributions. But if at the same time economies are implemented to an excessive degree on the expenditure side, which is the obvious thing to do if you do not want your budget deficit to grow too large, part of the spending impulse is in turn destroyed. The Netherlands is not the only country where views on this subject differ. An additional problem, however, for an open country such as the Netherlands, is that a large proportion of the demand stimulation ends up abroad, unless it is allocated to Dutch firms by discriminatory means. This is discouraged by the agreements which have been made within the EEC and GATT, amongst others. So the stimulation of demand only offers a workable alternative if it is carried out in co-operation with neighbouring countries. It goes without saying that the Netherlands is in favour of any efforts to tackle the economic situation, although the results of the consultations being held to this end have not been particularly glittering so far.

Whether effective action could be taken against the rising exchange rate of the guilder is another matter for examination. Measures are being taken in this direction by encouraging the export of capital, but so far without any decisive success.

I have indicated above that many developments can be identified as the causes of the present unfavourable situation, which means that a wide choice of approaches is available to bring about an improvement. Government, however, does not have the ability to turn the tide on its own; much of the load will have to be borne by the private sector, employers, and workers. This will also mean that the parties will have to make concessions to one another. The defensive attitude which both employers and workers tend to assume in order to defend their relative positions will have to be reversed if we are to be able to pursue a forward-looking policy together.

Acknowledgements

I wish to convey my thanks to the Staff of the Ministry of Social Affairs, who have co-operated in the preparation, translation, and processing of this contribution; in particular I wish to thank Dr. D. van der Werf, who was in charge of editing this chapter.

References

[1] Schumpeter, J. *Business cycles:* New York (1939).
[2] Kondratiev, N.D. The major economic cycles. *Voprosy Conjunktury* **1**, 28–79 (1925). English summary in *Review of Economic Statistics* **18**, 105–15, (1935).
[3] Driehuis, W. *Capital–labour substitution and other potential determinants of structural employment and unemployment.* Paper for the OECD experts' meeting on structural determinants of employment and unemployment, Paris (1977).
[4] Freeman, C. *The Kondratiev long waves, technical change and unemployment.* Paper for the OECD experts' meeting on structural determinants of employment and unemployment, Paris (1977).
[5] Centraal Planbureau. *De Nederlandse Economie in 1970.* 's-Gravenhage (1966). (Dutch). *De Nederlandse Economie in 1973.* 's-Gravenhage (1970) (Dutch).
[6] van der Werf, D. Developing the Rijnmond model. In *Relevance and precision: essays in honour of Pieter de Wolff.* Amsterdam (1976).
[7] Wetenschappelijke Raad voor het Regeringsbeleid. *Maken wij er werk van?* 's-Gravenhage (1977) (Dutch).
[8] Boersma, J. Inaugural speech, IEA-Conference, Noordwijk, Netherlands (18 April 1977).

Commentary
Duncan Davies

The key questions: political survival and long-term improvements

Professor Albeda's important chapter asks the key questions on the first page: 'Does technology, by a labour-saving bias, lead to a future of permanent unemployment? Must we look forward to unemployment as the one most fundamental problem during the last quarter of the century?' He then gives a most interesting economic analysis of the events of the past 50–100 years, and identifies causes of unemployment over this particular period. Crudely put, these are as follows:

(1) Growth in demand has not been stimulated so as to match growth in manpower productivity (what Albeda calls 'demand structuralism').
(2) Pay increases in prosperous industries have caused pay demands in impoverished industries that could be met only by subsidy, shrinkage, or other restrictive remedies ('cost structuralism').
(3) Labour surpluses cannot always be absorbed into the areas of shortage, because of geographical factors or mismatches of skill, etc. ('search structuralism').

It should be noted that these terms are not in general usage, although some of the thinking behind them is accepted. Deficiency of aggregate demand for goods and services, which in turn leads to a lower demand for labour, is usually tied into a consideration of the fluctuations in the economic cycle. Professor Albeda recognizes the part played by cyclical factors in bringing about the *current* recession in the world economy, but perhaps does not indicate sufficiently its significance.

Most of the rest of the chapter is, very properly, devoted to the political methods for tackling these factors. But in §16.4 'The non-market sector', he in effect answers his first key question affirmatively as far as industry is concerned, and turns his attention to expansion of the tertiary services sector as an essential contribution to the provision of enough jobs. He points out, however, that such expansion requires a successful industrial sector that can provide enough exports to pay for imports (after allowing for invisibles) and enough goods for home consumption to meet reasonable expectations. He also observes that the 'professionalization' of service activities has increased their cost, thus limiting their expansion according to his 'cost structuralism' argument (§16.2). Thus, expectations of consumption by all with jobs have been pushed up beyond the ability of goods, production, and exports to support them. Expectations of parity with the best wages have limited the extent to which jobs could be created so as to meet genuine social needs for care, cleaning, and so forth. Thus, the only way to balance the books is to keep a large and increasing number described as 'unemployed', this being one of the few classes in society lacking the economic power to stand out for parity of consumption with the highest paid. There is, of course, no economic obstacle to the employment of any number of the 'unemployed' in any service occupation that does not consume imports if they will perform service jobs for pay at the level of unemployment relief. It is rather important to distinguish clearly at this point between the *short-run* and the long-run implications of this analysis. Professor Albeda makes the distinction, but does not relate all his ensuing analysis to it. Increases in total productivity in any industry in a single country will, for a given level of demand, reduce the demand for inputs including that of labour. The industry in question may become more competitive as a result. All this is short-run, and the short-run implication is lower employment. However, if a larger share of the product market is obtained as a result of increased competitiveness, demand for labour in the industry might, in the longer run, actually increase. The same analysis applies to changes in technology, but with a different time-scale.

Alongside Professor Albeda's study of current and near future policy, it might focus the discussion if special attention is paid to his two fundamental questions restated as follows:

(a) Does technological investment now and for the next few decades reduce the total number of industrial jobs?

(b) If it does, then can we not substitute service jobs that carry general esteem and self-esteem for 'unemployment' that carries no general esteem, and erodes self-esteem and job skill?

This particular manner of statement has been chosen because it encourages us to look at the questions in terms of the analysis of particular technologies, resource supplies, and consumer needs on the one hand, and philosophy and morals on the other. This in no way undermines the politico-economic analysis, for this alternative view tends to focus attention on longer-run solutions that can only have value if political means can be found for attending to short-run survival — competitive and otherwise. However, an important preliminary comment must be made here; it is extremely difficult at a *macro*-level to unscramble the impact of technological investments on industrial jobs from other influences. Indeed, rather paradoxically, the level of manufacturing employment in the U.K. actually *rose* between 1976 and 1977 after declining steadily for some years, which is difficult to square with Albeda's question (a).

Technological investment, manpower productivity increase, market demand, and resource supply

There is no shortage of *micro*-evidence that, in the short run at least, technological investment reduces operating employment per unit of output; indeed, since the investment would presumably not be made unless the economic return was favourable, this result is virtually inevitable wherever materials and energy efficiencies are already high. Professor Albeda's Table 16.1 gives some results. After a period of rapid innovation and growth (e.g. in the chemical and allied industries over the past 30 years), competitive pressure, perhaps intensified by overcapacity, can generate annual increases in output per man at over 5 per cent per annum. The increase in demand can exceed this figure only when a new product is replacing older ones, creating a totally new market, or creating new markets for exports, processes that inevitably slacken over time. Thus, for mature products and processes, output growth of 2–4 per cent is common when the technology is holding its own (e.g. inorganic chemicals) and less when it is itself suffering substitution (e.g. 1 per cent for non-ferrous metals). Collapse of rapid growth is even more marked for materials such as steel, where substitution is accompanied by dependence on capital goods markets. Consequently, mature and efficient competitive industries will more often than not be shedding labour and, as technology and commerce stabilizes, reducing overhead manpower as well. As plant replacement becomes less common, construction industries will join the demanning process. Then, as time permits the studies underlying elaborate systems control, there are further manpower savings. A typical instance is the introduction of microelectronics into telephone switching to replace electromechanical devices. This can increase the manufacturing output per man by a factor of seven, and the system maintained per man by a factor of four.

It is sometimes suggested that increases in 'downstream' employment, made possible by cheaper materials from industries that are shedding labour, are sufficient to compensate for the reductions. The difficulty here is that the same 'downstream' job may be claimed by several 'upstream' industries, so that figures must be examined globally.

The generation of new industries has been a crucial source of employment, and indeed the calculation of productivity indices as output per unit of manpower (rather than per acre of land, tonne of oil input, or gallon of water) depends partly on this steady process in successful big and sparsely populated countries, such as the U.S.A. over nearly two centuries. Now, however, the range of cheap available goods covers most obvious human needs, so that new industries must usually displace something already efficient. This is one of the factors leading to high 'entry fees' for new technologies, comparable with the investment needs for very big plants. This has reduced the frequency of appearance of quite new industries, although obviously, spectacular new cases will continue to arise from time to time. Microelectronics will grow fast, and pave the way for products such as optical fibres, but these high-technology procedures usually employ few people in production (though increased numbers in design).

Next, increasing scarcity and the prices of raw materials and energy create two further limiting effects on employment. First, prices of final products rise because of the energy content. Second, the movement of funds to the primary producers restricts the availability of capital and the level of demand in the developed world. Lack of confidence erodes employment.

Finally, it is interesting to enquire whether any correlation can be detected – in either direction – between the level or rate of change of unemployment and technological capabilities in different countries. Fig. 16A.1 shows recent data, which are

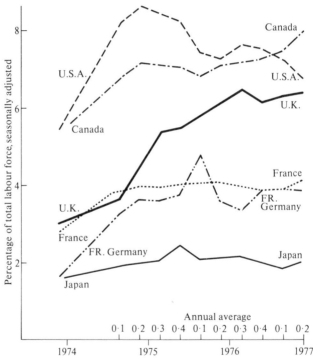

Fig. 16A.1. Standardized unemployment rates in major OECD countries since 1974. The national unemployment rates have been standardized to international definitions. [Source: *National Institute Economical Review,* May 1977 and August 1977.]

of interest. The most obvious point is the low Japanese figure, associated with the well-known job security in Japanese companies. However, the high figures for Canada and U.S.A. – both high technology countries – indicate that technology alone does not guarantee low unemployment.

None of these considerations 'proves' that there will be no technology-based recovery in employment over the next 25 years. But the fact that they are all explicably unfavourable, and more so than over the past 50 or 100 years, makes it reasonable to expect that overall manufacturing growth across the trade cycles, especially in energy-intensive or capital-intensive areas, may continue to decline. If so, industry may succeed in meeting demand with a steadily falling requirement of manhours – less people if the real working week (including overtime) does not come down sharply in length. If this happens, it too will be a source of further management problems. It would seem important to ask whether we generally believe that the answer to Professor Albeda's key questions is 'yes', or whether technological advance is likely to create enough new markets to maintain or increase employment. Whatever the level or trend in unemployment, competitive survival dictates that industrial nations attach enough esteem to industrial jobs to attract really able people to operations of increasing sophistication. For the United Kingdom, the short-term need is to increase such esteem. Yet if the number needed for industry is falling, it can hardly be acceptable in the longer term to reduce the esteem accorded to non-industrial service jobs. As time goes on, the philosophical attitude will have to change.

Before about 1950 or so, the major social objectives of longer, healthier, and more comfortable lives were greatly helped by increased production of goods and food: linkages were many. Now, however, further availability of goods relates less clearly to social priorities. Accordingly, it is reasonable to accord more esteem to non-manufacturers, but tradition will clearly die hard, and any big change may take several generations. The year-by-year balancing of the needs of industry for able people (but less of them) and for service and care trades to provide jobs (and more of them) will be very difficult. Moreover, as able people move more into service activities, these too will attract better technology and systems, and require fewer workers per unit of output. Only in the activities of care, compassion, tourism, the arts, and sport are the opportunities sure to increase (provided that enough goods and food are available to meet expectations).

The philosophical and psychological problems require consideration of the esteem attaching to leisured behaviour and activity. The unemployed, and the 'leisured classes' of bygone years, are united in not being required to do jobs, but divided by esteem – low for the unemployed, high for the leisured classes. Do we need (and can we afford) to take the emphasis off hard work aimed at goods, and place it instead on activity that is economical in scarce resources such as energy: football, musical performance, acting, teaching, pure science? But how can one combine two sets of values: 'hard work' while working in industry and the service activities; informed and constructive personal activity elsewhere? Without this, there may be difficulty in finding democratic solutions. But all of it is some way ahead in time.

Conclusion

Professor Albeda's key question has clearly raised wider questions in turn, no matter which answer one gives to it, and some of these are philosophical. It is indeed difficult to see how the problems raised by these questions can be settled other than by something along the lines that Albeda proposes — that is a consultative process including workers, unions, managements, and government. In the U.K. there is an embryonic framework for this process in two bodies: the National Economic Development Council, where all sides meet regularly to discuss economic issues, and the newly formed Manpower Services Commission (composed of union, management, and educational interests), which has started to map out an active manpower policy.

17

Long-term economic problems of advanced and developing countries

Richard S. Eckaus

17.1. Introduction

This is a propitious moment at which to assess the economic achievements of the last twenty years or so and the long-term economic problems that are now impending. Assessments of this type made ten or fifteen years ago would have been subject to a perhaps irresistible temptation to claim, optimistically, that control over our economic destinies, if not finally achieved, was within our grasp. At that time, the flush of general success in achieving substantial rates of overall economic growth was still upon us. True, it was recognized that there were many unresolved economic difficulties. There were persisting domestic issues of inefficiency in important sectors and inequity for some groups. There was then a faint but growing unease about some features of the international economy which seemed to harbour the potential for major disruptions. There was continuing, but not overriding, concern in the northern hemisphere with respect to the prospects of the developing countries, but in many of these countries, too, there had been unprecedented growth. There was still faith in the ameliorative powers of general growth, if not to resolve the problems of inefficiency and inequity, then at least to reduce greatly the obstacles to their resolution.

By 1978, however, we had lost some of our pride. The experiences of the last five years had demonstrated our limited capacity to manage our own economies successfully, to deal with our international economic interdependence effectively, and to assist the greater multitudes in the poor countries of the world to improve their conditions in an efficacious manner. We had had a series of humbling experiences: the economic crises set off, most proximately, by the drastic increase in petroleum prices in 1973, the rapid rise in commodity prices about the same time, growing disfunction in the workings of the international economy, a stubborn general inflation, the most severe recession since 1937 affecting most economies of the world in 1974 and 1975, and, perhaps most mortifying, a slow and uncertain recovery that, after more than two years, had still left a residual of high unemployment and unused capacity. In these circumstances, disquieting structural problems that had formerly been in the process of at least slow resolution or, perhaps, were submerged by general growth, have emerged and become pressing.

These recent experiences have demonstrated the limitations of the command which we have so far been able to exercise over our own future. It is not popular to say so, but economic policy, both domestic and international, in recent years has often been blundering and inept, and the truth about the essential nature of the problems faced has been obscured. This has, to no small degree, been the fault of the technical experts who have been doing the analysis and giving advice. But these faults have to a much greater degree undoubtedly been the responsibility of the policy-makers. In any case, with this new appreciation of our limitations, perhaps we can look more clear-eyed at some of the difficulties we now face. With necessary caveats, that is what I shall attempt.

As my colleague, Paul Samuelson, is fond of saying, 'A fool can ask more questions in five minutes than a wise man can answer in his lifetime.' In the space available, I can exhaust the lifetime of more than one wise man. I have to contain my ambitions in this respect. I do not dwell on those areas which other contributors to these volumes are addressing. Moreover, I restrict myself to long-term problems and for the most part abjure the current debate, however fascinating, over balance-of-payments deficits, differential rates of economic expansion, exchange-rate movements, and price inflation. The residual set of problems is still so large that I have to be selective. In the process, I unavoidably reflect my own anxieties; and so I apologize to those who would highlight a different set of issues and whose own worries are not reflected here adequately, if at all. However, I do not want to imply that the issues discussed precisely reflect my personal set of priorities. Rather they are convenient, obvious, and important examples of the character of long-term economic problems with which we are faced.

It is, I think, appropriate to concentrate on long-term problems in this commemorative volume. Long-term problems are easily put aside on the ground that their consequences arise only some time in the indefinite future, or on the ground that actions are required some time in the future. However, I would argue here that the correct way of appreciating such problems is that their causes and effects work out slowly, and the means of resolving them also operate only slowly. The implication of this characterization is that the attempts to deal with long-term problems must be started substantially before their effects become critical. Moreover, I would also argue that it is in the nature of many long-term problems that market processes cannot deal with them in a satisfactory manner. They are, therefore, recognized to be a responsibility of public policy, and to require some type of social planning.

17.2. A brief review of the macroeconomic patterns of the last twenty years

Table 17.1 presents the annual average growth rates of the OECD countries for five-year periods from 1950 to 1975. As can be seen ever from such overall data, these periods were quite different. The years of 'miracle' growth began in the early 1950s. The overall growth rates of the late 1950s were slowed by the widespread recession of 1957 and 1958. But before and after that recession, in the period 1955 to 1960, growth was often as rapid as in the first five years of the 1950s. In the early 1960s overall growth was faster than in the late 1950s, and in the smaller and

Table 17.1 Expansion of real gross national product or gross domestic product in OECD countries: annual average percentage

	Rates of increase*						
	1950–5†	1955–60*	1960–5*	1965–70†	1970–5†	1975–6*	1976–7*
Major countries	5.2	3.5	4.8	4.4	2.9	5.6	4.5
Total OECD	5.2	3.5	4.9	4.5	2.9	5.2	4.0

Source: Ref. [1]
*Growth rates of GNP.
†Growth rates of GDP.

less developed countries, faster than in the early 1950s. However, there was a general slowing of growth rates in the last half of the decade of the 1960s, although the pace accelerated at the very end of the 1960s and in the early 1970s in a differential manner in the advanced and developing countries.

The economic crisis which was set off in 1974 was virtually world-wide. The immediate cause was the extraordinarily large increases in petroleum prices imposed by the Organization of Petroleum Exporting Countries. However, the crisis came at an advanced stage of the economic expansion that had started in the late 1960s, slowed somewhat in 1971 and early 1972, and then developed again in 1972 and 1973 in many countries. This expansion was accompanied by a general increase in the rate of inflation. The petroleum price increases generated large balance-of-trade deficits in nearly all the non-oil-exporting countries. The adjustment of these deficits in the advanced countries took the form of managed recessions. The management was faulty, however, and the recessions were the most severe in the post-Second-World-War period, in the sense that they caused the largest absolute declines in levels of economic activity or reductions in growth rates, the highest rates of unemployment, and the largest amounts of unused capacity. Moreover, the recovery from that recession has been slower and more uncertain than has been the case in the previous post-Second-World-War recoveries from recession, and even now the pre-recession growth rates have been achieved in only a few countries.

Though the conjunction of recession and inflation was not unknown before this recent experience, it was thought to be a disease that affected only developing countries, primarily in Latin America or, if occurring in advanced countries, it reflected an advanced stage of disruption of the economy, as in a hyperinflation. Yet the record of the last three or four years has been quite striking in the relatively high rates of inflation that have accompanied recession in the advanced countries. This inflation rate in consumer prices climbed to 15 per cent for all OECD countries, before slackening to the current rates of roughly 8–9 per cent. Yet the differences in the inflation experiences within the OECD countries have been very great: in some countries, inflation rates as high as 20 per cent or more were recorded in 1974, 1975, or 1976, while in other countries the maximum rate of inflation never reached 10 per cent on an annual basis.

Accompanying the decline and recession in the various national economies since 1974, there have been major disruptions in the levels and patterns of international economic activity. In 1973 a system of managed floating of exchange rates super-seded the fixed rate system created by the Bretton Woods agreement after the Second World War. Before the new system had been thoroughly vetted, it was tested by the huge balance-of-payments deficits in the non-oil-exporting nations which were created by the increases in petroleum prices in late 1973 and 1974. The adjustments required clearly could not be handled solely by changes in foreign exchange rates. By and large the advanced countries opted for the adjustment process of the Bretton Woods era, and deflated their economies to change the relative demands for imports and exports. In effect, these countries used macroeconomic instruments to resolve a difficulty which, though widespread in its effects, was sectoral in its origins. Because each country tried to adjust separately, there can be little doubt that the overall problem was exacerbated.

Turning to the performance of the developing countries over the last twenty or twenty-five years, a retrospective view indicates that the pessimism that has become rather general is not fully warranted. Table 17.2 presents comparative growth rates of *per capita* gross national product since 1950 for developing countries and OECD countries. These data indicate that, on the average, a substantial group of developing countries have had faster *per capita* growth than many advanced countries. It remains true, however, that the growth rates of developing countries in some areas have remained sluggish.

Table 17.2 Expansion of real gross national product: annual growth rate of GNP per capita (per cent)

	1950–60	1960–70	1970–75
OECD countries	3.0	4.1	2.3
Developing countries			
Africa	2.4	2.2	2.8
South Asia	2.7	1.5	−0.4
East Asia	3.3	4.0	4.8
Middle East	4.0	5.7	7.9
Latin America	2.1	2.5	3.7
Southern Europe	4.6	4.5	4.4

Source: Ref. [2]

It is particularly striking that since 1970, with the exception of the countries of South Asia, growth in developing countries has been substantially higher than in the OECD countries as a whole. The recession starting in 1974 in the advanced countries has been much less marked in the developing countries. The predictions that the developing countries would suffer the most from the oil price crisis have not generally come to pass. This is not to imply that many of the developing countries have achieved self-sustaining growth and that their problems are now less severe than formerly; it has yet to be demonstrated that the means for achieving the recent relatively good performance are viable over longer periods. There are arguments that the foreign financial commitments assumed by the developing countries to maintain their imports and growth have 'mortgaged the future'. This issue will be considered in more detail below.

In drawing lessons from the experience of the last twenty years or so, it is useful to recall that there were periods in the 1960s during which it seemed, not that the economic problems of both the developed and advanced countries had been solved, but that the understanding needed to solve the problems was in hand. There was even some debate as to the possibilities and mechanisms of 'fine tuning' these economies. While that optimistic view may have been appropriate for those circumstances, it is clearly no longer warranted. The combined problems of inflation and unemployment and unused capacity of the late 1960s and 1970s have proved to be more difficult than the episodes of sluggish growth without inflation which occurred at intervals in the 1950s and 1960s. This has brought about a reappraisal of the conceptual frameworks with which economists approach overall economic problems and, more importantly perhaps, of the objectives which can realistically be set for aggregate economic policy in advanced countries. Thus one by-product of the 1970s has, perhaps, been a greater objectivity as to the limits of our understanding.

Yet the world does not stand still for scholars, technicians, or policy-makers to improve their methods. Whether we like it or not, old problems emerge with greater force, new problems develop, and occasionally, perhaps, some difficulties may fade away. So, with an awareness of the inadequacies of the methods we command, we must none the less face the issues that press upon us.

17.3. The focus on long-term problems and policies

There is no shortage of difficulties upon which we might concentrate; rather we have a multitude of choices. However, it appears most appropriate here to concentrate on long-term problems rather than to engage in the debate on current problems and their cure. While certainly not avoiding either relevance or controversy, this focus is more in keeping with the theme of this volume. To a considerable extent, the social implications of scientific change work themselves out over considerable periods of time and often generate changes that become major alterations in the characteristic features of the economy: the behaviour of the participants, both individual and institutional, the constraints of resources and production methods, and the organization of administration.

It is customary to think of long-term problems as requiring solutions some time in the future, with the exact dating not specified except that no current action is absolutely necessary. Meanwhile, the more immediately pressing 'short-term' problems can be resolved. It is true that the first task of a doctor and of a social policy maker is to help the patient survive the current crisis. But, even while that task is being performed, it is necessary to think about the future to avoid, if possible, any undesirable, though more distant, consequences of immediate actions, and to consider the difficulties that will arise if the patient does survive the immediate contigencies.

However, while there may be an overlapping, there is not necessarily an exact congruence between long-term and future problems. The latter will arise only in some distant period, and their causes as well as effects may be delayed; so may be their solutions also. But long-term problems are those for whose solution there are no policies that will act quickly. Or, more precisely, they are problems for which short-term adjustments, if they are possible at all, have relatively high costs, whereas slower adjustments are more likely to be feasible and/or have lower costs. This identification also implies that the consequences of long-term problems may be immediate. We can be living with them and may, currently, be attempting to deal with them. But it also implies that tactics that rely on responses expected to occur within a relatively short period will be more costly or ineffective than if the adjustments can be slower. Thus it is desirable to exercise foresight to identify the problems, attempt to understand them, and to begin to deal with them. The result of delay will be to make their resolution more difficult.

If one could be sure that the predominantly market-oriented economic systems of the North Atlantic countries would, of themselves, 'automatically' deal with long-term problems, then they could be put aside, much as the problems of the supply of food and clothing are left to the workings of the market mechanisms. Such a belief, however, would put us in the anomalous position of thinking that short-run fluctuations in domestic and international economic activity require intensive governmental policy-making and international co-operation, while the longer-term difficulties can take care of themselves. Or, rather, that they can be taken care of more or less effectively through the functioning of private markets.

Yet, to a considerable degree, long-term problems in the sense defined here are not issues of the kind with which markets can deal effectively. There may be considerable uncertainty associated with such problems, in the precise ways in which they manifest themselves and in the effects of proposed solutions. They require a long time-horizon for evaluation, and policy actions over an extended period of time. The adjustments involved are often not marginal ones but are on a large scale. Thus, on these grounds, the long-term problems are not of a kind that might be effectively resolved through market processes. In addition, many of the long-term problems clearly do not fall within the scope of market activity. As will be pointed out in the discussion of specific issues, many of the long-term problems have characteristics that determine in advance that markets will 'fail', i.e. will not operate optimally, in dealing with them. They have essential elements of a public nature. They involve production by natural monopolies. Some are partly the problems of other non-market institutions, the state, the university, professional groups, and even the family. Moreover, those long-term problems which are international in character may require the co-ordination of several governments. Thus it is not reasonable to expect that many of the long-term problems can be resolved through the functioning of market processes. Social decision-making and action is therefore required.

It would be a mistake in political understanding to imply that the obstacles to dealing with long-term problems are simply those of understanding the nature of the problems and the requirements of their solutions, though those are hard enough. To a considerable degree, the major obstacles are the lack of attention or will of the policy-makers, or their inability to exercise effective leadership. The latter may not be their own failure, but rather a feature of the systems in which they operate. It is beyond the scope of this paper to attempt a political analysis of this issue. However, it can hardly be denied that economic policy is characteristically made with a short time horizon. It often appears that the longest view which can be taken is the date of the next election, the next meeting of a parliament, or even the next public opinion poll.

While these political constraints cannot be avoided by technical analyses, perhaps such analyses, by emphasizing the character of the problems which are faced, can help to establish the necessary planning and policy mechanisms to deal with them.

Although the word, 'planning', has negative connotations in some circles, particularly when preceded by other such words as, 'social and economic', it is like speaking prose; something which is done all the time. In all the public decision-making with respect to public education, transport, and social welfare programmes, for example, planning for the future is implicit if not explicit. Public planning for the resolution of long-term problems is not a matter of ideology, but of pragmatic recognition of the locus of responsibility.

It may be noted that the contributions in this volume which deal with social and economic issues implicitly emphasize the importance of long-term problems. Changes in demographic structure are characteristically slow acting, but powerful in their implications for employment, education, and other features of society. Changes in the distribution of income are likewise typically slow to emerge, but also of profound significance for the patterns of work, consumption, investment, and for the operation of social and political institutions. These institutions also respond to shifts in political organizations and goals, whatever their deeper sources. While the range of long-term problems can excite the most jaded appetites for economic discourse, discipline is imposed by the constraints of time and space. Therefore, the general points will be exemplified and detailed in the following discussion of some special problem areas in the fields of domestic and international economic policy.

17.4. Policy implications of changes in energy prices

The rapid increase in petroleum prices in 1973 was a major shock to the economic systems of the non-oil-producing countries of the world. The consequent adjustments made in order to balance the international payments contributed to the most severe recession in economic activity since the 1930s. The recovery from that recession has been slow but is now definite. The task which, for the most part, still remains to be undertaken in the advanced oil-importing countries is the restructuring of their economies to the new conditions of relative scarcity. This restructuring is not just a matter of changing the setting on motors, turbines, and generators, which control the richness of the fuel mixtures, or adjusting the room heating thermostats. Those changes and adjustments can, and perhaps, should be made in the short run, taking into account the changes in efficiency of operation versus savings in fuel consumption in order to arrive at an evaluation of net savings. But it cannot be expected that such adjustments are optimal, either in the sense that they provide the most effective means of reducing energy costs when all other types of adjustments can be made, or in the sense that they create the most desirable pattern of consumption. After all, people may prefer rooms at a temperature of seventy-two degrees compared to rooms at sixty-eight degrees, and the former might be achievable with the same fuel as the latter if construction and insulation standards were changed.

The character of the long-term domestic adjustments to the changes in relative energy costs will be of several different types. There will be adjustments to develop domestic sources of energy to substitute for petroleum imports. There will be adjustments in power generating equipment of various sorts to use alternative fuels, and in the investment in new heat- and power-generating equipment. There will be consequent adjustments to the performance of the equipment using alternative fuels. And, finally, there will be a whole host of subsequent, but often exceedingly important adjustments in the patterns of household consumption and public consumption and industrial use of energy, as the consequences of the changes in relative energy prices move through the entire economic systems.

It is difficult, if not impossible, at this point to put reasonably precise quantitative estimates on the physical and economic magnitudes of all of the adjustments that can be expected. Part of the difficulty is in projecting the demand for and supply of the various energy sources in the future. On the demand side, adaptations will be determined by the patterns as well as overall levels of economic growth, the relative prices of the various energy sources, and the technologies available. On the supply side, the determinants are the relative responsiveness of supplies of various types of energy to prices, taking into account the uncertainties of resource availability, the gestation periods and costs involved in utilizing the resources, and the national policies of the resource-producing countries. Thus, in order to project energy use, many fundamental estimation problems must be overcome. If these can be resolved in some manner, the further difficulties of tracing all the direct and indirect consequences of the changes in relative energy prices remain. Thus it is not possible to project with any certainty the precise consequences of the changes which have occurred and which may yet occur. However, there is no doubt that the changes which have occurred are of the first magnitude, and that not all of the adjustments to those changes have yet taken place. Though precise quantitative estimates are not possible, it will be useful to exemplify the changes to begin to establish some appreciation of the character of the problems involved.

This is a not uncommon situation that arises in making policies for the future. Often even a rough projection is better as a basis for policy than no projection at all. For doing nothing is also a policy, and the consequences of doing nothing, of not making a projection, and not formulating policy on the basis of the projection will be worse than acting on the basis of even a rough guess which is made in such a way as to minimize the most expensive errors.

To begin to establish the character of the adaptations, it is helpful to look first at projected changes in energy use. These are shown in Table 17.3, which presents data for all OECD countries on energy production from various sources, imports, and consumption for 1972 and projections for 1980 and 1985. These are taken from a 1974 study by the OECD, and naturally embody a set of specific assumptions with respect to overall growth and prices. Since the data are intended only to exemplify the magnitude of the adjustment problems involved, the details of the projections will not be discussed here.[1]

Making a mental adjustment for possible overestimates, as suggested in a more recent OECD study, the changes involved are clearly substantial: the growth in supplies of indigenous total energy must achieve roughly the rate of 4 per cent per year from 1972 to 1980, and 5 per cent per year from 1980 to 1985. Moreover,

Table 17.3 Projections of OECD energy production, imports, and consumption compared with 1972 (Million tons of oil equivalent-10^{13} Keal)

	1972	1980	1985
Coal			
Indigenous	670.3	880.2	1050.9
Imports	21.0	9.7	−14.8
Consumption	669.8	889.9	1036.1
Oil			
Indigenous	692.6	1089.4	1380.0
Imports	1224.3	1084.4	1071.6
Consumption	1917.3	2173.8	2451.6
Natural gas			
Indigenous	747.0	983.0	1100.4
Imports	7.1	103.2	144.4
Consumption	744.5	1086.3	1244.8
Nuclear	34.8	325.6	756.1
Hydro and geothermal	96.9	124.5	161.5
Total primary energy	3463.3	4600.2	5650.5
Indigenous	2241.8	3402.8	4448.9
(per cent of total)	64.7	74.0	78.7
Imported	1253.3	1197.4	1201.1
(per cent of total)	36.2	26.0	21.3

Source: Ref. [3]

Table 17.3, even if taken at face value, does not reveal the magnitude and the difficulties of the adjustments involved. For example, achievement of the OECD projections requires substantial effort in various forms of conservation. And within the European countries of OECD, many of the changes must be more substantial than indicated by the averages.

The total additional costs which result from the energy price increases were estimated in the 1974 OECD study cited [3] as 4 per cent of gross domestic product

[1] Table 17.3 is taken from an OECD publication[3]. The estimates embody assumptions about resource availabilities, supply and demand elasticities, prices, and growth rates, which are spelled out in the study. This 1974 study was followed by a 1977 OECD study that projected energy requirements at 6.9 per cent less in 1985 than the previous study, due mainly to a lowered forecast of GDP growth rates in OECD countries. Thus the table should not be taken as definitive or representative of the most recent OECD position. It serves here only to indicate the order of magnitude of the changes involved.

in all OECD countries in 1980, but as high as 5.2 per cent for Japan and 4.7 per cent for the European countries of the OECD. The OECD study also contained estimates, admittedly even more approximate, of the additional investment requirements necessitated by the rise in energy prices. The estimate for the period between 1974 and 1985, corresponding in assumptions to the previously cited figures, is U.S. $540 billion; with reasonable conservation efforts, the total could add U.S. $300 billion. The projected result is that investments in primary energy production and conservation could rise to 9.4 per cent of gross investment in 1977, 10.2 per cent in 1980, and 13.5 per cent in 1985, as compared to an average of 7.7 per cent in the early 1970s.

These estimates, though rough, are none the less heroic attempts to assess the magnitude of the adjustment problems resulting from the increase in energy prices. Yet they still omit many of the changes that can be expected to occur under the impact of the various price and tax incentives and direct regulations to further alleviate the impact of the price increases. The changes will take place in both the consumer and producer sectors. Consumers are already reacting to changes in relative prices, and can be expected to continue to do so. For consumer durables which are linked closely with energy use, the changes may be slow, as consumers are 'locked in' to patterns of consumption associated with durable goods. Thus, the effects of high energy prices are not seen quickly, but will work themselves out over a longer period. For example, the effect of the increase in petrol prices on the operation of private automobiles can be expected to induce a shift to public transportation systems. The change will take place slowly, due to the substantial investment existing in private automobiles, which reduces the pressures of adjustment. However, as the existing stock of automobiles is replaced, the substitution effects will become more important. Offsetting that, will be the improved mileage performance characteristics of new automobiles. In any case, any shift to public transportation systems will require new investments in such transport.

Similar kinds of slow adjustment can be expected to occur over longer periods in other consumer durables. Perhaps the most important consumer durable is the housing stock, and about 27 per cent of the total primary energy is used for residential and commercial purposes in all OECD countries. It is tempting to argue that there is little, if any, flexibility in housing to adjust to relative change in the price of energy. For some types of housing that may well be the case, but for other types, it just clearly is not. For dwellings in which the ratio of roof area to living area is small, as in apartments, heat leakage through roof, and therefore roof insulation, may be relatively unimportant as compared to heat leakage through walls, doors, and windows. But this can also be modified through investment, and will be if the incentives to do so are strong enough.

The adjustment to higher energy prices is also occurring in industry and will continue. In this sector, which absorbs roughly 50 to 60 per cent of total primary energy, there is also a locked-in effect whose constraining power will diminish with time, the scrapping of old equipment and the shift to less energy-intensive processes when substitution possibilities exist. It has been argued that this type of substitution is absorbing a large part of total investment, accounting in part for the difficulties that some countries are having in reducing unemployment. While an ingenious argument, it is difficult to decide definitively. However, it does make the point that the potential substitution effects may be quite important from a macroeconomic viewpoint, as well as for energy usage.

One of the more significant types of substitution may occur through the change in the composition of industry in particular countries, with the more energy-intensive industries tending to move to countries with relatively low energy prices, just as the aluminum industry moves toward low electricity prices.

Similarly, there may well be changes in the location of populations as a result of the higher energy prices, which increase the costs of transport so that the shift of populations from central city areas to the periphery may be reversed. This will have further consequences for the investment requirements in urban infrastructure.

Perhaps the most comprehensive type of adjustment to higher energy prices is in the foreign trade sector. Nearly every projection for the OECD countries, even with major efforts at conservation or substitution against energy use, concludes that there will be further increases in oil imports in many countries. This will require corresponding increases in exports and/or long-term capital inflows from the oil-exporting nations. Another alternative, presumably unacceptable, is that there be a permanent deflation of economic activity in the oil-importing nations in order to reduce demand for petroleum products. Even if the alternative of increasing exports rather than reducing imports is taken, that will require shift in the use of resources and, most likely, new investments in export industries.

If the alternative of increasing exports is taken in order to pay for the higher-priced petroleum imports, then international demand for a country's exports at prevailing international relative prices must be high enough to induce the increases necessary to pay for the higher-priced petroleum imports. If this is not the case, then a change in relative prices is required and that, in turn, is most likely to take the form of a change in foreign exchange rates, although a sufficiently comprehensive system of taxes and subsidies could, conceivably, provide an alternative. In either case, the change in relative foreign and domestic prices would be intended to be powerful enough to induce the changes in resource use desired throughout the economic system.

The foregoing arguments describe the character of substantial adjustments that are required in response to the major relative changes in energy prices. The arguments also suggest that many of the necessary adjustments can be expected to result from the responses of household and productive enterprises. However, the arguments also indicate implicitly that private adjustments will not be optimal in terms either of social or individual goals. This should now be made explicit, and the general consequences for long-term policy-making described. The arguments are of three types: 'microeconomic', and concerned with the conditions of production and use of energy, 'macroeconomic', and focusing on the implications for overall economic performance, and 'distributional', that is having to do with the effects on the relative shares of income received.

Taking up the microeconomic aspects first, one of the characteristics of energy use is that it has associated features that prevent market mechanisms from operating optimally from the standpoint of social goals, or even from an individual point of view. There are so-called 'external effects' in some uses of energy. These are effects which are not fully transmitted through markets. Of course, it cannot be expected that markets would work well to take account of such effects. For example, the use of private automobiles creates pollution and congestion, and the size of the automobile and other characteristics chosen by one individual may affect the choice of others. All these choices will influence petrol consumption. Interactions such as these, though they are not mediated in markets, are none the less real and important. Since markets 'fail' to take them into account, there is a rationale for the exercise of public policy which recognizes and attributes their real effects. This can be done by direct regulation as, for example, pollution controls on automobiles or controls on automobile speed and size. Or public policy may be exercised through pecuniary incentives, with taxes and subsidies, or through progressive taxes on automobiles according to size or weight or petrol consumption.

Another of the features of energy supply and use is that the industries are often characterized by decreasing costs. In this case, the conventional rules for pricing are not appropriate. This is the case with respect to electricity generation, public transportation, and highway use, at least until the limits of capacity are approached, and space heating. In such circumstances optimal pricing would require, not that the services be valued at the cost of the marginal unit produced, but at a price which, when collected from all users, would cover the total costs of production. Such pricing procedures typically cannot be imposed unless there is either government control or regulation.

However, in some types of energy use, it is quite difficult to develop a means for charging persons according to their usage in a manner that would provide effective incentives for balancing the energy use in relation to other types of goods and services depending on costs and satisfactions. For example, space heating is provided to apartment dwellers essentially on demand, and there is no sliding scale of charges for differential temperature settings. In this case there is no control by prices on energy use and direct regulation, which may be used, may also be unsatisfactory. Similarly, in the use of private automobiles for transportation, the price of petrol is only a part of the total costs, and not always such a significant part that it alone can efficiently allocate all the transportation services provided by the streets, roads, and associated services.

There is an analogous characteristic of research and development in the energy field. While the increase in relative energy prices makes the savings from greater efficiency in production or use more important, that does not translate directly into increased incentives to develop technical improvements to accomplish the goals. That is because it is likely to be difficult, if not impossible, for individuals or companies to capture a sufficient part of the benefits of such improvements. More than that, it may not be desirable from a social point of view that all, or even a major part, of the benefits from technical innovations be captured by individuals. That provides a rationale for public support for energy research and development through government or other institutions.

Still another feature of the microeconomic adjustments occasioned by the increase in energy prices is that they involve long time horizons, and information is unfortunately often sparse and inexact. This is another of the conditions in which private market adjustments cannot be expected to produce optimal results.

Turning to macroeconomic aspects of economic policy required by changes in relative energy prices, a number of these will be discussed when the subject of international adjustments of balance-of-payments disequilibria are taken up. There is no doubt that the change in relative energy prices has been one of the major sources of such disequilibria. The policies followed to correct the disequilibria, in turn, have set off major changes in overall levels of macroeconomic activity in most of the countries of the world, and the changes in energy prices have contributed to, if not generated the inflationary pressures of the last several years.

The need for macroeconomic policy does not have to be argued. The need for *improvements* in macroeconomic policies is subject to more debate as, undoubtedly, policy-makers in some countries believe that, on the whole, the programmes that they have followed have been as efficacious as could reasonably have been expected. None the less, the overall record of macroeconomic policy of the past several years must be counted a disappointing one. It might even be argued that, on the whole, the macroeconomic approach to the difficulties created by the increase in energy prices has been inappropriate, inasmuch as aggregate tools of policy have been used to deal with problems which are essentially those of sectoral adjustment. However, this may have been unavoidable as the public policy tools that can be used to facilitate sectoral adjustments have not been generally available. Thus policy-makers have had to use 'second-best' or, perhaps, 'third-best' or even less good instruments in order to accomplish anything at all.

Taking up the distributional consequences of energy price increases, it may be noted that, although the workings of markets undoubtedly have important consequences for the personal distribution of income, it is true that, to a considerable extent, the factors determining the distribution of income are non-market-determined. Inherited wealth, sickness, old age, differentials in the quality of education, and social environment are influences which are not generated by market forces though they are mediated in markets to determine the distribution of income. It is also true that a distribution of personal income which is determined solely by market forces would not be accepted as socially desirable. That, in turn, provides the rationale for public intervention. The increase in energy prices, though a sectoral phenomenon, has widespread impact in production and, because of the importance of energy in household budgets, on real incomes as well and therefore on income distribution. Unfortunately, the macroeconomic policies which were followed to adjust to the changed relative scarcity of energy and the resulting balance-of-payments problems must often only have exacerabated the skewness in the distribution of income. Other policies are therefore required, not only to offset the effects of the energy price increases, but also to offset the effects of the macroeconomic policies on which such reliance has been placed.

This is not the time to attempt to discuss in detail the character of the microeconomic, macroeconomic, and distributional policies that should be developed to deal with the problems occasioned by the increase in energy prices. Rather it is the purpose to argue that long-term policies are necessary and, incidentally, to point out that in spite of a good deal of rhetoric on the subject, little progress has been made in formulating such policies.

For example, one of the most frequent complaints of policy-makers is that the data base necessary for the formulation of policies is woefully inadequate. Yet that complaint has continued for several years without the difficulty being rectified. Another obvious indication of the inadequate rationality and foresight being used in making policy is the limited degree to which expertise has been organized by government policy-makers. There have been some notable examples of makeshift arrangements for the development of long-term comprehensive policies, in which the organization of the expert group seems to have been determined as much by bureaucratic claims as by the desire for effective analysis. While there has been public deception in this respect, experienced analysts cannot fail to recognize inadequate staff organization.

The OECD publications, *Energy prospects to 1985* [3] and *World energy outlook* [4], each have a list of recommended energy conservation measures for each OECD country, and compare those with the policies actually being implemented. While there may well be grounds for disagreement with respect to some of the proposals, most are relatively straightforward. Yet the discrepancies between what is desirable and what has been done are still substantial four years after the crisis was provoked. Thus, it remains true that in the energy sectors there is a need for the exercise of rationality in policy-making with a long-term horizon.

17.5. Urban problems and policies

There is a risk of losing an audience by dwelling on urban problems. In the post-Second-World-War period, these problems have received more general recognition than ever before. Their study has flourished, and the public has been treated to continual warnings as to the dire condition of urban areas and the desperate need for the implementation of effective urban policies. Yet urban problems remain one of the areas most desperately needing foresight in long-term planning and the implementation of long-term policies.

In spite of the continuing preoccupation with urban problems, it is still not possible to obtain an overall view of their dimensions. There is information about the age distribution of the housing stock in a number of countries, which provides some indication of the condition of a major type of durable good. Tables 17.4–17.6 present such distributions for France, Germany, and the United Kingdom. Though

Table 17.4 Age composition of housing stock in France

Year	Per cent of total built before 1945
1960	83
1970	71

Source: Ref. [5].

housing is expected to be long-lived, the age distributions suggest that such a large proportion of the stock is so old that the standards of housing that it provides must be relatively low. This tends to be confirmed by surveys of housing quality which report in 1976, for example, that one-half of the housing stock in Germany did not meet, 'current standards', and that six million of the dwellings in France did not meet 'minimum' standards of comfort. In the United Kingdom in 1969, two million of the eighteen million dwellings were judged to be structurally unsound; two million needed extensive repairs and redecoration, and 2.5 million were without bath and adequate hot water.

Table 17.5 Age composition of housing stock in U.K. 1969

Age	Number	Per cent of total
over 90 yrs	4 million	22
50–90 yrs	3 million	17
30–50 yrs	4 million	22
less than 30 yrs	7 million	39
Total	**18 million**	**100**

Source: Ref. [6].

Data such as the above are only suggestive with respect to urban problems, as there is no distinction between urban and rural dwellings, yet there are *a priori* reasons to believe that rural dwellings may be both older and in generally poorer condition than urban housing.

Table 17.6 Age composition of housing stock in Federal Republic of Germany: 1972

Age	Number	Per cent of total
> 54 yrs	5.8 million	28
24–53 yrs	3.3 million	16
9–23 yrs	8.0 million	39
< 7 yrs	3.5 million	17
Total	**20.6 million**	**100**

Source: Ref. [7].

Other aspects of the condition of the capital stocks that provide urban services cannot be documented with even this degree of specificity. For example, there appears to be agreement among experts that the water and sewage structures are approaching their capacities in many of the older cities, or in their central sections. These structures were typically built before the turn of the century, especially in Europe, and are sometimes more than a hundred years old.

There are old cities with good public transport systems and old cities which never had good public transport systems. The public lighting systems, being relatively easy to update, are usually adequate. On the other hand, the standards of personal safety provided by public security services are generally regarded as having deteriorated.

While the problems of the cities of Europe are different in important ways from those in the United States and other OECD nations, there is no country in which there is not a recognized urban problem. Moreover, it is also recognized that this problem is primarily one which must be resolved through the implementation of public policy, rather than through individual action and market-mediated decisions. This is in spite of the fact that the proportion of dwellings owned privately is high in most of these countries.

The need for public policy arises because of the character of the services provided in urban areas and the technical conditions under which they are provided. These goods and services often have externalities associated with them, or they are public goods, or they are generated by technical processes or conditions such that there are increasing returns to scale, or decreasing costs with increasing output. Therefore, it cannot be expected that markets will perform adequately in the production and distribution of such goods. As a result, public authorities are created to provide such services, or private companies are regulated in the conditions under which production and distribution take place.

In addition, it is often the case that some urban services are provided under state subsidies, which vary with the income or other circumstances of the individual using them. The consumption of such services may potentially command a relatively large share of the income of individuals or households with low incomes. Control of the prices of such services therefore provides a relatively effective means, though not necessarily the optimal means, of contributing to the real income of low income groups.

Finally, it is recognized that the conditions of life in urban areas are such that 'subcultures' are often created, with values and life behaviour patterns which are rather different from those of the society as a whole. These conditions, though perhaps originally thrust upon the individuals involved, sometimes represent defensive measures and means of mutual support for the members of the subcultures. Yet they also perpetuate a pattern of exclusion from opportunities for social and economic mobility. Even where there is relatively little evidence of subcultures with permanent members, it is recognized that urban society has not always evolved patterns of life that provide mutual support to its members. Therefore, it is necessary to develop explicit institutions to perform the functions that are often intrinsic features of rural village or small town societies.

The general conditions described above provide the rationalizations for the active role of public policy in improving the urban environment. While these rationalizations are not controversial, they have seldom led to an effective urban policy. Before taking up the question of why this is so, it will be useful to examine some of the urban problems in more detail, in order to establish that they are essentially long-term in nature, and that it is less costly to take actions to resolve them slowly than to attempt to resolve them quickly, even if that is possible.

To a considerable extent, the existence and use of old and substandard housing in urban areas is a reflection of the poverty which still characterizes substantial parts of the population. Just as the diets and clothing of the poor people are 'substandard', so is their housing. However, the substandard characteristics of much of urban housing also reflects the inability of the markets for private housing to deal with housing problems efficiently, because of externalities involved, or as the phenomenon is characterized in this sector, because of 'neighbourhood effects'. The value of a housing unit is determined not only by the quality of the unit itself, but by the local environment in which it is set. Thus the same effort and expense invested in improvements in existing housing stock or new construction by individual owners or tenants will have different returns, depending on the location of the dwelling. But the individual owner or tenant can do little if anything to affect the neighbourhood environment. As a result, owners and tenants do not respond optimally to the prices which they face for housing, and the prices may not capture the full costs and benefits of housing of different qualities. All this is well known, and motivates public housing programmes or public programmes of support for private housing which try to embody the 'neighbourhood effects' that individuals' actions cannot themselves create. This reasoning also provides the rationale for direct regulation through zoning.

There are similar elements in the conditions of urban transport. However, in addition, there are undoubtedly economies of scale in this sector. So the natural monopoly feature of this activity generates another motivation for public action.

There are both economies of scale and externalities in public security services. Fire and police services provided individually and to particular locations also benefit other individuals in the neighbourhood in the deterrence of crime and prevention and/or containment of fire. But fire and police services provided jointly undoubtedly would have a higher cost than the same level of services. The same arguments apply to the various sanitation and waste collection and disposal systems which are so important in urban areas because of the large volumes involved and the need for timely disposition of them.

There are other services, such as education and health, which will only be mentioned, that are an integral part of urban management in some areas and, in other areas, are a part of regional or national systems. While each has unique characteristics, they also possess features which dictate that market processes will, typically, not produce socially satisfactory results. There are still other determinants of the character of urban society that include the factors already mentioned and other conditions as well. For example, the incentives and disincentives provided for various types of cultural activities may be only slightly within the influence of contemporary public policy. Though individuals may and should take initiative in endeavours of this type, public support for such activities and institutions is often essential. In general, these arguments and conditions are well recognized; this is not an area, such as international exchange policy, in which experts disagree not only as to what should be done, but whether anything should be done. Why then, has so little been accomplished on urban problems that have been so clearly identified and which are agreed to be long term in the sense specified here? The obvious answer is in the difficulties of financing the policies required. Perhaps there is something in the argument that the electorates themselves are willing to trade off short-term advantage for long-term costs. Yet the costs may not have been clearly and dramatically identified, and the difficulties may in turn be laid to some extent at the door of the policy-makers.

17.6. International co-ordination of economic policies, foreign exchange rates, and international finance

After more than twenty years of relative stability in the structure of international exchange and finance, enormous changes have occurred since the early 1970s. The post-Second-World-War patterns of foreign-exchange-rate determination, and corresponding patterns of adjustment of balance-of-payments disequilibria were established by the Bretton Woods agreement, which provided for fixed but adjustable parities. The system that Bretton Woods created broke down in the early 1970s as the domestic economic adjustments which it implied became unacceptable. A new system, nominally one of flexible exchange rates, but in reality one in which there is a kind of 'managed floating' of rates, emerged and is now itself under increasing pressure, again created by difficulties of domestic adjustment.

The international financial system has also transformed itself in recent years in ways which were unexpected. The Eurocurrency market, which had grown steadily but slowly in the 1960s, expanded much more rapidly in the 1970s, and the direction of lending changed substantially toward the developing countries.

These changes in the system of international payments and finance, while perhaps resolving old problems, have created new ones. The system of managed floating of exchange rates has permitted new kinds of competition of national policies and speculative reactions. There is increasing concern about the virtually uncontrolled and possibly volatile system of international finance. The difficulties of co-ordinating economic policy have not gone away and have not been resolved. These problems, although they have usually been treated as if they were short term in nature, having to do only with the variations in economic activity, often have long-term issues at their heart. So they are also areas in which long-term planning is necessary.

The fixed but adjustable parity rate system established by the Bretton Woods agreement broke down under speculative attack on both the hard and soft currency countries. The nature of the defence of fixed parity rates reduced the risk of speculation to such an extent that the defence itself could not succeed. In 1971, 1972, and 1973 there were a series of realignments in currency prices, and then general movement toward a full floating of currencies. Subsequently there were a series of international meetings whose objective was the formulation of an agreement on a new exchange rate system. These meetings did not produce a consensus, and almost by default, the new system of floating rates emerged. This was advertised as, and perhaps originally it was thought that it would be, a system of flexible rates in which relative prices of currencies would be set by the unimpeded play of market forces for the foreign exchange. In actuality, however, the new system is one of 'managed floating'. There has been active intervention in exchange rates by governments. Their objective has not been to maintain a fixed rate but, rather, to manage a flexible rate that is consonant with its domestic economic policy.

In the regime of fixed exchange rates, the mechanism by which balance-of-payments adjustment for a particular country was expected to occur was through changes in its levels of overall economic activity. For example, a deficit country would have to reduce its effective demand for imports, which would require reduction in effective demand for all goods. The domestic consequences would be increases in unemployment and unused capacity. So elimination of the international imbalance required adjustment of the entire domestic economy.

By comparison, one rationale that was put forward for flexible rates is that they would permit adjustments in balance-of-payments positions without forcing overall changes in the levels of domestic economic activity of the deficit country. In this case, a change in the price of a country's currency would alter the effective price of its exports and imports, which would then bring about changes in trade and payments that would restore equilibrium. For example, the price of the currency of a deficit country would fall relative to the price of other currencies because the supply and demand conditions would not clear the market at the going rate; there would be an excess supply which would force the exchange rate down. This would stimulate exports and reduce imports and under proper conditions eliminate the deficit. The increase in exports, with domestic supply prices unchanged, would increase export earnings. The decrease in imports will reduce import spending, if imports are sufficiently sensitive to prices. Even if the latter were not the case, there might still be an improvement in the trade balance due to the increase in export earnings. Thus it was argued that, with flexible rates, the adjustments necessary to remedy a disequilibrium in the balance of payments would occur primarily in foreign trade, rather than in the overall levels of activity in the domestic economy. This, in turn, was expected to reduce the dangers to international trade and finance of incompatible domestic macroeconomic policies in the major trading countries.

There has been enough experience since the move to a system of managed floating of exchange rates to provide the basis for an assessment of its efficacy. This assessment suggests that the attempt that was made after the breakdown of the Bretton Woods system to establish some sort of new international exchange agreement should be renewed. Again it is an initiative that should be taken with a long time horizon in mind, rather than under the pressure of attempting a 'quick fix', as seemed to be the case in the meetings held in 1972.

It is now clear on both theoretical and empirical grounds that flexible exchange rates do not in fact isolate the adjustment of balance-of-trade disequilibria from the domestic economy. In order for trade imbalances to adjust via flexible rates, there must be changes in exports and imports. These changes, in turn, create changes in effective demand for domestically produced goods and services, and in employment and use of capacity. There is as much temptation to engage in competitive exchange depreciation as ever there was to engage in competitive, beggar-thy-neighbour policies of trade, export promotion, and import restriction. Moreover, it is rather easier to do this in a regime of managed floating of exchange rates, because it can be carried out piously in the name of free exchange markets rather than appearing as a conscious attempt to export unemployment.

There is another effect of exchange rate flexibility which makes it into an active instrument of domestic macroeconomic policy as well as for balance-of-payments adjustment. That is the effect on the domestic price level. Appreciation of the exchange rate reduces import prices which, especially in countries relying heavily on foreign goods, reduces the domestic price level. The effect is direct when the imports are final goods and services and indirect, but just as certain for imports of intermediate goods. Countries importing a relatively small share of their requirements will be less attracted to this instrument of anti-inflationary policy. But countries in which this share is substantial are attracted, and have actively used this means for reducing inflation. It may even be the case that exchange rate depreciation can also assist in controlling inflation when that depreciation is made by use of a crawling peg. Since the close relationship between domestic inflation and exchange rate depreciation is widely appreciated, the crawling peg may be accepted as an indicator of limits to future inflation. In a situation of expansive inflation, such limits may serve to dampen unrestrained inflationary expectations and thus restrain prices and wage increases.

Thus the new system of managed floating of rates is by no means one that isolates domestic economic policy from the adjustments required to achieve balance-of-payments equilibrium. The relationships are now generally appreciated and widely exploited. There appear to have been competitive depreciations via managed downward floats to resolve balance of trade problems. There are apparently conscious policies of undervaluation to generate recoveries from recession and further growth via export expansion, rather than domestic expansion. Currency appreciation has been used to reduce inflation through the cheapening of imported goods.

Exchange rate flexibility is a new tool, which can be used together with other macroeconomic tools, to manage the domestic economy and the foreign trade balance. Therefore, there is as much need to co-ordinate economic policies in this new system of managed floating as ever there was under the Bretton Woods system.

Yet the obstacles to co-ordination of economic policies among countries are still the same as they were under the Bretton Woods system. There are different goals of economic policy in the various countries, and different opportunities for trade-offs among those goals. For example, countries in which the labour movement is especially aggressive in defending real wages may find less scope for depreciation to eliminate a balance of trade deficit, or deliberate undervaluation to generate an export-led recovery, than countries with more docile labour movements.

However difficult it will be to obtain the objective of international co-ordination of economic policy, and it would be grossly unrealistic to pretend that, in fact, it could ever be fully attained, the attempt must be made. The problem is a long-term one, for the adjustments necessary for co-ordination cannot be achieved quickly. It is not enough for agreements to be made in principle when there is a general level of high economic activity and balance-of-payments equilibria. Unless the essential barriers to carrying out such co-ordination are reduced, or devices found to get around them, they will rise again to bar co-ordination when the stresses of balance-of-payments problems or domestic recession recur.

Analogous issues arise in the field of international finance. There has been a rapid expansion and diversification of the international financial activities in the advanced countries which has provided important services to the international community, as well as making profits for the participating banks.

The amounts and rates of change in international bank lending from 1972 through 1976 are shown in Table 17.7. The total amount has grown more than

Table 17.7 Estimated bank lending in international markets (billions of U.S. dollars)

	European Group of Ten	Canada and Japan	United† States	Total*
1972 – Amount outstanding	149.6	24.0	29.2	203.8
1973 – *Per cent increase*	*41.6*	*22.9*	*68.9*	*43.0*
Amount outstanding	211.8	29.4	49.3	291.5
1974 – *Per cent increase*	*16.5*	*17.3*	*67.1*	*24.8*
Amount outstanding	246.8	34.5	82.4	363.7
1975 – *Per cent increase*	*20.5*	*−0.9*	*34.6*	*21.6*
Amount outstanding	297.3	34.2	110.9	442.4
1976 – *Per cent increase*	*18.7*	*14.6*	*40.2*	*23.8*
Amount outstanding	353.0	39.2	155.5	547.7

† Includes branches in offshore centres.
*Includes double-counting due to redepositing among reporting banks in the amounts of (U.S.) $217.7 billion in 1976, (U.S.) $182.4 billion in 1975, and (U.S.) $143.7 in 1974.

two and a half times during this period, with the greatest rate of change occurring in the participation of US banks and their offshore branches. These amounts, even after adjustment for double-counting due to redepositing, are substantial by any standards. The motivation in recent years for the large-scale movement into international lending by banks in advanced countries was undoubtedly the accumulation by them of excess funds. These came partly from the 'recycled' earnings of the OPEC countries which could not be fully spent currently, and accumulated as their foreign exchange reserves, which they chose to hold in large part in the banks of Europe and the United States. The excess bank reserves have also originated as a result of the slow pace of economic activity in the advanced countries on account of the recession and modest recovery of the last several years. This has resulted in a demand for credit in the advanced countries which has been substantially less than the internally generated liquidity.

The excess supply of potential credit in the advanced countries in recent years has been used for a number of purposes. Depending on relative interest rates and other loan conditions, firms have often found it advantageous to borrow abroad rather than at home, even for domestic investment. International loans have also been used for investment in second and third country activities. Borrowing and investing in the same foreign country provides a means of hedging at least partially against the risk of exchange fluctuations. As international lending becomes a more familiar activity for a larger group of banks, and the formula for the organization of and participation in loan syndicates also become more familiar, bank credit is increasingly used for third country activities as well.

One of the major changes in the pattern of international private lending in recent years has been the expansion of such credit to governments and public or semi-public authorities in developing countries. The excess liquidity in the banks of northern hemisphere countries has fortuitously met the growing demands of the developing countries in the last several years. The relative growth of these demands has, in large part, been the result of the quite different patterns of adjustment to the petroleum price crisis which began in 1973, as compared to the adjustment patterns that have prevailed in advanced countries. The latter countries, in general, deflated to bring about the necessary balance-of-payments adjustments, and in an attempt to control inflationary pressures. By contrast, the developing countries often resorted to large-scale foreign borrowing in order to sustain the import levels required to maintain their growth patterns. The tactic was successful in achieving the objective, and accounts in part for the relatively better overall economic performance of the developing countries in recent years as compared to that of the advanced countries.

Although the details of the distribution of this type of lending are not known, there is general agreement that it has become a major use of international credit. However, the large-scale lending to developing countries has led the banking systems in advanced countries into unfamiliar financial territories. The standards of creditworthiness and sound banking practice in lending to countries in the process of development are not well established. As a result, there is growing uneasiness about the substantial commitments of the banks engaged in such lending. The lack of comprehensive information as to the patterns and amounts of lending to developing countries in itself is a source of uncertainty and, therefore, of uneasiness. Since the country loans are larger than a loan to any single firm, it is feared that a 'default' or a 'moratorium' on payments by one of the borrowing countries might raise problems for banks participating in the lending.

On the other hand, it is also claimed that the risks of lending to developing countries are spread rather widely through the use of syndicates. In addition, it is argued that there are, in fact, only a few countries in which there is or has been any danger of payment problems arising; although there are a few spectacular cases of countries or public authorities going into arrears on its bank loans, there is no general danger of widespread default. Of course, as noted, given the magnitude of the lending in some of the individual cases, one such default by itself could set off major shock waves. For example, according to a report of the Foreign Relations Committee of the U.S. Senate, the threatened default of Pertamina, the state oil company of Indonesia, in 1975 would have involved an estimated total of $6.5 billion. The proximate cause of the possibility of default was created when Pertamina was unable to meet payments on a $40 million short-term loan made by a banking syndicate headed by a regional U.S. bank in Texas. Had a default been declared on that loan, the cross-defaulting provisions on other credit extended to Pertamina would have brought its total outstanding bank credit into default. The default was finally avoided through negotiations in which the Indonesian government guaranteed the Pertamina bank debt. Another current example of potential difficulty is the case of Peru, where the amount of credit in danger of default has been estimated at $1 billion for U.S. banks alone. Roughly one-half billion dollars of loans are also estimated to be in default to Zaire. And growing worries are expressed about several other countries, involving, perhaps, several billions of dollars.

These examples suggest, but do not, of course, prove, that it cannot be claimed that the risks of non-repayment by a developing country on an originally agreed schedule are small. But actual default, which would bring legal machinery into play for collecting claims, has been strenuously avoided. Extraordinary steps have been taken to avoid this exigency. Rollovers and bail-outs have been arranged. In the Pertamina case, the U.S. government participated in finding devices that would encourage the conventional sources to maintain the extension of their credit.

There may be other important offsets to the concern about the growing volume of lending to developing countries. It has been suggested that such lending is actually less risky than lending to private corporations. If the latter should default and suffer bankruptcy, liquidation may occur with a resulting loss of principal as well as interest. In any case only a company's assets stand as surety against its obligations. By comparison, loans to governments, or government-guaranteed loans to public authorities are essentially secured by the future foreign exchange earning power of the country. In spite of loose talk, there is no national bankruptcy.

Another argument adduced to diminish the character of the risk involved in lending to developing countries is that they benefit from the general inflation, so that the real value of the necessary repayments is continually reduced. Whether they will also suffer from the devaluation of their currencies, and what the net effect will be, is more difficult to project.

Still another major use of bank credit which has emerged since the movement to managed floating of currency exchange rates is the participation in the forward exchange markets. The floating of foreign exchange rates has created a large-scale spot and forward market in foreign exchange which is often quite volatile. There are, of course, sound business reasons for banks which engage in international lending to participate in both of these markets in order to obtain required foreign exchange, and to hedge its transactions against future fluctuations in exchange rates. However, there is also evidence that there is an amount of active speculation by some banks. These banks take uncovered positions which are not the result of their credit operations but rather are intended to achieve relatively short run gains. While there is no statistical information which records these positions, there have been some unedifying examples of bank difficulties resulting from active speculation in foreign exchange markets. It is resonable to presume that the large-scale operations of some banks are reproduced on a smaller scale in other banks. There is little information on such operations, but the total use of funds for this purpose may be substantial.

Thus international bank lending has evolved from a relatively small-scale activity concentrated primarily among the advanced countries to provision of financial services on a large- and world-wide scale. This has certainly helped overcome some of the major problems associated with the 'recycling' of OPEC surpluses. It has contributed to the efficient mobilization of capital around the world. It also has certain dangers associated with it which will be taken up below.

The growth in the other major sector of international credit, bond issues, both

Eurocurrency and foreign issues in national markets, has also been remarkable. The total amount of new issues has grown more than two and a half times from 1974 through 1976, with the growth of Eurocurrency issues being faster than the issues abroad for national purposes. While international institutions continue to participate in this type of borrowing and, at an increasing level, in Eurocurrencies, issues by other private and public institutions have grown most rapidly. Table 17.8 presents a general picture of these developments.

Table 17.8 International bond issues (millions of U.S. dollars)

		Eurobond issue	Foreign issues	Total
1974	International institutions	2070	3410	5480
	Other private and public issues	2450	4380	6830
	Total	**4520**	**7790**	**12310**
1975	International institutions	1480	3980	5460
	Per cent increase	*−28.5*	*16.7*	*−0.4*
	Other private and public issues	8720	7850	16570
	Per cent increase	*255.9*	*79.2*	*142.6*
	Total	**10 200**	**11 830**	**22 030**
	Per cent increase	*125.7*	*51.9*	*79.0*
1976	International institutions	2960	4960	7920
	Per cent increase	*100.0*	*24.6*	*45.1*
	Other private and public issues	11 970	13 050	25 020
	Per cent increase	*37.2*	*66.2*	*51.0*
	Total	**14 930**	**18 010**	**32 940**
	Per cent increase	*46.4*	*52.2*	*49.5*

Source: Ref. [8].

The expansion of this type of borrowing and lending has been due in part to the reduction in the general rate of price inflation, and the premium of bond yields over short-term interest rates. Both of these changes have increased the attractiveness of bonds to potential buyers. The private international banks themselves have been major purchasers of bond issues. On the issuing side, some banks have also taken advantage of this market, as well as some major deficit countries and private firms. The increasingly close relation between the domestic bond rates in the major financial centres and the Eurocurrency rates indicates the growing acceptance of this type of financial instrument in national as well as international financial markets.

The international bond market is another aspect of the integration of the various national financial markets, and undoubtedly has important advantages in facilitating the flows of funds and capital. Yet it is also true that both the international bond market and international bank lending are relatively unsupervised by any authority. The international lending by banks within the U.S. is subject to the various reserve requirements and loan conditions which regulate their activities within the country. On the other hand, the offshore and overseas branches of United States banks are not subject to these regulations, and the total lending by the offshore branches in 1976 was only about seven per cent less than the total international lending of banks within the U.S. There is some supervision of the loans by overseas branches of U.S. banks because the parent bank is ultimately responsible. Yet this supervision by the Comptroller of the Currency cannot be so rigorous as for loans by the parent banks.

The regulation of banking practices in European countries is characteristically less rigorous than within the U.S., and those banks also have recourse to the use of branches which are relatively unsupervised. Therefore the growing ease of transfer of funds through the Eurocurrency market, while having important advantages, has also the effect of reducing the degree to which financial systems are regulated. Similarly, with respect to the issue of Eurocurrency bonds and borrowing in some foreign currencies, there is likely to be less careful control than for domestic borrowing.

Thus we are in the position of having financial markets of already great importance and continuing to expand rapidly whose participating institutions, while not out of control, are not completely within the control of any authority. Even the information on the range and scale of their participations is not complete. Certainly it is not centrally collected and dispersed. And there are suggestions from the available evidence that the informal means of communication among the institutions does not protect them by keeping them all informed as to the commitments of borrowers.

While the international financial markets represent important improvements in intermediation, they are also major sources of vulnerability to financial disruption of both international and domestic economies. There is some limited experience on the general effects of difficulties within any one bank in its international operations. This indicates that the repercussions are not isolated, but spread quickly through the international system and penetrate domestic financial markets. Even bankers who do not like to discuss such eventualities, talk of the importance of the various central banks serving as lenders of last resort in case of weakness in any major international bank. There are continuing suggestions by bankers that the International Monetary Fund or Bank for International Settlements take on some of the functions of a central bank.

Major problems have so far been avoided, but it would be unwise to expect that to continue. Though private interest counsels caution, there is a lack of an organized flow of information and a set of regulations that would help in reducing the vulnerability of the system as a whole. No doubt, if there should be a crisis of some type, the central banks would attempt to isolate and control it. It is natural that there should be resistance to regulation in any 'normal' period, because it means that some apparently profitable opportunities will be lost by someone. The advantages to the international financial system and the international economy as a whole, however, are more important if one considers the dangers of unregulated activity. Therefore, rather than improvising in an emergency it would be preferable to engage in long-term planning, which would reduce the probabilities of such an emergency and provide methods for dealing with it should one arise.

17.7. The interrelated futures of the developing and advanced countries

The last twenty years or so virtually encompass the period during which the problems of the developing countries have been recognized on a world scale, and have become the object of explicit policy within the developing countries themselves and as a partial responsibility of the advanced countries. While the poverty in the developing countries is immemorial, foreign imperialism and/or the dominance of traditional hierarchies forestalled a concerted attack until after the Second World War. At the time, attention was drawn to this poverty by clamorous independence and revolutionary movements.

As the scholars and policy-makers began to focus on development problems, two contrasting views emerged. One view, associated with former colonial administrators but with technical experts as well, and finding support among experienced businessmen, was relatively pessimistic. The capital requirements and the associated savings, the transformations in social structure required to generate entrepreneurship and achieve productive efficiency were all regarded as beyond the achievement of the developing countries, at least within any foreseeable future. The implicit forecast was that, at best, there could only be slow change while gradually the conditions necessary for growth were created, and then only could there be optimism about potential growth. Another more optimistic view was expounded first by national leaders in the developing countries and then became almost an article of faith: that the release of energies associated with independence or revolution would brush away the obstacles to development. However, a similar optimism was expressed by some scholars who revised the earlier calculations of capital and savings requirements, and saw more clearly the existing reservoirs of entrepreneurship and skills in many of the developing countries.

This latter optimism provided a rationale for the initiatives undertaken in the advanced countries to supply resources and technical assistance to the developing countries to assist in their development. In one form, the optimism made it possible to claim that within a relatively short time the goals of self-sustaining growth in the developing countries would be achieved and the assistance programmes substantially reduced, if not terminated. Associated with both contrasting opinions of optimism and pessimism, there was also the view that the relationships between developing and advanced countries, while complementary in important ways, were primarily those of the dependence, by the former on the latter, for capital and technology.

Although the dependence of the advanced on the developing countries as suppliers of primary products was recognized, in the 1950s and early 1960s an influential school of thought expected that dependence to decrease. This was expected to be the result of increasing substitution of synthetic for many natural materials, and a declining rate of growth in demand for other natural products. It was in large part from this prediction that the policy of large-scale import substitution was deduced as a means of making up for the expected slow growth in export earnings in developing countries.

At one level, the contrast of expectations with the actualities of the last twenty years is simple: one only needs to inspect the overall growth rates. At a deeper level, however, it is necessary to have a theory which permits the evaluation of what constitutes viable growth and the qualities of different patterns of growth. Table 17.2 permits such comparisons, with impressive results. Only in the last five years has it been true that there has been an absolute decline in *per capita* output in any major part of the developing world. That occurred in South Asia and has been due mainly to an exceptional combination of circumstances: of drought and/or flood, political disruption, high food and energy prices, and disruption of the world economy. With this single exception, since 1959 the *per capita* rates of increase of gross national product in the developing countries range from noticeable to substantial. Remembering that population growth rates range from two to three per cent in the developing world, the overall growth rates in gross national product have not only been large relative to past experience but relative as well to the average growth rates in advanced countries. The latter comparison is also shown in Table 17.2. To compare the realized expansion in developing countries with past experience, it is only necessary to appreciate the power of compounded growth. For example a one per cent per annum *per capita* growth rate would result in a more than sevenfold growth in 200 years. That is something that clearly has not happened in most of the developing world.

Yet the growth rates in developing countries have outrun what was expected when attention began to be concentrated on the developing countries of the world. At that time five per cent growth rates in gross national product were regarded as 'an almost unmeetable challenge' [9]. Moreover the growth rates achieved country-by-country are typically above the specific projections that were made as late as the early and middle 1960s.

Without going deeply into an evaluation of the performance of developing countries, it is nevertheless possible to move quickly beyond a simple comparison of growth rates. An inspection of the record of each of the developing countries over the period 1950–75 shows that, with only a few exceptions, each of the countries whose average annual growth exceeded three per cent benefited from some exceptional circumstance. In most cases, the country has been the beneficiary of relatively large amounts of economic assistance *per capita*, primarily from the United States, and motivated usually by the national security of the United States. Thus it can be claimed for only few developing countries that the processes of endogenous growth have led to growth rates of gross national product of above three per cent.[1]

Yet, while appreciating the remarkable character of what has been achieved in the developing countries, it must also be recognized that the problems remaining are of virtually the same magnitude of those which have been overcome. The countries which since 1950 have achieved growth rates of three per cent or more have populations which constitute less than 20 per cent of the populations in all the developing countries. Thus 80 per cent of the populations in these countries are in

---

[1] This observation helps explain the fact that the projections of growth in developing countries in the 1960s have turned out to be underestimates, on the whole of actual achievements. The original projections understandably could not take into account United States security policies.

states in which the average growth rate has been less than three per cent overall, and roughly less than one per cent on a *per capita* basis.

These relatively slow growing countries are also among the poorest in the world in terms of their *per capita* annual incomes, which average less than $200. If the inequality in the distribution of income within as well as among the developing countries is taken into account, it follows that, in spite of the important achievements over the past several decades, the problems of poverty in the developing countries are far from being overcome.

This is not the occasion to attempt to probe deeply into the causes and cures for the difficulties of development. It remains true that for most developing countries internal programmes and initiatives are more critical than influences from abroad. Yet there are many areas of political and economic policy for which the actions of the advanced countries are of great importance to the developing world. It is these areas which are the focus of the intensely expressed desire for a *new economic order*. While there may well be differences of opinion as to the precise content of the goals of a new economic order and their desirability and feasibility, there is little doubt as to the sentiments which are expressed in these goals. The calls for a new economic order reflect the widely-held belief in the developing world that the international economic system does not promote its development and provide an equitable distribution of the benefits of international commerce.

Perhaps the most common response to the proposals for a new economic order from the advanced countries has been that international commerce is free and competitive and, therefore, provides maximum total benefits to all the participants. That type of response is inadequate in two respects. First, assuming the premise of competition, it ignores the issue of the distribution of benefits from international commerce. That is just the question that excites opinion in the developing countries. Second, the assumption of competition in international trade and capital movements is, from the viewpoint of the developing countries, patently inaccurate. They believe that they typically face monopoly power in their purchase of materials and equipment and in their borrowing in international financial markets and, frequently, monopoly power in selling to these markets. They find that there are a variety of constraints placed on their exports to the markets in the advanced countries, and even on sales to markets in other developing countries that depend on their use of technologies licensed from advanced countries. These basic sources of dissatisfaction with the prevailing international economic system have not yet been resolved on an intellectual level, and virtually no progress has been made on a policy level.

Whether or not there continue to be international conferences on the new economic order, the views which are held in the developing world of the prevailing international system will not go away and will provide the basis for the policies of the developing countries individually. Moreover the possibility of joint actions on the basis of these beliefs cannot be set aside.

The complaints of the developing countries are sweeping in their scope. The evaluation of these complaints requires careful analyses which have not yet been done. But some simple facts will suggest the grounds for the complaints.

It is often the case that manufacturers in the developing countries face import restrictions by the advanced countries that are designed to preserve old industries in the face of growing competition from abroad. This is the case in the European Community and in the European Free Trade Area, as well as in North America, Japan, and Australia. Such import restrictions also apply to some commodity exports of the developing countries, such as sugar. The developing countries also often face only a few sellers of the technologies which they desire and are subject to restrictive provisions in the use of those technologies. While they have been successful in borrowing from commercial banks in a number of advanced countries, they have found that in periods of financial exigency they confront the International Monetary Fund, dominated by the interests of the advanced countries, and without understanding of their growth objectives. The criticisms that have been levelled at the developing countries, over macroeconomic policies which have generated inflation and balance-of-payments problems, as well as over detailed investment programmes which are inefficient, are thought to be one-sided and to reflect standards which are not applied in the advanced countries themselves.

It would be desirable for the advanced countries to respond positively to the complaints which have been made by the developing countries, rather than contenting themselves, as has been mainly the case so far, with rejecting proposals for large-scale cancellations or postponements of debt, or fixing of prices of internationally traded commodities. But positive responses will require long-term planning. In order to open markets in advanced countries which are preserved for domestic industries in advanced countries, adjustments in those industries are necessary. Since such adjustments require displacement of labour and of established financial positions, they naturally face vested interests which oppose and delay. Thus the adjustments cannot be carried out quickly, and will not be carried out at all unless there is long-term planning. This is also the case with respect to meeting the demands for effective competition in the sale of technology. Changes in this area also can only be made slowly, and must be started long before any progress is likely to be observed.

The programmes of financial and technical assistance to developing countries require reappraisal in order to put them in a long-term context. The experience of the last twenty years or so demonstrates that development is not a set of short-period problems but rather contains difficulties which will be with us for many more years. One of the sources of the inadequacies of the existing programmes is in the fact that they have not been regarded as long-term, if not permanent, in nature. It may not have been politically feasible in some countries, perhaps in the United States, to admit that this is the case and to create a permanent programme of development assistance. Yet since it was not tried, one cannot be certain.

Programmes of financial and technical assistance by advanced countries for the developing countries are only one aspect of their interrelated futures. There can be no doubt that the developing world will increasingly be the source of many primary commodities. In this respect the trend is clear. On the other hand, as incomes rise, the developing world will become, even more than now, important markets for the manufactures of the advanced countries. In addition, the developing countries with relatively abundant labour will become an increasingly important source of supply to advanced countries of those manufactures in which labour costs are important.

With reasonable stability in world trade, the mutual long run advantages of trade can be achieved, but only if existing obstacles to it are cleared away and new ones not added. This requires a vision of the future of international trade and finance, and concerted policy to realize the potentials. Without such long-term policies, the frictions between the advanced and developing countries will increase and the progress of both will be impeded.

17.8. Conclusions

There is no single set of substantive conclusions for a discussion such as this one. A few examples of domestic and international long-term problems have been set out. The treatment of these examples was not intended to be, and has not been exhaustive with respect to their substance. The issues were taken up to illustrate the nature of the long-term economic problems which require that immediate action be taken if the problems are to be dealt with most efficaciously.

There is a relative disregard of long-term problems in advanced countries. The issues discussed here are a small sample of the many that are neglected. By comparison, though it does not follow from the discussion above, problems of this sort are considered more carefully in the developing countries, yet, because of resource and other technical constraints, they may not be dealt with more effectively. However, the preoccupation in the developing countries with long-term development has created a climate in which it is regarded as quite normal that attention should be given to such problems, that plans be made for their resolution and resources be allocated for that purpose.

It is not that this sort of thing is never done in advanced countries, but it is done less conscientiously. There are, less commonly, institutions, public or private, such as exist in the developing world, which regard long-term problems as their special responsibility and which are allowed a claim on the national budget.

There are several reasons which may be advanced to explain the relative neglect of long-term problems in advanced countries. Perhaps the most obvious reason is ideological. As noted, long-term problems typically require public policy for their resolution. In the advanced countries, even those with a substantial dose of socialism, there is little public planning. In these countries, socialism has, for the most part, taken the form of the nationalization of some industries. Lacking that, social policy has dealt with the most pressing requirements of individual welfare such as education, health, unemployment, and poverty problems.

Neglect by technical experts may be part of the reason for lack of attention to long-term problems. In advanced countries, it is true that it is only recently that the tools have started to be developed which would permit the effective analysis of long-term problems. However, if one examines the specialization in planning problems among economists now, it is clear that this is found mainly among those who devote their attention to developing countries or the socialist countries of the

Eastern Bloc. The reasons may be partly ideological, reflecting a distaste for issues requiring public action. Or the reasons may not be ideological but simply the persistence of traditions of professional interest after the ideological rationale or the technical inadequacy has disappeared.

Perhaps the explanation for the neglect of long-term problems lies with the citizenry itself: they simply cannot be interested in the resolution of such problems and will not support the political leaders who place them high on the agenda for action and the use of public resources. This might be explained by the public goods nature of such problems. The reasoning may be that action now is certain to involve sacrifices. Postponement may permit the costs to be shifted to someone else. Yet against this argument must be set the willingness of the peoples in many of the developing countries to support programmes to deal with long-term problems.

And finally we come to the policy-makers – the politicians. As noted earlier, their time-horizons seem generally to be breathtakingly short. Yet we do not know whether that is first cause or final effect: whether they have failed to exercise leadership, or whether they are responding to what they have been taught is important by their technical experts or to their own perceptions of the inclinations of their electorates.

These questions are not empty intellectual exercises. The long-term problems are important, and somehow efforts must be mobilized to deal with them. Those efforts could be more effectively mobilized if there was understanding as to the obstructions to their appreciation. Thus, at this point it is necessary to end with the customary plea of the academic for greater efforts at understanding of a neglected set of issues. There is no self-interest in this plea, however, but rather the interest of the community at large.

References

[1] OECD publications.
[2] Morawetz, D. *Twenty-five years of economic development 1950–75*, Table A1. The World Bank, Washington, D.C. (1977).
[3] *Energy prospects to 1985*, Vol. I, p.55. OECD, Paris (1974).
[4] *World energy outlook*. OECD, Paris (1977).
[5] *Report on the development of the social situation in the communities in 1976*. ESC/EEC/EAEC, Brussels (1977).
[6] Stone, P.A. Resources and the economic framework. In *Developing patterns of urbanization* (Ed. Peter Cowan), Sage Publications, Beverley Hills, Calif. (1970).
[7] *Statisches Jahrbuch*. Bundespresseamt, Bonn (1976).
[8] *Forty-seventh annual report*. Bank for International Settlements, Basle (1977).
[9] Lewis, W.A. Objectives and prognostications. In *The gap between rich and poor nations* (Ed. G. Ranes), p. 415. Macmillan, London (1972).

Commentaire
Edmond Malinvaud

Apportant à ce tome scientifique après le professeur Eckaus le message d'un économiste, je crois nécessaire de prendre d'abord un certain recul par rapport aux deux thèmes majeurs qui retiendront notre attention : la nécessité de politiques à long terme et les conditions d'un nouvel ordre économique mondial. Ainsi s'explique la première partie de mon intervention.

Moyens et fins de la croissance économique

Les sociétés humaines disposent de certains moyens pour leur développement : leur population, des connaissances techniques, des ressources naturelles en leur possession ou qu'elles doivent acquérir. Elles poursuivent certaines fins plus ou moins explicitement reconnues : satisfaire les besoins essentiels des hommes, réaliser au mieux leurs aspirations profondes. L'activité économique met en oeuvre les moyens en vue de la réalisation des fins.

Les économistes concentrent naturellement leur attention sur les lois caractérisant comment les sociétés s'adaptent ou devraient s'adapter au potentiel de croissance dont elles disposent ainsi. C'est sur cette question que les spécialistes des autres sciences s'en remettent aux économistes. Mais notre conférence ne peut manquer de consacrer quelques instants aux moyens desquels l'humanité disposera pour son développement futur et aux fins qu'elle assignera à son activité.

Pour exprimer les enseignements principaux de la science économique moderne, le professeur Eckaus a retenu un certain nombre de problèmes qu'il présente comme autant d'exemples : il n'a donc pas visé l'exhaustivité. Je trouve néanmoins surprenant qu'il ne mentionne pas celui de l'alimentation. En fait je trouve surprenant que ce problème ne semble avoir aucune place dans une conférence concernant le développement futur de l'humanité. Pouvons-nous tenir pour acquis que les ressources alimentaires suffiront? Sinon, ne devons-nous pas nous attendre à ce qu'il y ait là un élément susceptible de jouer un grand rôle dans les conflits que le monde risque de connaître?

Si le professeur Eckaus n'a pas considéré la question alimentaire c'est sans doute qu'elle n'a pas été sensiblement renouvelée par les recherches économiques récentes. Les économistes n'ont pas la responsabilité de faire des prévisions démographiques ni de dire quels progrès techniques seront réalisés en agriculture. Ils savent que l'équilibre futur entre population et ressources alimentaires est aussi incertain qu'il y a trente ans, ni plus ni moins. Ils savent que de grandes possibilités existent encore pour que la production agricole augmente fortement en utilisant les techniques aujourd'hui connues. Mais ils connaissent aussi la difficulté des transformations socio-économiques que ceci exige. Il s'agit d'un domaine dans lequel les politiques à long terme sont indispensables, mais difficiles à bien concevoir et mettre en oeuvre. Il n'est pas question d'en traiter plus à fond en quelques phrases.

Le second grand problème que peuvent poser les moyens de la croissance économique est évidemment celui des disponibilités en ressources énergétiques utilisables. Le professeur Eckaus le considère dans sa partie IV, traitant des *prix* de l'énergie. Quant à moi, je trouve cette partie un peu trop optimiste du point de vue auquel je me place ici.

Parler des prix de l'énergie et des adaptations qu'ils imposent plutôt que de s'interroger sur les réserves en énergie primaire et le meilleur rythme de leur utilisation, c'est déjà accepter implicitement que le relèvement des prix est certes durable mais aussi suffisant : à l'horizon utile il ne semble pas nécessaire d'envisager une plus grande rareté de l'énergie que celle résultant des prix relatifs actuels. Une telle éventualité est sans doute la plus vraisemblable ; mais on ne doit pas, me semble-t-il, éliminer la possibilité d'une éventualité moins favorable.

En effet, les données prospectives citées par le professeur Eckaus non seulement se limitent à 1985, ce qui constitue un horizon court pour l'étude de la croissance à long terme, mais aussi pourraient bien surestimer les possibilités d'expansion de la production et les possibilités d'économie et de substitution dans la consommation.[1]

Certes, le professeur Eckaus insiste à très juste titre sur les nombreuses formes que peuvent revêtir ces économies et substitutions et sur le fait que le relèvement des prix est encore beaucoup trop récent pour que nous puissions apprécier son plein impact à ce sujet. Les non-économistes ont toujours tendance à ignorer ou à surestimer les possibilités de substitution, alors que les économistes ont eu de nombreuses occasions de constater dans d'autres domaines leur ampleur et le rôle qu'elles jouaient pour permettre l'adaptation des sociétés à des changements de leur environnement.

Maix l'existence de ces possibilités ne suffit pas à rendre sûr le choix d'une politique à long terme pour l'utilisation des réserves énergétiques, les seules parmi les réserves non renouvelables à sembler susceptibles d'opposer par leur insuffisance un obstacle significatif à la croissance économique. En effet, les succès que rencontreront les efforts actuels et futurs pour la mise au point d'énergies nouvelles, la recherche de nouveaux gisements ou la réduction des besoins énergétiques ne sont pas connus ; l'avenir à deux ou plusieurs décennies est très incertain. Or dans le cas particulier de l'utilisation des réserves non renouvelables et par opposition à tous les autres problèmes économiques, l'incertitude de l'avenir lointain réagit à plein sur les décisions actuelles.[2]

Considérant en 1978 la croissance économique future, on ne peut s'abstenir de penser aussi aux doutes qui s'expriment de plus en plus souvent sur les vertus de cette croissance. Si des hommes de plus en plus nombreux estiment que le type de développement vécu depuis la dernière guerre mondiale n'assure pas la réalisation des fins que l'humanité voudrait se donner, alors on peut craindre d'en voir les répercussions sur la production : l'effort au travail peut devenir moins intense, l'esprit d'entreprise devenir moins fréquent. L'éventualité d'une telle modification des attitudes dans le monde développé doit être prise très au sérieux.

Je peux citer à ce sujet deux passages significatifs d'un nouvel état d'esprit et empruntés à un ouvrage récent [1]: «On croyait à un développement harmonieux. . . . , où tous les développements économique, social, culturel et politique devaient

---

[1] En contradiction avec le tableau de la page 233. R. Eckaus reconnaît d'ailleurs à la page 235 que maintenant on s'attend à une croissance encore prolongée des importations de produits pétroliers dans les pays de l'OCDE.

[2] Cette particularité théorique importante me semble insuffisamment prise en compte par la plupart de mes collègues économistes.

se faire de façon synchrone. Il est clair maintenant que cette conception n'était pas nécessaire, car on a pu voir presque toutes les combinaisons possibles» (C. Mendès, pp. 190–1). «La crise du développement. . . , est surtout la conscience obscure et inquiète d'un épuisement de la volonté, de l'imagination et des mythes qui ont inspiré le progrès» (J.M. Domenach, p.20).

Necessité de politiques à long terme

Le message principal de le professeur Eckaus est exprimé à la fin de son résumé : «Les problèmes économiques à long terme ont des particularités telles qu'ils ne peuvent pas être bien résolus par des actions privées opérant par l'intermédiaire des marchés. Des politiques publiques sont requises». Il en montre la nécessité sur quelques exemples bien choisis, exposant dans chaque cas comment l'analyse économique du problème a fait apparaître en quoi les actions privées, livrées à elles-mêmes, conduisaient à des conséquences inadaptées.

A ce message je m'associe sans aucune réserve. Je conseille une lecture attentive de la contribution du professeur Eckaus, que je ne pourrais pas paraphraser sans en réduire la force. Il faut savoir en effet que la réflexion économique, au cours des deux dernières décennies a beaucoup nuancé la confiance que la majorité des économistes témoignait vis-à-vis du «système des marchés» ou du «mécanisme des prix». Le «laissez-faire» n'a pratiquement plus d'adeptes aujourd'hui, de même d'ailleurs que la planification quantitative intégrale.

La place de l'intervention publique dans le domaine économique ne devrait donc plus être l'objet d'affrontements dogmatiques, mais devrait être examinée sereinement dans chaque cas, en fonction des particularités du problème ou de la situation, comme l'a fait le professeur Eckaus dans sa contribution.

Une fois reconnue l'opportunité d'une intervention publique, il reste à en déterminer les modalités. On a généralement le choix entre deux types d'interventions. D'un côté, la puissance publique peut prendre en charge directement certaines opérations ou encore elle peut fixer des obligations s'imposant aux individus ou aux entreprises. D'un autre côté, elle peut laisser jouer les timulants matériels du marché mais en y subventionnant ou taxant certaines activités.

Le choix entre l'action publique directe «sur les quantités» et l'action indirecte «par l'intermédiaire des prix» doit faire intervenir des considérations très variées, dont notamment les possibilités de fraude ou d'évasion et les coûts de la gestion administrative de l'une et de l'autre action. Les économistes théoriciens en ont étudié la nature en faisant abstraction de telles difficultés de mise en oeuvre.

Si la puissance publique était parfaitement informée de toutes les conditions qui régissent l'activité économique, le choix n'aurait pas d'importance. A toute action sur les quantités on pourrait substituer une action par les prix conduisant rigou - eusement au même résultat, et vice versa. Mais dans la réalité la puissance publique ne dispose que d'informations très partielles de sorte qu'elle apprécie imparfaitement l'effet final de chacune de ses actions.

L'étude théorique a montré que dans ces circonstances certaines actions ont moins que d'autres des effets finals sensibles aux incertitudes de l'information disponible au niveau central ; ces actions, qui doivent dès lors être recommandées, interviennent dans certaines cas sur les quantités dans d'autres sur les prix. Ici encore le dogmatisme serait donc hors de propos.

La nature des informations disponibles au niveau central et au niveau individuel de chaque entreprise, ou même de chaque ménage, joue aussi un rôle crucial dans toute réflexion sur la planification décentralisée, c'est-à-dire sur la concertation précédant le choix des plans (à long terme notamment). Il est significatif que l'étude théorique de cette planification ait intéressé depuis quinze ans beaucoup d'économistes et pas seulement, comme le prétend le professeur Eckaus, ceux travaillant sur l'Europe de l'Est et les pays en voie de développement. Encore une fois nous constatons ici que la nécessité d'interventions publiques bien conçues et adroitement appliquées est maintenant admise par la profession des économistes ; peut-être leurs efforts pour aider la mise au point de ces interventions sont-ils encore insuffisants ; ils ne sont pas négligeables pour autant.

Mais nous rencontrons la question fort pertinente du professeur Eckaus : pourquoi l'adoption de politiques à long terme tarde-t-elle autant? Pourquoi, dans chacun des exemples considérés, après avoir compris la nécessité d'une politique publique, doit-on constater tantôt qu'aucune n'est encore en application, tantôt que elle existante reste insuffisante?

Je m'associe pleinement à cette question. Je suis malheureusement conscient de n'avoir que peu d'éléments de réponse à lui proposer' D'ailleurs un économiste n'est sans doute pas mieux placé que d'autres pour le faire.

Le professeur Eckaus cite fort justement le coût des politiques à long terme. En effet, l'étude précise de chaque cas montre vite que les actions adéquates obligent à accepter des sacrifices alors que leurs résultats bénéfiques (parfois simplement le fait qu'elles empêchent des évolutions dommageables) n'interviendront que plus tard, par définition du long terme. Ces coûts jouent induscutablement un rôle.[1]

J'attribue aussi un poids important au fait que l'adoption d'une politique suppose toujours un accord entre des parties qui ont des intérêts souvent divergeants. Le cas est particulièrement net en matière internationale ; mais il se présente aussi pour les politiques nationales. Même quand la nécessité d'une action est reconnue par tous, chacun peut avoir intérêt à dissimuler l'avantage qu'il en tirerait, ceci afin d'obtenir que les modalités de cette action lui soient finalement particulièrement favorables (c'est une difficulté bien connue des économistes qui la rencontrent pour tout ce qui touche les «consommations collectives»).

Je pense avec le professeur Eckaus que, si les coûts sont refusés et si les accords ne se réalisent pas, la raison fondamentale doit en être trouvée dans la mentalité et les choix personnels des citoyens. Car si une volonté claire existait à la base, elle obligerait en fin de compte les hommes politiques à se faire les avocats des sacrifices et des concessions.

---

[1] On pourrait penser à prétendre que joue aussi contre les politiques à long terme l'incertitude sur leurs résultats bénéfiques qui, concernant un avenir non immédiat, sont évidemment plus ou moins aléatoires. Mais l'étude des décisions en face de l'incertitude ne semble pas justifier une telle thèse. Une incertitude croissante de l'avenir pourrait suivant le cas exiger tantôt plus tantôt moins d'interventions publiques.

La carence du citoyen peut elle-même recevoir deux explications alternatives ou éventuellement concurrentes. Peut-être le citoyen est-il mal informé ;certes faire constater et faire comprendre est long, comme le savent bien les professeurs. A un moment historique où certaines réorientations doivent s'opérer, cet aspect des choses n'est pas négligeable.

Peut-être le citoyen néglige-t-il délibérément l'avenir. Devant une telle interrogation on ne peut manquer de faire une nouvelle référence aux finalités de l'activité économique et au diagnostic de J.M. Domenach selon lequel on assiste à un «épuisement de la volonté, de l'imagination et des mythes» qui seuls peuvent pousser les hommes à entretenir des projets à long terme et à savoir accepter les sacrifices que leur réalisation impose.

Un nouvel ordre économique mondial?

Fort justement le professeur Eckaus consacre près de la moitié de son rapport aux problèmes internationaux. Parmi ceux-ci il donne la plus grande place à l'étude du système monétaire et financier ainsi qu'à celle des difficultés qui se présentent en la matière depuis 1971.

On pourrait se demander si ce dernier problème relève vraiment du long terme. Il est fort actuel et sa solution est urgente pour enrayer la crise financière que le monde connaît, crise dont les effets réels sont évidents. Comme le professeur Eckaus l'expose, un contrôle des marchés financiers internationaux devrait exister et jouer un rôle analogue à celui du contrôle effectué dans chaque pays par la banque centrale.

C'est uniquement par pessimisme, un pessimisme malheureusement justifié, que nous perdons l'espoir d'une création rapide d'un tel contrôle. Nous pensons alors utile d'adopter au moins un plan à long terme susceptible d'aboutir à l'institution d'un système monétaire adéquat et de réduire d'ici là les inconvénients résultant de son absence. Certains esprits estiment d'ailleurs, bien que ce ne soit pas évident, que la solution du problème monétaire est intimement liée à l'instauration d'un nouvel ordre, mondialement accepté, dans les relations commerciales internationales. Or les difficultés, auxquelles je dois maintenant consacrer quelques instants, paraissent encore plus redoutables.

Comme il se doit, le professeur Eckaus aborde le sujet par l'examen de la situation et des aspirations des pays en voie de développement. Il nous rappelle que ces pays considèrent n'avoir pas reçu leur juste part des avantages que le commerce international comporte et des possibilités de développement qu'il suscite. Ces pays veulent tirer à l'avenir un meilleur profit des échanges et pensent pouvoir y arriver. Devant une telle attitude, le professeur Eckaus recommande aux pays industriels de répondre positivement. Cette politique exigera de leur part des reconversions difficiles qui ne peuvent pas être immédiates mais qui doivent pouvoir s'effectuer dans le cadre de plans à long terme.

On doit adhérer pleinement à une telle analyse et à un tel projet. Mais il ne faut pas se cacher que la réalisation sera difficile.

Les économistes ont souvent à étudier des situations dans lesquelles les intérêts mutuels sont partiellement concordants mais aussi partiellement antagonistes. Ils le font aujourd'hui en se référant à la théorie des jeux qui eut une vogue exagérée vers 1950 mais qui n'a pas été oubliée par la suite et a même fait l'objet de progrès discrets mais substantiels depuis lors.

La solution d'une situation conflictuelle n'est pas pensable tant que les antagonistes surestiment ce qu'ils peuvent obtenir. Or pays en voie de développement et pays développés ne semblent pas aujourd'hui prêts de conclure parce que les uns ou les autres (probablement les uns *et* les autres) se trompent sur les avantages qu'ils retireront des échanges internationaux.

Comme le professeur Eckaus l'écrit, une analyse précise des plaintes des pays en voie de développement n'a pas été faite. Sans doute n'ont-ils pas souffert au cours des dernières decennies d'une exploitation aussi marquée que certains le prétendent. La modification des terms de l'échange en leur faveur ne peut leur être vraiment favorable que si elle ne fait pas disparaître l'échange, comme ce serait le cas si elle était trop forte. Le cas des exportateurs de pétrole peut induire en erreur les autres pays, qui sont beaucoup moins bien placés.

Les opinions publiques des pays développés n'ont pas compris non plus l'étendue des sacrifices qui devront, semble-t-il, être acceptés. Elles sousestiment aussi bien les modifications à attendre dans les termes des échanges que l'importance des réorientations que leurs systèmes productifs devront subir.

Aussi, l'information réciproque doit-elle être activement promue et l'amorce de concertation que constitue «le dialogue Nord–Sud» être développée si possible, ou en tout cas maintenue. C'est bien l'ordre ou le désordre international qui risque de se révéler déterminant pour l'avenir à long terme de l'humanité.

Reference

[1] Mendés, C. (Ed.) *Le mythe du developpement.* Le Seuil, Paris (1977).

18

Science, technology, and international relations

H. Brooks and E. B. Skolnikoff

18.1. Introduction

The nations of the NATO alliance collectively represent the majority of the world's scientific and technological resources and, seen as a group, have been the dominant supporters of research and development. Much of the result of that R and D has served to transform first Western societies, and ultimately others, as new technologies have moved across borders or have been independently developed by other countries.

Of course, the effects are not confined within borders, but have been major sources of change internationally as well. Some of the international effects are the result of the changing pattern of relationships among states owing to differential capabilities to exploit science and technology for military or economic purposes. The military and economic dominance of the U.S. in the west, and the Soviet Union in the east, arises from those differences, though the decay of that dominance, at least in its economic aspects, is an important change that will affect the coming twenty years. More striking is the disparity between north and south in ability to develop, assimilate, and use technology effectively for economic purposes, a disparity increasingly a focus of attention in international politics. Other international effects grow out of the greatly intensified mutual dependency relationships that have accompanied the continuous industrialization of technologically advanced countries. Under the general and rather ill-defined term 'interdependence', it is evident that resource, food, information, capital, manpower, and other sectors are ever more closely intertwined across borders, and that even the most advanced technological nations find themselves increasingly dependent on others to provide the primary resources for the operation of their economic systems. Technology itself has become a basis for dependency relations, a point to which we shall return.

Another class of international effects includes those that flow from the basic international or global nature of some technologies, or the implications of the extended use of technologies that may have local effects individually, but global effects in the aggregate. Space communications or weather-forecasting systems are examples of the former; transnational and global environmental issues or population growth are examples of the latter. In this class can also be included the international policy issues that arise from applications of technology, such as concerns over controlling the danger of nuclear weapons proliferation associated with civilian nuclear power, or attempts to control the spread of advanced conventional weapons. The NATO alliance itself attempts to use science and technology for more than national purposes through the Science Committee and through attempts to share and standardize military technology in NATO.

Given these international effects from the application of research and development, which in sum have profoundly altered the nature and substance of international relations, it is striking that R and D support and programming remains primarily a national endeavour. There are exceptions, of course, found more often in scientific than in technological subjects, especially those associated with global phenomena such as climate, or oceanic circulation. But, by and large, R and D is a national enterprise. That is, the planning and setting of R and D objectives and decisions on the allocation of resources, especially in applied research and development, are determined largely within a national framework, with all that implies about the nature of those decisions. The exceptions are important, for they may establish newer trends, or at the least show what is and is not possible, or what the barriers are to greater collaboration or joint planning.

Clearly, determining R and D objectives on the basis of national criteria is not likely to reflect adequately international or global needs. But other questions are also raised by this situation. The onrush of technology has given rise to problems of unprecedented complexity and risk. How can all nations affected by nationally developed technologies, or exposed to their potential risks, participate in the decisions regarding the management and regulation of these technologies, or in assessing their potential risks before irrevocable decisions are made? Are our political institutions adequate to cope with technological issues and their consequences, even in the most technologically advanced nations, let alone in those with less technical competence? And in the face of complexity and uncertainty, how can solutions be devised for international issues when these solutions are likely to have widely varying implications for various affected countries?

With this backdrop, we shall explore several international issue areas closely related to science and technology that are likely to be of particular importance to the countries of the NATO alliance over the next twenty years. Our focus is not only on a description of the parameters of the issue, but more critically on the political and institutional problems inherent in dealing with those and similar issues before us today, or likely to arise in the future.

18.2. Issue areas

Consequences of differences in technological performance among nations; transfer of technology
Competence in science and technology, and even more the capacity to convert it to economic application, is not evenly distributed among nations, even among those with similar resource and economic endowments. Differences in size, resources, wealth, and historical pattern of development are part of the explanation of the differences, but other factors also seem important: religion, form of economic and

political organization, perhaps ideology. It is inappropriate here to explore the causes, but the fact of existing differences among nations and even among regions is important. Moreover, it is not simply differing capacities for the conduct of R and D that is relevant, but also the ability of a nation's economic system and infrastructure to assimilate the results of R and D into the formulation of policy and the creation of goods and services.

Relations with Communist countries and with developing nations are both heavily influenced by these differences in technological performance, especially in relation to the transfer of technology between NATO countries and those in either category. This question of transfer of technology has come to pose important immediate and future policy issues for NATO countries; it deserves discussion in the context both of East–West relations and of North–South, for these settings pose quite different policy concerns.

East–West transfer of technology

The Communist countries of Eastern Europe, and especially the Soviet Union, have long had the goal of catching up to and surpassing the West in scientific and technological capability. They have devoted extensive resources to this effort, and have created a variety of incentives to encourage their best talent to enter scientific and technological fields. The motivation can be seen simply as a recognition that a strong science and technology are necessary for military strength *vis-à-vis* the West, but other motives are also apparent: political influence deriving from technological performance, requirements of economic growth, and internal pressures for consumer goods available in the West.

Notwithstanding substantial progress, and strong basic science competence in a variety of disciplines, the Communist countries have remained behind the forefront established in the West. And they have experienced, particularly in the Soviet Union, continued difficulty in translating scientific knowledge and laboratory technology into producible and marketable products and services. There appear to be many reasons, most stemming from the form of economic organization of the society, the difficulties encountered in applying an ideology committed to planning to the scientific and technological enterprise, often extreme compartmentalization among government entities created by the incentive structure, and other difficulties that can be largely encompassed under the headings of economic organization and management. Notwithstanding these difficulties, the Soviet Union has been able to create a formidable modern military force, though it is generally accepted that Soviet weapons systems have been designed to compensate for the country's technological shortcomings rather than encompassing the latest technology known in the West. Sometimes, however, this results in weapons systems, especially for ground warfare, which are superior in design to Western counterparts.

The result has been a continued, and in recent years, intensified interest in the East in acquiring Western technology in order to bypass the weaknesses in their own system. Access to technology has become one of the important political goals of Eastern European communist countries, and clearly one of their principal motivations for moving toward *détente* with the West. The extensive array of bilateral

S and T agreements between the Soviet Union and Western countries is one product of this policy goal. However, the meagre results of technology transfer from the West have undoubtedly been disillusioning to the Soviet Union. More substantial changes than simply gaining access to Western technology are required, changes that, as will be noted below, raise important political problems for the Soviet leadership.

Communist China also, starting from a much more technologically backward position, has shown increasing interest in acquiring Western technology for more rapid economic, and presumably also, military development. The solidification in power of China's present rulers, who have demonstrated a much more pragmatic approach than their predecessors, appears to have accelerated this trend. [1]

The interest of Communist countries in Western technology has raised several policy concerns, and disagreements, among Western countries. There is first of all the obvious question of the impact on the security of NATO and the West of making technology available, especially to the Soviet Union. The question is not a simple one to answer, for it is bound up, among other things, with the ability of the Soviet Union to reproduce the latest Western technology even if it had it, its availability through 'leakage' and clandestine channels in any case, and the different needs of Russian military systems because of the patterns of design, as noted earlier, that incorporate less sophisticated technology, usually with larger numbers of less multipurpose weapons.

The Western allies established shortly after the Second World War (Japan is now also a participant) an informal process for controlling technology transfer called Cocom, within which an agreed and frequently revised list of technologies is the basis for embargo of specific technological products and systems from sale to Communist countries by any member country. Gradually, the list has been shortened as some of the technologies have been developed in the Soviet Union, and others were seen to be of little military significance, or as other motivations of trade and detente became more important. Even the shorter list is a source of considerable friction today, as most European countries see new trade opportunities with the East as more important than the security risks which they rate as small. The U.S. has tended to be more conservative, in part because of the U.S. enabling legislation that gives the Department of Defence an effective final voice on any export licence.

The disagreements have recently become sharper, carrying the danger of the breakdown of the control machinery altogether. Even now, there are indications of increasingly lax enforcement of controls on the part of some Cocom members. This has led the U.S. to undertake a substantial review of the whole issue, with results as yet uncertain.

Clearly, no technology can be prevented indefinitely from reaching the Soviet Union, nor should the transfer of technology be seen only in negative terms. Not only are there trade opportunities, but also potential political benefits to the West. Not the least of the latter, though the hardest to measure in policy terms, are the changes required in the Soviet system to make effective use of technology, and to have productive scientific and technological relations with the West. Unless those changes are made, for example, breaking down communication barriers within the system and to other countries, allowing greater independence of judgement and access to Western ideas at more places within the system, and similar moves, it is unlikely that the gap will ever be closed. If those changes are made, the evolutionary political process within the Soviet Union is likely to be more desirable from the Western standpoint.

Moreover, the next generation of Soviet leaders, certain to succeed to power in the near future, must be well aware of the myriad of major economic issues facing that country, many deriving from their mediocre technological performance. A forthcoming attitude on the part of the West to help in technology could be an influence towards moderation on other policy fronts. In addition, the Soviet Union has natural resource endowments likely to be of interest to Western nations in the future. It is in the Western interest to facilitate the development of these resources possibly in exchange for technology.

The latter political arguments may have even more force in the case of China since that nation now appears to be embarking on a massive effort in science and technology; nations prepared to participate in the early stages of that effort may well develop the greatest policy and economic interaction over time as that huge nation grows in the years to come. These relations cannot be seen independently of those with the Soviet Union, of course, but that fact only adds to the political significance of establishing a technological policy for China.

Notwithstanding some of the political and economic benefits that might be achieved in more relaxed transfer of technology and the erosion of present controls, it remains true that some technologies at any given time might, for security reasons, be better kept from Communist countries. Is it possible to identify such technologies – so-called critical technologies – for which agreement is possible among Cocom countries to accept and enforce controls? The problem is complicated by the fact that many technologies have multiple uses, commercial as well as military. For those, safeguards of various kinds can be, and have been, devised and accepted, though no safeguard can ever be perfect.

There is no clear-cut answer to the question. One proposal is to focus not on technology products, such as computers, but on process or know-how, such as manufacturing techniques for integrated circuits. That seems reasonable, but also hard to enforce in practice. Are there other candidates: gyroscopes, for example?

The answer is important, for without a reasonable identification of critical technologies, and a pattern for phasing out embargoed items as the technology advances, it is quite likely that controls will erode entirely, without adequate consideration of the significance to Western security.

It is worth observing that one of the difficulties inherent in coming to grips with this issue is the inadequacy of the institutional base for its resolution. Cocom is some 30 years old; many member countries have little or no formal domestic machinery to consider the issues, and because of their domestic situation, are more willing than the U.S. to let economic considerations dominate their views. In the U.S. the formal policy machinery is elaborate, slow, entrenched in the bureaucracy and at too low a level to give effective representation to non-military considerations. Moreover, the legislation now gives the Department of Defence a dominant voice which tends to lead to worst-case analyses that play down political and economic values. Political attitudes from the Congress and the public are also likely to be felt on the conservative, restrictive side.

The result is a policy process inadequate to deal effectively with changing situations in which technology has become a much larger factor in international politics, and is available as a subtle tool of policy. Instead, it is likely to become an oversimplified source of friction among NATO allies, with benefit only to the Communist countries.

Although security considerations have generally dominated discussion of technology transfer to the Eastern Bloc, concern has arisen also with potential commercial competition further down the road. The ability and willingness of Communist countries to subsidize competition with other countries (as currently in the case of shipping rates) gives rise to the fear that technologies transferred to the East might some day be sold in the West at cut rates if a sufficient political or economic advantage could be obtained by doing so.

North–South transfer of technology

The relations between the rich and poor countries have emerged as a major political problem of the present decade, for the first time vying for attention with East–West relations. This is a product of advances in science and technology in two senses. First, the income gap between the developed and underdeveloped countries has been increasing partly because of the differential technological performance of the two sets of societies, so varied in other respects as well, partly because of differential population growth rates. Second, modern communications and travel have made the gap in living standards more apparent to people on both sides, and have stirred feelings of undeserved injustice among the élite of the poor countries, and of guilt among the élite of the rich countries. The present average gap in income *per capita* is estimated at 12 to 1, whereas it was probably only about 3 to 1 at the end of the nineteenth century [2]. The gap continues to grow despite the fact that the overall percentage growth rates in the LDCs have slightly exceeded those in the developed countries for most of the last twenty years. However, population growth has absorbed a large proportion of the economic growth of the LDCs, and this too is a consequence of the application of western public health and nutritional technology in the poor countries. Indeed it could be argued that the gap has arisen from an imbalance between the rate of transfer of public health technologies and of other technologies for improving living standards and controlling fertility.

The problem of the gap between rich and poor has thus come to be seen by both sides as mainly a problem of technology transfer. According to Orville Freeman, president of Business International, 'technology in the broadest sense – including material, managerial, marketing, organizational and other skills, as well as advanced technical information such as secret know-how – is at the heart of the difference between developed and developing countries [3]. It is recognized that the mere sharing of wealth without transfer of the knowledge and skills that make wealth productive is not very useful, and even self-defeating. Partly in consequence of this perception, the modes of control and the terms of transfer of technology have become an important ingredient in the North–South debate.

There is a paradox in the fact that the most successful mode of technology transfer from North to South, outside of agriculture, has been the Multinational Corporation (MNC), but this is also the most deeply resented of all developed country institutions. It cannot be argued that MNCs have been an unalloyed benefit to the LDCs, or that if nations would rid themselves of their unnatural preoccupation with sovereignty, the multinational trading system, and its efficient division of labour, would solve all of the problems of development. The adverse impacts and misbehaviour of MNCs may have been exaggerated in recent years, but the problems are real.

The LDCs especially see proprietary technical knowledge as a wilful obstacle to technology transfer, and express the view that such information should be made available to the poor countries either free or at a very low price. They argue that most technology sold to LDCs has had the cost of innovation amortized in the developed country markets, and therefore the marginal cost of supplying it to poor countries is very small. The MNCs argue that the technology transferred is an expensive commodity and that the time in which the cost of development can be recovered is severely limited by obsolescence. The LDCs counter with the argument that their markets provide an extension of life for technology which would be obsolescent in developed country markets, so that revenues received from LDCs constitute a bonus not originally factored into the economics of the innovation [4].

Another source of conflict between DCs and LDCs about technology transfer has to do with potential penetration of certain kinds of goods from LDCs into DC markets. Especially during a time of lagging growth in the DCs, there is increasing fear that the transfer of production technology will create competition that will result in labour displacement in the DCs. There are actually few cases to date in which the volume of trade from LDCs is large enough compared with developed country markets to cause appreciable injury to domestic industries in the developed countries [5]. Nevertheless, labour in the DCs is very conscious of the competition from Japan and from the middle-income rapidly developing LDCs, such as Taiwan and Brazil, in areas such as textiles, consumer electronics, and ship-building. As a result, there is growing suspicion of too energetic efforts to transfer production technology which might create efficient competitive capacity in the third world. This is likely to prove an increasingly troublesome political problem in the next decade, even if high rates of growth resume in the DCs. High growth rates would facilitate structural adjustments in the labour market of the DCs. However, some of the labour displacement is likely to occur in high technology industries, such as electronics, which still belong to the growth sector of the DCs. Unless the DCs continue a high rate of innovation to create *new* industries and services, rather than just refinements of existing technology, their economies will probably not be able to stay ahead of the rapid diffusion of new productive capacity around the world. The target of UNCTAD calls for the LDCs having 25 per cent of manufacturing capacity by the year 2000, compared with 7 per cent today [6]. Although the target is unrealistic, it implies a determination of the LDCs to make many goods which are now exported from DCs. It is essential that greater effort be devoted to inventing adjustment mechanisms that insure that structural changes which benefit the world economy as a whole do not result in costs which are borne excessively by particular groups within DCs. This is necessary not only for reasons of equity, but even more to avoid a resurgence of protectionism and the development of intransigent political opposition to the transfer of technology and to the opening of DC markets to LDCs [7].

The conflict between DCs and LDCs regarding the role of MNCs can be understood in terms of the limitations of classical market incentives in relation to broader development goals. These broader goals include the creation of employment, the wide diffusion of technical sophistication and know-how within the working population, and the extension of the benefits of development to the poorest segments of the population and the least developed regions of the country.

The line of least resistance often leads LDCs to adopt capital-intensive production technologies when in fact cheaper, more labour-intensive technologies would be both more efficient in strictly economic terms and much more supportive of development goals, including employment creation and diffusion of technological know-how. This happens because the transferring agent tends to think in terms of the relative factor costs appropriate to the host country, while the recipient government officials in the host country see elements of prestige for themselves, and the opportunity to bypass difficult problems of labour training and the initially inefficient use of local inexperienced engineers and technicians.

This leads directly into the debate over 'appropriate technology' and the accusation that most technology transfer from the DCs to LDCs is 'inappropriate' to the setting in which it is implanted. The term is somewhat misused insofar as it implies that high technology is always inappropriate and low or 'soft' technology is always the most appropriate. The aim should rather be to use the full range of instruments made available by modern technology; the important point is that criteria of selection should be set in terms of overall development goals rather than merely economic efficiency in the narrower sense of maximum addition to total national income. The menu might then include anything 'from improved traditional technologies to others which are modern but small-scale, labour-creative, and indigenously developed; to others which are second-generation or "obsolete" technologies imported from developing countries; and to others which do not exist anywhere yet but which must be developed *ex ovo* to suit special needs' [8]. It is often pointed out that the People's Republic of China (PRC) has not hesitated to adopt the most capital-intensive and sophisticated modern technologies for its capital goods industries, while using employment-creating small-scale technologies for basic consumer goods production. [9]

A policy of advocating 'appropriate technology' can also carry the connotation that the DCs in fact have the motive of preventing the kind of national economic development in the LDCs that is the hallmark of a modern industrial state. Other policy recommendations, such as urging emphasis on agriculture or decentralized energy production, are often received in the same way. Development is a political, and not just an economic, process and must be seen that way by all the actors.

Another point where politics and market incentives may conflict with development goals is in technological training of local labour and technicians. Timely delivery of needed goods and services argues for heavy reliance on imported managerial and technical skills, and yet the optimal transfer of technology argues for a more gradual and costly process in which the necessary skills are acquired by local people. In the long term, economic development should be regarded as primarily a problem of human resource development, rather than capital or infrastructure development, and its success should be judged by the degree and speed with which the host country can dispense with foreign tutelage. Yet this sort of organic development is not the cheapest form and tends to be in conflict with the efficiency and quick return on investment goals natural to private foreign investors, and often also to development bureaucrats in the host countries, donor countries, or international agencies.

In this connection, it is interesting and somewhat disquieting to observe that one of the most successful forms of technology transfer has occurred in the military field. This is probably because it is automatically assumed that sophisticated transferred weaponry will have to be operated and maintained by nationals, and therefore every effort is made to incorporate training as an integral part of the transfer process. Partly as a result of this fact, many LDC regimes see the acquisition of modern weapons as a route to rapid modernization and national technical self-sufficiency. In actual fact the rate of transfer is probably exaggerated, and dependence may follow weapons transfers for longer than is believed.

A more difficult problem with foreign technical assistance, especially via private capital, has to do with its distributional effects in the host country. Rapid economic growth has frequently resulted in increasing rather than decreasing income inequalities within LDCs, and has benefited least the poorest part of the population. However, rapid economic growth is sometimes associated with increased equalization of incomes as well as with increased inequality. Brazil and Mexico seem to be examples of countries where rapid development has exaggerated the gap between rich and poor, while Malaysia and South Korea are examples of the opposite trend.[10]

The policy levers necessary to effect development with improving income equality are not fully understood. Apparently a high level of education in the general population is a helpful factor, as is the existence of strong credit and savings institutions. Since there is evidence that increasing equality of incomes is a factor in reducing fertility, especially among the poorer fractions of the population, it is a particularly important development goal. In fact there seems to be positive feedback here in that it is easier to achieve greater equality if the poor have smaller families; the world over, in developed and underdeveloped countries alike, there is a high correlation between poverty and family size.[11]

There is general agreement that the most successful technology transfer system to LDCs is in the field of agriculture, where the network of nine semi-autonomous international research centres supported by the international consortium of donors known as the Consultative Group on International Agricultural Research (CGIAR) has a record of substantial achievement in the short six years of its existence (though some of the constituent research centres go back much earlier than that)[12]. These institutes are becoming the nodes of a world-wide network in which advanced research facilities sponsored by developed countries and international agencies provide basic technologies, national research institutes provide adaptation to local conditions, and members of the network test solutions in experimental plots in a wide variety of circumstances. There are now five formally established international collaborative programmes based on this networking principle [13]. Although agriculture ordinarily requires more local adaptation than industrial or infrastructural technologies, the agricultural research system provides a potential model for other technical areas. [14]

Technology transfer in agriculture has not escaped the same kinds of criticism that have been levelled at other kinds of technology transfer. In particular, it has frequently resulted in greater benefits to the better-off and more entrepreneurial farmers; it has increased vulnerability of the food system to stresses such as drought or new pests, it has sometimes displaced labour, and its adverse ecological consequences may be severe in the long run. Even in agricultural research, it is important that developmental goals other than merely increasing productivity be kept in the forefront of attention [15]. Analysis has shown that the adverse consequences of the Green Revolution are not inherent in the technology, but avoiding them depends on careful attention to social supporting systems as well as technology [16].

In summary, the key role of science and technology, and of technology transfer, in closing the gap between rich and poor countries is acknowledged by both sides, but the beneficial impact of S and T is not automatic and depends on the formulation of development goals and incentives much broader than the expansion of economic growth and productivity, and the setting of priorities for technology transfer which are consistent with these goals.

National R and D versus transnational problems

It is evident that R and D is supported by public or private sources for purposes defined by those sources. In industry, it is natural that R and D should be aimed at the needs of a company, as seen by that company. In government, it is equally appropriate that ministries support R and D related to their missions as defined by law. The problem relevant for us lies in two factors: (1) in some areas, especially related to international concerns, there is no ministry with the requisite responsibility, or where there is, it is too often a ministry with little experience, competence, or interest in R and D; and (2) ministries in functional areas tend to operate in a political framework that necessarily gives more weight to domestic issues and interests than international ones, or to domestic effects of international issues.

Both of these factors are perhaps more in evidence in the U.S. than in other NATO countries, though they are relevant to all. The situation leads to several undesirable effects.

(1) Some international issue areas of ultimate relevance to national objectives and concerns do not receive timely R and D support.
(2) R and D in such issue areas is often fragmented and of sub-critical size, spread among several ministries, no one of which sees it as a major goal.

(3) It is difficult to give adequate attention to foreign policy considerations ar
to the international context when determining R and D objectives and all
cating resources in functional areas defined in domestic terms.

(4) As a result, policy-makers are often confronted with products of R and
that have major international impact without having had an opportunity t
influence the course of development of those products.

A series of examples will illustrate these effects, and point up more sharply th
institutional weaknesses from which they derive. How those weaknesses may t
improved is another matter.

Development R and D

The overwhelming majority of the world's R and D is carried out in the develope
countries, more than one-third of it on military-related topics, and most of the r
mainder is of primarily domestic relevance in each country supporting the R an
D. A small percentage of the total is devoted to problems defined exclusively o
primarily in terms of the needs of developing countries. In the case of the NAT
countries, such research is often related to specific economic needs of form
colonies, especially in agriculture and health. A precise figure is impossible to dete
mine, but a fair estimate is that it is under 5 per cent of the total global R and
budget.

Of course, that understates the amount of R and D actually relevant to th
LDCs, for many issues are common to both DCs and LDCs; much industriall
oriented technology development is applicable to both; LDCs cover a wide spe
trum, including many with problems similar to the DCs; many important glob
issues transcend country categories; many of the products of basic research are o
significance to all; and even defence R and D, unfortunately, is increasingly pert
nent to the needs of LDCs, as perceived by the leaders in these countries.

On the other hand, if the question is asked another way; how much of th
world's scientific and technological capability is devoted to the needs of the world
poorer countries, the answer is a pitifully small fraction.

Economic growth and eventual elimination of poverty in the poorer countries o
the world have been repeatedly articulated as important goals of the industrialize
nations, for reasons of self-interest if no other. Damping of population growth, wis
management of environmental and resource problems, and greater political stab
lity will, among other factors, depend on visible progress toward these goals, if no
their attainment. It makes little sense in that setting for the world to devote so littl
scientific and technological effort to problems that are peculiarly those of LDCs.

Much of this R and D cannot and should not be done in industrialized cour
tries, for practical as well as philosophical and political reasons. To be effective, t
work on the right problems, to be sensitive to local needs and preferences, to pro
duce solutions that fit and are likely to be adopted, to keep up and adapt, all re
quire R and D defined and carried out locally. In turn, this implies much greate
attention to the building of the scientific and technological infrastructure in LDC
than has recently been the practice.

But, this does not mean that *all* R and D relevant to LDCs needs to be carried out *in* LDCs. Many areas of basic research can more effectively be carried out in existing laboratories; many problems of LDCs are generic and can be more quickly investigated in experienced laboratories with adequate resources and skills already deployed; many technological problems require general solutions before locally adapted applications are possible. Perhaps most important is finding ways to commit scientists and engineers in industrialized countries to work on problems of development in a sustained way that allows for the cumulative benefits of 'organizational memory' and continuous attention to a problem area. Long-term availability of financial resources is essential, not only to make such commitment possible, but also to make such a commitment respectable in the eyes of the disciplinary peers to whom technical people look for recognition.

Some might still argue that all that is needed is transfer of existing technological knowledge to the LDCs. Though some of that is useful and will always need to be fostered, one clear lesson derived from experience over the last decades is the ineffectiveness of such a transfer unless adequate receptors exist to choose, adapt, finance, and develop knowledge to fit local environments and needs. Technology is seldom sufficiently codified to be transferred as an intact package; it requires adaptation, including adaptation to a unique social, economic, and political, as well as technical, environment. Also it tends to change that environment, often quite rapidly, so that in fact mutual adaptation of technology and the environment is a continuing and dynamic process.

The selection and adaptation of imported technology by LDCs requires a significant and growing indigenous technical capability in the host country. This technical capability must embrace basic science as well as technology, for without the insight and self-confidence created by an indigenous scientific community an LDC will lack the ability to control its own technological development.[17]

An intriguing example that exemplifies all of these arguments is agricultural research. Enormous strides in agricultural productivity have been made in Western countries, very substantially as a product of extensive R and D, especially in the U.S. The same R and D approach has recently been applied through the creation of international research establishments *in* developing countries also with impressive results. But the needs of agriculture in LDCs have as yet barely been touched.

Only a small part of the extensive agricultural research establishments of the West have concerned themselves with problems outside their own countries. This is not surprising, for they were created first to serve domestic needs, and for many years assistance to foreign producers meant assistance to potential competitors. Domestic agricultural lobbies have traditionally been politically powerful and had little interest or incentive to encourage, or even allow, R and D to assist other nations.

The situation, in the U.S. at least, is now changing. A National Academy of Sciences report[18] laid out a long and critical agenda of agriculture research needed for developing countries. The Department of Agriculture is seeking its own funds to allow its research laboratories to work on non-U.S. problems. The Congress has created a new part of the Foreign Assistance Act (so-called Title XII) to provide funds for American land-grant universities to work on agricultural problems outside the U.S.

In effect, the U.S. is coming to recognize that its own R and D capabilities can legitimately — in the sense that it is in the U.S. interest — be applied to international as well as domestic problems. This will be of limited value unless there is a concomitant growth in R and D capacity within LDCs, but the two developments are complementary, and promise considerable benefit if both are carried out. Because of their former overseas colonies, many NATO countries possess R and D capabilities and scientists knowledgeable about conditions in tropical countries to a much greater extent than the U.S., and such capacities as they have are better focused.

Nuclear power and proliferation

Nuclear power and proliferation represent an example of a quite different kind. Here, technological development of nuclear power plants proceeded until recently with little detailed reflection of foreign policy considerations except for the overall general (and important) objective of exploiting the technology for foreign policy purposes. Essentially, that influence simply resulted in a policy of 'the faster, the better'.

But there was no serious attempt to influence the development process in ways that would have injected foreign policy criteria into, for example, the choice of technological designs for reactors or fuel cycles. By hindsight, it is clear that the world and the West would be better off today if reactor designs, and the pace of development, had taken account of proliferation concerns early in the process, rather than largely after development was completed and huge investments had been made in the reactor systems which were favoured in terms of domestic economic criteria.

In fact, the situation was even worse in that nuclear reactors were exploited, by the U.S. in particular, in the Atoms for Peace Programme, with inadequate attention to possible longer-term dangers, particularly those associated with extensive dispersion of plutonium in weapons-usable form among many political jurisdictions.

It is interesting that now, when it may well be too late, many nations participating in the International Fuel Cycle Evaluation Programme are, in fact, attempting self-consciously to set R and D objectives for major systems based on foreign policy criteria. The question is whether other nuclear fuel cycles can be chosen as candidates for R and D based on their relatively greater resistance to proliferation. It is perhaps one of the first major efforts to use such political criteria early in the R and D process as a central criterion of technical choice.

But the political criteria depend heavily on subjective judgements about probable human behaviour. These judgements are bound to be much less clear-cut than traditional development criteria based on objective and quantifiable measures of system performance. Subjective values injected at that early level of the R and D process may simply cause distortion and ineffective diversion of resources, rather than development of alternate technologies. Most of the technical alternatives now being considered were considered and rejected earlier on the basis of economic or technical criteria. On the other hand, the effort to set political constraints on development objectives early in the process has rarely been made, and certainly has not been analysed. Both deserve to be done.

Trans-national environmental issues

Environment issues have emerged as significant in three ways:

(1) Certain industrial effluents cross national boundaries and create problems in adjacent countries, sometimes greater problems than in the country of origin. This has been a long-standing problem in river basins that are heavily industrialized and which traverse several countries, e.g. the Danube or the Rhine. More subtle river basin problems have arisen between the U.S. and Mexico, where irrigation activities in the U.S. have given rise to mineral concentrations in the waters flowing into Mexico such that the value for irrigation and other economic purposes is significantly degraded. A more dramatic effect has been the transport of sulphates arising as conversion products of SO_2 in smoke stack plumes from heavily industrialized areas, such as the Midlands in Britain or the Ruhr Basin in W. Germany. These sulphates are claimed to have appeared as acid rain in the Scandinavian countries and other parts of Europe, with adverse environmental effects, for example, on the fish populations of fresh water lakes and streams [19]. The long distance transport of sulphates is likely to become increasingly important in the future as the production of electric power from coal grows. The extent of their transport and their possible adverse health effects are still poorly understood, but are likely to become an increasingly divisive political issue between neighbouring countries.

Another growing transnational problem results from the location of nuclear power plants and reprocessing facilities close to international boundaries. Issues of safety and radioactive emissions across these boundaries are raised, as in current problems between West Germany, France, and Switzerland.

(2) A number of human activities have now reached a sufficient scale to affect the global environment. Examples which are of current concern are the general pollution of the world's oceans by hydrocarbons and by DDT residues. Until the banning of atmospheric nuclear testing in 1963, radioactive fall-out from nuclear tests was also a global problem of rising concern. Recently the U.S. has started the banning of fluorocarbons in many uses because of their likely future effect in reducing the ozone content of the stratosphere. Many other problems, such as CO_2 from the burning of fossil fuels, heat release due to energy production, release of dust to the atmosphere as a result of desertification, changes in the earth's albedo due to deforestation, agriculture, water, and irrigation projects, and other human activities are somewhat further in the future. These will be discussed in more detail below.

The characteristic of such problems is that the activities of any one country make only a fractional contribution to the global problem, and yet the overall management of the problem requires equitable restrictions on the activities of all countries. Where the economic consequences of such restrictions are small and relatively short term, management can probably rely on voluntary co-operation among a few countries, but where the adverse economic consequences may be severe, as in the case of restrictions on the burning of coal or the reprocessing of nuclear fuels, the negotiation of an international regime presents much greater difficulties. Fortunately it appears that most of the global problems which may require severe economic restrictions are far enough in the future to allow a good deal of time for evolutionary institution-building. Furthermore, the problems for the most part originate from the activities of advanced industrial countries, so that agreement on controls and the development of alternate technologies is easier than if the LDCs were significantly involved as well.

(3) The application of environmental standards and regulations affects the relative costs of goods that move in international trade, and thus can affect comparative economic advantage. This could lead, and no doubt has, to competitive relaxation of environmental controls for trade advantage, and creates the need for some kind of agreement on standards across national boundaries. Yet complete uniformity of emission standards may be economically inefficient; there is something to be said for a division of labour in environmental pollution as in many other economic activities. Such a division of labour would argue for the movement of highly polluting industrial activities away from regions where they are already heavily concentrated, and to regions and climates where the possibilities of rapid dispersion or environmental assimilation are high. The dilemma is how to achieve a sensible balance of abatement and relocation of pollution, while avoiding the operation of a sort of Gresham's law of environmental quality between countries. This problem has already been tackled by OECD, and seems to be manageable among the developed countries, which are still the main source of the problem. Management may become more difficult as industrialization spreads among the less mature economies, in which growth has much higher priority than environmental quality.

It is hard to foresee exactly how this issue will develop. Both the Eastern Bloc countries and the LDCs are increasingly sensitive to the charge that they are becoming 'pollution havens', and show signs of moving towards more restrictive policies. For example, a new development plan for Malaysia requires environmental assessments of all major projects, and Ghana has recently required the filing of environmental impact statements for all new industries [20]. There is increasing recognition that it is the rural poor who frequently suffer most from certain kinds of development-induced environmental degradation.

Co-operative development of global technologies

The term 'global technologies' is used here in two senses. The first sense refers to large-scale technologies whose socio-political or environmental effects have a major transnational component, but are largely developed from a national base, though their deployment often involves multinational institutions, including MNCs. Examples include satellite communications, resource monitoring and weather satellites, multinational data networks, seabed mining, and new fish harvesting techniques. The second sense refers to technologies whose scale is such that only multinational development and deployment may make economic sense. There are as yet o clear-cut examples of the latter category. A possible future example might be satellite solar power. [21]

It is also possible that if fusion is ever successful as an energy source, its scale will be such that the first commercial reactor may well be multinational. The argument has been made that fission breeder reactors might be more appropriate for co-operative international development and deployment, especially if considerations of security against proliferation argue for deployment in secure energy 'parks' with colocation of reactors, chemical reprocessing, fuel refabrication, and radioactive waste disposal. [22]

It is interesting to observe that, in the wake of Hiroshima and Nagasaki, consideration was given to developing fission power on an international basis in the form of the Acheson-Lilienthal plan [23]. The argument was primarily that fission was too dangerous a technology to be left to private or even national exploitation, an argument to which we now seem to be returning.

Although the Soviet Union, France, Britain, Germany, the United States, and Japan each have separate national programmes to develop liquid metal fast breeder reactors (LMFBR) there is close co-ordination and exchange of information among the five main non-Communist programmes, and a number of the projects are actually joint ventures. Experimental reactors in one programme are often used to test components for other programmes. However, all the national programmes have essentially similar technical objectives, and one may well ask whether it would not have been better if the countries involved had gotten together to develop several competing breeder concepts, rather than having all concentrate on a single technical path [24]. What might have made sense would be to assign a number of multinational consortia to develop different concepts, with agreement that the winning development would be available to all participants for final commercialization. To make such a proposal, perhaps, reveals its political unrealism, but shows also how far we are from a truly global technological system.

For a time Europe tried to develop nuclear energy on a multinational basis through Euratom, but as soon as nuclear power began to look economically competitive, national commercial development took over, with Germany and France each developing a strong manufacturing industry with export capability. Isotope separation and reprocessing industries, however, are being built up on a partially multinational basis, primarily because of the scale of the investments involved, and the fact that one enrichment or reprocessing plant is best scaled to serve a large number of reactors. On security grounds, it has also been proposed that enrichment and reprocessing be restricted to internationally operated and governed facilities. [25]

Probably the most extensive technology development and deployment that has taken place on a multinational basis is in the natural resource field. This has been especially striking in North Sea oil development, where many of the drilling platforms have been built by consortia of national and private oil companies. A race appears to be on to develop technology and operations for recovery of manganese nodules from the floor of the North Pacific. The largest consortium, Ocean Mining, Inc., includes participation by German, Japanese, Canadian, and American companies, with the German Ministry of Science and Technology contributir financially to the supporting research [26]. Because of the risks and scale of mar natural resource developments, we may expect the continuation and expansion c joint ventures involving both governments and private companies in this field.

Interfacing of national technological systems

There are many technologies, such as air traffic control, communications, nav gation systems, weather observation and forecasting networks, postal systems, ai line reservation systems, and financial and banking systems which require clo interfacing of national systems. All of these systems are essentially describable : information technologies. For the most part, the interfaces and the potential r sources of international conflict lie in the flow of information rather than in tr sensors which gather the information. By and large, these interfaces have worke rather well, as long as the technological environment remained rather stable an evolved slowly. Recently, however, the explosive development of information tecl nology has begun to place serious strains on the system. This has been particularl true because technology has broken down previous sharp distinctions betwee different functional areas, e.g. communications versus data processing, informatio systems versus banking, postal versus electronic communications. [27]

The situation has been further complicated by public concerns about privacy c records of individuals, and differences in national practice have become sources c conflict as economies of scale lead to storage of credit, reservation, and other typ of personal information in central computer banks. Credit information abou citizens of one country stored in another country and transmitted freely acro borders in response to computer interrogation has been a particularly sore poin Several European countries, such as Sweden, have passed rather strict laws limitin the transport beyond their borders of certain kinds of information which migh violate privacy or confidentiality. This subject has been a source of disagreemer between the Canadian government and U.S. firms offering various kinds of ir formation services.

In addition, many national PTT systems have limited the use of their services b private information networks because of the fear that the shift of revenues fror carriers to terminal facilities would erode their revenues. When large amounts c information can be conveyed in short signals and interpreted by elaborate pr cessing at terminal facilities, the carrier loses revenue to the owner of the termin: facilities. A few bits of data, supplemented by information already stored at th terminal, can substitute for an extended voice conversation. All of these problem have been exacerbated because most private international information network are based in the U.S., and are seen elsewhere as a new source of American technc logical dominance [28]. American control of the flow of information, even thoug provided in a manner open to all, is seen as a potential threat to national sovereignt a threat that could become actual in time of international stress.

The problem is complicated by a perceptible difference in philosophy betwee the U.S. and other countries, a philosophy which can be seen as reflecting thei different interests. Americans advocate the free flow of information, with competitiv forces determining the relative investment in different kinds of services, and the maximum economic benefit from economies of scale, regardless of how the systems involved may straddle national boundaries. Americans tend to be joined in this attitude by some of the smaller countries of Europe, such as the Netherlands and Belgium, which cannot aspire to self-sufficiency in information technology, but can benefit by becoming 'free ports' for international trade in information. An open system favours American interests because the U.S. for the most part dominates the technology, and is in the best position, because of the size of its domestic information market, to take advantage of scale economies. The Europeans, on the other hand, fear loss of autonomy, and see their national interests potentially threatened by decisions made on purely economic grounds outside their own control.

Perhaps even more serious is the conflict between the U.S. and LDCs over control of information flow, a conflict in which the LDC viewpoint is strongly supported by the USSR, which feels the control of information to be vital to the maintenance of its system. The conflict has flared in many different forums. Perhaps the most interesting, because it involves new technology, is the debate over the possible future deployment of direct broadcast satellites, i.e. satellites capable of relaying radio and TV programmes from a country of origin into the homes of the citizens of another country. This is viewed as a fearsome form of 'cultural imperialism', especially advantageous to America's dominance of the world scene, and intruding massively between a government and its citizens.

However, conflict has emerged in somewhat more prosaic arenas, particularly between the US and Canada. Canadian TV has been dominated by American programming, because most Canadian cities are so close to the U.S. border. In response Canada has gone heavily into cable TV and prohibited American advertising on programmes received over cable in Canada.

The problems outlined above have only just begun to emerge into the political arena. What makes them particularly serious is the extremely dynamic character of information technology. This makes it very difficult for the slow process of international negotiation to deal with the situation. Not only are the problems changed by the time negotiations are completed, but the flexibility of the technology makes it relatively easy to 'design around' any specific regulations, and also makes general regulations difficult to formulate. We believe that the management of all types of global information flows will be increasingly on the agenda of international negotiation in the next two decades. There is, perhaps, no area in which the principle of national sovereignty and self-determination comes into sharper conflict with the principle of freedom of information and the unfettered flow of ideas. So great is the potential for one country blanketing another with information at extraordinarily low cost, that the most dedicated devotee of freedom of information could well ask whether there are some legitimate limits to the applicability of this principle in the international arena.

Resource issues

Cornucopians and Malthusians.
Up until the early 1970s it was assumed that natural resources would not be a significant constraint on development, at least provided that population growth could be brought under control, a relatively free international trading system continued, and scientific and technological progress continued to be implemented in international investment. [29]

Perceptions began to change rather rapidly in about 1972. The emergence of the 'Limits to Growth' debate was immediately followed by the 1973 oil crisis, a world-wide commodity price explosion, and the shortfalls of world grain harvest and shrinkage of grain reserves in the period 1972–4. Serious people began to ask whether the DCs would skim the cream of non-renewable resources before the poor countries could develop to the point where they could use them. Would the LDCs find the cupboard bare just as they were ready to sit down at the world table? Would technological progress permit the transformation of potential resources into economically recoverable reserves faster than resources of existing exploitable quality were exhausted? Would the DCs apply the technology needed to exploit lower quality substitutes fast enough to leave some of the higher quality resources for the poor countries? For example, in the field of energy would the DCs convert rapidly enough to capital-intensive long-term energy sources such as breeder reactors, solar generated electricity, or thermonuclear fusion in order to leave cheaper and less capital-intensive coal and oil reserves available for the development of the LDCs? Or, conversely, will the LDCs try, and be able to, by-pass dependence on existing high quality resources by 'leap-frogging' existing technology and adopting radically new 'soft' technologies much less dependent on internationally traded resources like oil, coal, uranium, or phosphates? [30]

More fundamentally, is continued rapid economic growth in the DCs desirable because it creates markets for the exports of the LDCs and the surplus financial resources necessary for foreign aid and development investment, or is DC economic growth undesirable because it pre-empts exhaustible natural resources which would otherwise be available at low cost for the rapid development of the LDCs? Today we find the world full of contradictory rhetoric on this point. A group of sixteen world economists recently advocated the first alternative in the following terms:

'We cannot stress too strongly the importance of improving economic performance in the industrial world. Slow expansion of output and persistently high unemployment endanger our own societies. They worsen the already difficult problems of the developing countries. They threaten the hard-won progress toward an open international order.' [31]

In contrast is the statement of a group of British and American scientists that 'if present trends are allowed to persist, the breakdown of society and the irreversible disruption of the life support system on this planet, possibly by the end of the century certainly within the lifetime of our children, are inevitable.' [32]

Although the rhetorical contrast exhibited above to some extent results from a difference in time perspective, one still has to face the question of when and how the transition from the all-out growth orientation of the first quotation to 'a condition of ecological and economic stability that is sustainable far into the future' is to come about. This question has apparently never been addressed within a realistic policy framework by either side. The most authoritative analyses appear now to agree that neither non-renewable nor renewable resources *need* be a constraint on development for the balance of the century, but what happens after 2000 is much less apparent. [33]

Sovereignty and international resources

A major political phenomenon of the 1970s has been the assertion by the LDCs of the right of absolute national sovereignty over natural resources. This, of course, comes into sharp collision with the interests of many of the most highly industrialized countries, the maintenance of whose economies is vitally dependent on uninterrupted access to raw materials, almost all of which are under the sovereignty of other countries, many of them LDCs. Japan and Western Europe in particular live by importing natural resources and exporting finished products embodying a high added value. Basic materials, including fuels, constitute less than 9 per cent of world GNP. The rest is 'value added', i.e. essentially information incorporated in the fabrication and assembly of products and in their distribution, and in the utilization of various products in providing services to the consumer. But the production system is very inelastic in the short term with respect to the availability and cost of energy and materials. This was dramatically illustrated by the world-wide recession following the abrupt rise in the price of oil, and historically by the near-disaster to the U.S. economy in the Second World War, when sources of natural rubber were abruptly cut off. By the same token, however, the industrial system may be very elastic with respect to long-term changes in the price and availability of materials and energy [34]. This is largely because of substitution possibilities and the development of hitherto subeconomic resources. Thus even the rubber crisis was overcome by forced development of a synthetic rubber industry [35].

While it is still true that about 60 per cent of all basic materials are produced in developed rather than undeveloped countries, the proportion coming from LDCs has been increasing, at least until recently [36]. Moreover, it is not only LDCs that put political restrictions on the export of basic materials for non-economic reasons. Recently, for example, both Canada and Australia have placed restrictions on the export of uranium ore as a non-proliferation measure. The United States placed an embargo on soya bean exports for a short time in order to stabilize skyrocketing prices. The new U.S. nuclear export legislation places conditions and potential embargoes on export of enriched uranium based on policies toward inspection of reactors and reprocessing of reactor fuel. Actions such as these create exaggerated fears about future access to vital raw materials.

Within the LDCs, there has been a growing trend to nationalize basic natural resources. Even in the absence of outright nationalization, many LDCs attempt to recapture what they see as economic rents by means of higher taxes [37]. Neither nationalization nor high taxes necessarily cut off supplies immediately, but in the long run they discourage exploration, investment, and even maintenance of the productivity of existing investments. Between 1971 and 1976 capital expenditures by U.S. affiliated companies in mining and smelting industries abroad dropped precipitously, with the greatest drop in countries showing hostility to foreign capital, particularly in South America. In 1970–73, 80 per cent of the spending on exploration for non-fuel minerals was in four industrialized countries: the U.S., Canada, Australia, and South Africa [38]. This is likely to have the indirect effect of making industrial raw materials more expensive in the long run than they would be in a freer market. On the other hand, there may be other cases where nationalized raw materials industries overexpand in an effort to capture a larger share of the world market, thus resulting in large losses [39] to the investing government.

Probably in the case of most minerals it is possible to achieve protection by stockpiling, although this can be expensive. In general, the stockpile has to be large enough to buy time for the development of alternate sources or substitution technologies. In this respect oil is probably quite unique in the small ratio of stocks to flows. There is seldom more than a few months' supply in the 'pipeline' and even stockpiles are impractical beyond about 90 days. By contrast the uranium 'pipeline' is usually two or three years' consumption, and stockpiling can extend this rather inexpensively, since fuel is a much smaller part of the total cost of energy. For reasons such as this it is generally believed that the formation of cartels analogous to OPEC is less likely with non-fuel minerals.

For many raw materials, subeconomic resources exist which could be developed if prices rose somewhat. In some such cases it may be worthwhile to 'stockpile' technology which could be used to exploit these resources in case of cut-off of cheaper supplies; the primary purpose would be to reduce the lead time for bringing new sources into production. Such stockpiling of technology would include exploration for higher cost reserves in politically more secure areas, and a greater effort at inventorying potential areas where higher cost raw materials might be found. If such measures were taken by groups of nations (such as NATO) rather than by one country alone, the cost of stockpiling both technology and the resource knowledge gained from exploration could be shared. There is a question whether private investors would be motivated to do this if the contingencies to be hedged against were largely political. At all events the whole issue is worthy of study on a multinational basis by a group such as NATO or OECD.

An interesting suggestion is that certain essential metals such as chromium could be 'stockpiled' in various non-essential economic uses such as coins, decorative metalwork, car bumpers, etc., whence they could be readily reclaimed in an emergency. [40]

An important question is whether conflict over access to raw materials could become an important source of international hostilities at some time in the future. The case of Japan prior to the Second World War is often cited as an example of a nation that went to war partly to assure access to vitally needed raw materials. It seems probable that the only mineral that could lead to this kind of conflict would be petroleum. On the other hand, if there was severe erosion of the present international market in other basic raw materials as a result either of political interventions or a rise in protectionism, it is possible the political climate would change so that hostilities were a much more likely outcome of raw materials conflicts. The present degree of interdependence is such that it is unlikely any one resource would lead to hostilities, with the exception always of oil as noted.

Another possible source of conflict leading to hostilities is exploration for resources in hitherto inaccessible frontier areas, such as the arctic, the southern polar regions, or the hinterlands of Brazil or Siberia [41]. In each case these explorations might lead to the development of political 'presence' in hitherto unoccupied regions which would stimulate fears on the part of neighbouring countries. For this reason there is something to be said for multinational development and exploration in such areas. If properly organized, such development would downgrade the geopolitical significance of the activity, and could even permit participation of the threatened neighbour.

The global commons

Another resource issue has to do with the status of the 'global commons', mainly the oceans. The outcome of the long and laborious Law of the Sea negotiations is still in doubt. One outcome that presently seems irreversible, however, is the extension of the economic zone to 200 miles offshore. The implications of this development are still not completely clear. In the short term it will certainly lead to conflict as countries are partially excluded from traditional fishing grounds and seek new areas of exploitation. On the other hand, sovereignty over fishing areas may lead to wiser management of fisheries, since countries that overfish will be less able to move to new areas. The aspect of the Law of the Sea negotiations most in doubt is the regime of the deep oceans, which will govern mining operations for manganese nodules and eventually other minerals.

Global environmental issues

A prime characteristic of global environmental issues is the long time that may elapse between the commitment to using the technologies that give rise to them and the appearance of significant effects. Another characteristic is the high degree of uncertainty that usually attends any predictions of future effects. The uncertainty is compounded by the fact that it occurs at several different levels.

These characteristics are well exemplified by the problem of the buildup of CO_2 in the atmosphere resulting from the combustion of fossil fuels, especially coal. The first level of effect is the direct chemical or physical result of the activity, in this case the actual CO_2 content of the atmosphere. Since the late 1950s this has been carefully monitored in several locations, and there is virtually no uncertainty about the 14 per cent buildup above pre-industrial concentrations which has occurred so far. [42]

When it comes to future buildup, uncertainties increase because of the lack of an adequate theory of the partitioning of the CO_2 generated between the atmosphere and other sinks. Some theories indicate that the fraction remaining in the atmosphere will increase, and hence the rate of buildup will accelerate towards the turn of the century.

The next level relates to the climatic effects of predicted CO_2 concentrations.

This requires elaborate computer modelling of the earth's heat budget and of global atmospheric circulation in response to temperature changes of the earth's surface. The prediction of 1.5–3.0 °C average warming from doubling of CO_2 concentration is a fair possibility. The prediction of 4.5–9.0 °C warming in the polar regions with about 7 per cent increase in precipitation is more speculative. Changes in the global atmospheric situation and their effects on temperature and precipitation distribution are extremely uncertain. [43]

But there is still another level of uncertainty that has to do with the ecological effects of these climatic changes, even if they were certain. In addition there is the speculation that the polar warming may melt the continental ice caps of Greenland and Antarctica over a period of approximately 1000 years, causing a rise in ocean level of over 100 feet. But to translate all these effects into predictions of agricultural productivity in different regions of the world and effects on human settlements is even more difficult because it involves speculations on the ability of man to adapt to these changes, which will occur gradually, given his technological and management capabilities not in the twentieth but in the twenty-first century and beyond. It will also depend on the degree to which technological planning anticipates the problem during the balance of the twentieth century. It is important to emphasize that the curtailment of CO_2 production in the twenty-first century will not suffice to mitigate any adverse effects which appear then, since it is estimated that it will take 500–1000 years for the cumulative CO_2 concentration in the atmosphere to return to approximately its present value. Add to all these speculations the possibility that CO_2 may simply compensate for a long-term trend towards climatic cooling that might be occurring naturally, and one sees the difficulties of prediction.

Many other long-term environmental effects have received attention recently. Among them is the possible contribution of N_2O derived from fertilizers and fossil fuel combustion in adding to the 'greenhouse' effect produced by CO_2 [44] . Another much discussed effect is that of fluorocarbons, which both deplete the stratospheric ozone layer and add to the greenhouse effect[45]. The point to be emphasized is that we have only just begun to study global chemical cycles and the effects of man upon them. There are cycles for phosphorus, nitrogen, sulphur, and carbon, as well as water; within about 50 years or less man's activities will cycle amounts of these elements which are comparable to the amounts circulating in the natural cycles. However, the complexity, and indeed interdependence, of these cycles make it impossible to predict how they will be altered or what the human effects of these alterations may be. The characteristic times and uncertainties are similar to those which exist for CO_2.

Large-scale reprocessing and refabrication of nuclear fuels for a major worldwide nuclear power industry could have equally major effects, though most experts point out that radiation and its effects have been much more extensively studied than other chemical cycles, and they think they know better how to control them. Recycling nuclear fuel will nevertheless release to the atmosphere small amounts of ^{85}Kr, ^{129}I, ^{3}H, ^{14}C, as well as various heavy radioactive elements[46] . These involve dose commitments extending to thousands of years, and though the effects are tiny they are not zero, and their duration raises important ethical issues. Once these isotopes have been released to the environment there is no way of reversing the effect.

Another class of issues has to do with the effects of human activities on the reflectivity and other optical properties of Earth's surface. Earth's 'albedo' determines the surface temperature at which Earth equilibrates with incoming solar radiation. In the future, changes in albedo could produce effects comparable to those predicted for CO_2. This is a subject about which very little is known, but it clearly deserves as much study as some of the other global environmental effects that are receiving more attention at present. [47]

18.3. Institutional and policy aspects — problems of governance

Competence of international machinery
For many of the issues discussed above, some form of international regime is not only desirable but necessary. Increasingly, issues cannot be dealt with on a purely national basis but require a mechanism for international consultation at the least, and often much more substantial international agreement, control, management, and allocation of resources. The easy assumption is that various formal international governmental organizations must be invested with requisite responsibilities and authority, or new organizations created if none exist.

In fact, international regimes can and do take many forms, including informal private industrial management of some international issues, dominant but uncodified control by one or a few countries, formal international organizations built around specific functions, subjects, or regions, and variations and combinations in between. Each method of dealing with an issue, whether self-consciously planned, or a product of historical evolution, has its own strengths and weaknesses. For most issues there is likely to be found a complex variety of coexisting and mutually dependent mechanisms. An examination of the working even of formal intergovernmental bodies would show several informal mechanisms that operate only by implicit rules and yet are critical to and determine the functioning of the official body.

There is nothing intrinsically wrong with this *mélange* of formal and informal international mechanisms; in fact, it cannot be avoided. But many problems stem from it, some that pose pressing questions for the future.

One is simply the growing requirement for functions to be carried out in an international environment that are clearly 'governmental' in nature — i.e. functions normally thought of as being within the province of public bodies: allocation of resources, regulation and enforcement, adjudication, systems management, and others. These functions, in turn, require governmental bodies with political legitimacy, with officially recognized responsibilities, and the necessary authority to fulfil those responsibilities. But nations are always reluctant to yield sovereignty to international bodies, for it means some loss of control over decisions that directly affect important national interests and domestic constituencies. The results of delegation are frequently not predictable in advance, and the competence of international bodies to carry out the responsibilities they do have is open to question.

In practice, some delegation of responsibility cannot be avoided, and a variety of means are attempted by each nation (not always consciously) to make it more acceptable. For example, it is easier to yield authority to 'expert' bodies which can make decisions according to comprehensive decision rules largely negotiated in advance. Thus there tends to be pressure by those nations with the largest stake in an issue, to create or define jurisdictions as narrowly as possible. But this runs counter to the implications of increasing interaction between issues, for example energy, monetary systems, and environment, which can frustrate the effort to delimit jurisdictions. This approach has been referred to as 'piecemeal functionalism' [48]

Expert organizations also have the characteristic of emphasizing the value and influence of 'neutral competence' in the subject (which is also a source of greater organizational efficiency). But in practice, this has the effect of allowing advanced nations to retain control over international bodies invested with substantial responsibility, since they tend to have a monopoly of the necessary competence. Obviously, those without that competence, but still with a political stake in the issue, are not as pleased by this approach, and point out that expertise is not as neutral as first appears, because experts have many unconscious value assumptions which they bring to their expert judgements. [49]

Transnational networks among government ministries and with international organization secretariats can also be an effective means for nations with particular competence in a subject to maintain *de facto* control, or at least reduce uncertainty about future outcomes.

Limiting membership, or establishing weighted procedural rules, is another way to preserve influence and reduce the uncertainty inherent in delegation of responsibility to international organizations.

All of these techniques and others (such as creating new organizations to bypass existing bodies when the latter do not meet the test of adequate performance or influence) are natural responses to the reluctance of nations to yield sovereignty and sacrifice freedom of action. But aside from the difficulty that these techniques lead to fragmented institutions not able to reflect the interaction among issue areas, there is the more serious long-run problem of participation and legitimacy. It is possible for advanced nations to attempt to maintain control of the progressively more important international policy process through wealth and superior knowledge, but eventually other accommodations must be made. The problem in the international arena indeed has a counterpart in the domestic politics of advanced countries, where new constituencies and affected interests in the domestic polity are increasingly challenging the legitimacy and questioning the claims to neutrality of expert bodies.

Nations affected by a particular technology or technical policy domain are increasingly unwilling to leave control of that technology, or a determining voice in the subject, to the countries who command the expertise. The International Atomic Energy Agency (IAEA) started with clear dominance of the advanced nations, particularly the USSR and US. As nuclear technology has spread, as energy issues have grown in salience, as proliferation has become a more central issue, other nations have demanded a restructuring of the IAEA to give them a greater voice. Some of this has already happened; a further dilution of advanced-nation influence is in the offing. Similarly, the US was forced to sacrifice much of its dominant position in Intelsat as space communications technology became of greater economic interest to member countries. Controversy has arisen over US insistence on free dissemination of resource data from the Landsat satellite system, and the long-range institutional framework for that system has yet to be established. Questions of form and nature of participation are certain to be prominent [50]. In the long run, perceived economic or political stakes in an issue will tend to outweigh command of expertise as a basis for participation, even in supposedly expert bodies.

In many international organizations, especially global UN bodies, strong-willed coalitions of Third World countries have effectively changed the priorities of the organizations toward a central focus on development-related questions.

Of course, developments such as this in UN agencies are also significant reasons why advanced nations seek means of maintaining their influence over international bodies, or prefer to create new organizations more likely to reflect their interests and maintain greater efficiency and predictability, at least for a number of years.

Unfortunately, as decision-making power devolves increasingly on organizations in an international environment, the question of legitimacy of those organizations must be faced. Those with only limited membership, or dominated by a few countries, or without governmental sanction, are not going to be acceptable bodies for those decisions that directly affect nations not represented. Yet full participation by all interested countries may mean ineffectual organizations, with results that are frequently the unpredictable outcomes of coalition bargaining among countries with quite different stakes in a particular issue. There is thus a growing conflict between the requirements of political legitimacy and the necessity for efficiency and internal consistency in decision-making on highly technical issues.

The dilemma has been stated in stark, either–or terms. Undoubtedly, the overall international system must be a mixed one, with some matters reserved, when necessary, for truly global organizations, while others are handled in smaller groupings or perhaps even more self-consciously by transnational coalitions of national bureaucracies. A useful concept is that of 'appropriate scale', that is, always to seek the minimum size organization that can deal with the issue. [51]

Problems of planning for long-term impacts

There are many problems inherent in governance, especially for modern technological societies. Three stand out as being particularly related to science and technology, and magnified in difficulty by international interactions: the difficulty in planning of taking account of long-term impacts, risk evaluation, and the special problems arising from the need to deal with technological information. First, we deal with problems of planning for the long-term.

Man has always faced long-term consequences of his present actions, but ours is the first generation that has taken seriously the impact of its decisions on its descendants. This is in part the result of a change in consciousness, but in part a result of the fact that the 'systemic' effects of decisions extend much more distantly in both time and space than was true in the past. In the past, the relatively limited area of effect of an action never led man to consider large-scale aggregate effects (such as deforestation or widespread use of the automobile), nor would he have had the tools to evaluate them had they occurred to him. Now the situation is quite altered, for it is evident that man has the power through both individual actions and the cumulative effect of millions of small actions, to alter his global environment in substantial, even catastrophic, ways.

This is obvious with regard to nuclear weapons, the management of nuclear wastes, overfishing, the production of CO_2 from fossil fuels, and a wide variety of transnational environmental and resource depletion issues. Most of these are international in their impact. All of the issues that arise out of the aggregation of millions of individual decisions in particular raise the special problem for the planner of anticipating effects 30–50 years in the future, sometimes much more. In such cases it is not sufficient to consider only most probable outcomes, or to use conventional economic tools or market signals. Instead, ranges of possible outcomes must be taken into account, and the vulnerabilities associated with relatively improbable events.

However, the international nature of the issue – at least its effects – raise special questions about who should have a voice in the decisions; where should responsibility lie? And the problem is further complicated by systemic effects, that is, effects in one area that rebound in important ways to alter other aspects of the international system (much as the temporary disappearance of the anchovy off Peru in 1972 was a factor in the massive commodity inflation in the U.S. the following year).

For some issues, it is even more serious. Those are the ones *requiring* present choices with uncertain, but potentially catastrophic effects that will not be known until a point at which it may be too late to counteract the effects. Such choices arise, for example, in the energy field where, as discussed earlier, the only present (probably) viable alternative to oil and nuclear power is coal, yet the long-term climatic effects of expanded burning of coal are potentially catastrophic, though still highly uncertain as to timing. Once heavy investments have been made in coal as a source of energy, it will be economically traumatic to later transform the energy system rapidly in response to evidence of serious climatic changes [52]. Yet we have no choice but to expand use of coal in the intermediate term until other alternatives are available.

There is no clear appropriate response to these planning problems involving very long-term effects. In part, it is a need for more research to improve the capability to estimate effects; in part it is a need to invest in development of alternative technologies, even when a complete economic justification cannot be made (thus making it a governmental responsibility).

In part, also, it comes back to the development of international machinery that can more effectively represent the interests of those likely to be affected by what is seen as a purely national decision, that can provide a forum for negotiation of decision rules for action as new information becomes known, and that can have the responsibility to ensure that new information is, in fact, made known whatever the consequences for an individual country. Such capabilities exist only in embryonic form internationally at present, and certainly are now of only passing relevance to the nationally-based planners wrestling with these issues.

Risk evaluation

As we have seen, the industrialized world is increasingly deploying technologies whose potential risks are very hard to assess, and yet whose consequences cross national boundaries. One of the earliest such cases was atmospheric testing of nuclear weapons, with fall-out widely dispersed. An example recently in the news was the accidental re-entry into the atmosphere of a Soviet satellite containing a small nuclear reactor, which resulted in the deposit of radioactive debris in the Canadian arctic. The prospective deployment of ever larger objects in space also arouses concern as to risks of re-entry of such objects, with massive fragments falling in inhabited areas. Other earlier examples are the wide international concern raised by U.S. plans to deploy copper needles in orbit as part of a global communications system, and the conduct of test nuclear explosions in the upper stratosphere.

In another domain, the growing oil tanker trade raises questions of accidental oil spills contaminating beaches and estuaries. The prospect of a very rapid increase in multinational trade in liquified natural gas (LNG) raises questions of safety which require assessment, since gas clouds released as a result of accidents could conceivably cause fires or explosions at long distances from the site of the accident. As discussed earlier, an increasing number of industrial and agricultural activities may result in the release of chemicals that may ultimately damage the ozone layer. There has been talk of recent Soviet plans to reverse the flow of rivers emptying into the arctic ocean, with possible risk of triggering major climatic changes in the northern hemisphere.

The question raised by all these examples is that of appropriate mechanisms for assessing such risks before individual nations proceed unilaterally with activities whose effects on other countries or on the global commons (oceans or atmosphere) may be serious and irreversible. So far we have relied mainly on the publicity given to the unofficial assessments by the world scientific community to restrain governments. Frequently, however, the world scientific community cannot mobilize the resources to perform definitive assessments. These usually require assembly of new data from many sources, and often require experiments, elaborate calculations, and field measurements.

There is not always agreement on such unofficial assessments, and national governments often refuse to act to terminate activities to which large public commitments have been made, especially in the face of technical disagreement. A good case in point is the reaction of Japan and several European countries to the efforts of the U.S. government to secure a world-wide pause in plans for reprocessing nuclear reactor fuel, and for deployment of commercial nuclear reactors employing plutonium in the fuel. To some extent, the so-called International Nuclear Fuel Cycle Evaluation Programme (INFCE) [53] is one of the first examples of a formal international assessment of risk. Though proposed by the U.S., it was deliberately put under the management of the IAEA in order to emphasize the international character of the assessment [54], and perhaps to set a precedent for similar assessments of other global technologies.

Can we continue to depend on *ad hoc* arrangements for conducting timely assessments as problems arise, or should we look for a more formal mechanism, a kind of institutionalized international early warning system? An alternative might be to negotiate a very general set of rules within which nations initiating the deployment of technologies with potential transnational effects would prepare and submit formal technology assessments to an international body. The ultimate responsibility would thus still lie with individual nations, but the international submission process would serve to put moral pressure on nations not to proceed with plans having adverse consequences. The system here would be somewhat analogous to the Environmental Impact Statement (EIS) procedure used in the United States [55]. The advantage of such a system is that it would not be seen as infringing sovereignty, and could thus be instituted rather easily, possibly under the auspices of UNEP, OECD, NATO, or others, depending on the subject. For it to be effective it would have to be able to mobilize finances to support important assessments on a priority basis. However, such effectiveness as that of the U.S. EIS system, probably has arisen from the access of public interest groups to the courts, and there is a question of whether it would work on an international scale without such a mechanism of judicial review of proposed international actions. This is an area in which NATO, which includes most of the nations likely to develop or deploy global technologies, might take the initiative in creating a system, possibly under the auspices of its Committee on Challenges of Modern Society.

Another type of approach which has recently received attention is exemplified by a proposal that foreign projects partially or wholly financed by the U.S. government be subject to an EIS procedure under section 102(c) of National Environmental Policy Act just as is the case with domestically oriented federal actions affecting the environment [56]. An interesting legal problem would arise if the suits of environmental groups in the U.S. resulted in the application of the EIS procedure to multilaterally funded development projects to which the U.S. was a contributor.

Policy-making involving technological information

A key difficulty that technology adds to the problems of governance is that raised by the need to deal with technological information in the policy process, especially information related to high technology. Increasingly, as has been repeatedly stressed above, central issues in domestic and international affairs are raised by, or revolve around technological subjects. The technological aspects must be an intimate part of the policy process, yet to do this effectively or even satisfactorily poses severe difficulties in practice. The problem is partly one of structure, but also is an inherent problem for which there is no fully appropriate approach. As difficult as it may be in a domestic policy process, it is even more difficult to solve in an international framework, in which there is no recognized channel of accountability between decision-makers and affected constituencies.

There are several interrelated aspects that can be summarized briefly under the headings of accessibility, monopoly, and uncertainty.

Accessibility. One of the most obvious problems of effectively including technological elements in a policy process is simply the esoteric nature of the material. Technologically-untrained participants must usually rely on intermediaries to provide the information and, most important, to evaluate its significance for policy purposes. This is, of course, not a new problem of government. The pitfalls of relying on experts have long been seen and debated. They take on added dimensions when technology is involved, because generalists so often assume they cannot understand technological expertise at all, and are more likely to defer to the expert.

Moreover, the apparently highly rational nature of science and technology, coupled with their inaccessibility, can lead to technocratic elements in governance: policy-making in effect by experts who do not understand the broader political and social implications of their recommendations. The problem is further complicated as experts increasingly line up on opposite sides of controversial issues of high technical content, often with hidden political agendas that transcend their expert knowledge.

This is not only a difficulty within governmental decision processes, but also affects any attempts to broaden participation in decision-making to allow or encourage affected groups in society at large to take part. The inaccessibility of information, the difficulty of understanding the details of the technology or its implications, makes much harder any attempts at greater participation, and the public is further confused by sharp public disagreements among experts. This contributes, in fact, to the sense of alienation — the feeling that a critical element of change in the society that will affect everyone — cannot really be understood (or controlled) except by a small elite, which appears less and less trustworthy as its members fight among themselves.

Internationally, the same elements are evident, for the competence to participate fully in deliberations and negotiations involving high technology aspects varies widely among nations. Technology is seen as the key to growth and prosperity, and also as a threat to national independence and social structure. But many nations find themselves dependent on others for the information required even to know where their interests lie with regard to many technology-related issues.

The problem is, of course, a most serious one, though frequently misunderstood. Intermediaries certainly are required, but what is usually needed most in a policy process is an understanding of the political and other implications of the inevitable technical uncertainties. That is, the important interaction is between the policy choices and the unfolding technological development. It is *not* simply a matter of laying out the known facts, but rather of representing in the policy process the implications of present knowledge, and of developments in the future that depend on the decisions that are made now. Understanding of that interaction necessarily must involve experts, but the significant issues can be made accessible to the generalist.

The problem is more difficult internationally because of the wide variation in expert knowledge, let alone the ability to interpret and understand the interactions between policy and technological uncertainty. The realization that an increasing number of issues involving high technology must be dealt with internationally makes this a problem of growing seriousness for all the affected parties.

Monopoly. A concomitant aspect of the problem of accessibility is the fact that high technology information is necessarily not widely held or understood. Virtual monopoly situations arise domestically or internationally, which give particular leverage (explicit or implicit) to those who command the knowledge. And for some subjects, that may mean a quite small number of companies or countries. This problem is already apparent in the field of transnational information networks, and is beginning to arise in connection with the Landsat satellite system for resource monitoring. It also arises in connection with many projects for natural resource development.

The result is that, for some policy issues, the choices seen in the policy process may be unnecessarily constrained because the alternatives cannot be developed adequately or authoritatively. Given the advanced nature of many important technological developments, and the large capital resources often involved (e.g. computers, some large weapons systems), this is a situation very difficult to reverse or even to ameliorate.

Uncertainty. Contrary to popular assumption, a major characteristic of the technological aspects of important policy issues is uncertainty. Rather than technology providing a clear, factual background for policy, it more often presents as unfocused, disputed, and unclear a picture as the more obviously subjective aspects of the same issue. The reason is clear: for most issues of any significance, the perspective looks to the future, to what science and technology may be or may mean, rather than to what already exists. Most technical controversies that enter the policy arena involve predictions of the consequences of various actions, rather than delineation of existing facts. Thus, neither the future character of the technology nor its effects can be known with certainty, and, it turns out, both are usually the source of as much disagreement as are any other aspects of the future.

In fact, the seeming certainty of science and technology is a problem in this regard, for they frequently present an illusion of precision to the layman, when none exists.

It is this uncertainty that underlies the problems of risk evaluation and long-term planning, and the just discussed dangers inherent in excessive influence of experts. For decisions cannot wait for uncertainties to be resolved. Judgements must be made about likely futures, and as in any subject, judgements are likely to be influenced by personal preferences. Thus, if the policy process relies on a very few sources of judgement, the results are likely to reflect the biases, conscious or unconscious, of those sources.

The obvious conclusion of opening the policy process to wider participation and to competing judgements raises questions noted above about adequacy of access to information. And it raises important questions about the institutional means by which conflicts among experts may be resolved, and about the costs of extended and expensive processes of policy-making.

These difficulties, once again, are as relevant internationally as they are domestically, sometimes in exaggerated form. In particular, the problem of dealing with competing evaluations of the evolution of large-scale environmental or other phenomena, and of the measures to deal with them, looms very large. The scientific judgements are usually quite uncertain and tentative, while the economic and political implications of either action or inaction are enormous. Even the relatively innocuous salmon dispute in the North Atlantic, where technical information appeared reasonably definitive, took many years to settle because of the economic impact on Denmark of restricting salmon fishing off Greenland.

Uncertainty contributes enormously to the complexity of technological issues, adds yet a further load on political institutions, and makes even more difficult the problem of effective participation by those nations with lesser competence in technological areas. Technological dependence on advanced nations is unavoidable, yet that also implies dependence that is much broader than technological alone.

International institutional means of dealing with some of these problems, and especially that of uncertainty, are even less in evidence than they are domestically. In particular, better means of providing independent assessments of important science and technology-related situations, both as a backdrop to policy deliberations and as an aid in dispute settlement, are badly needed. Domestic capabilities are somewhat more advanced, though in most countries the institutions for this purpose fall far short of the need.

In this respect there is an essential and probably insufficiently recognized need for non-governmental institutions that can provide multiple independent assessments of major technical–political issues as free as possible of national or parochial biases. For such organizations to be effective they must have a realistic understanding of political and economic constraints and of impacts on various constituencies, but be less bound by them than would official bodies accountable to particular constellations of governments or economic interests. It is especially important to develop non-governmental research and policy analysis institutions which are capable of integrating several different issue areas within a common framework. To date the most influential NGOs have been those which operated within fairly narrow subject domains, with the result that their policy recommendations have reflected a parochial view of the world, which is not readily translatable into action by decision-makers who, perforce, must have a wider perspective.

Disenfranchised electorates

It is interesting to note, and perhaps no more than that, to what extent the advances of science and technology have created a world of disenfranchised electorates. As the effects of the application of science and technology have become global in scope, they have by implication had impacts on literally billions of people who have had no say in those scientific and technological developments, nor in their application.

It is certainly not a new situation for decisions taken in one country to affect 'innocent' citizens of another. Yet the very scale of the effects arising from the application of science and technology, and the parochial nature of the decision processes involving the deployment of large-scale applications, appears as a change of kind and not just of degree.

To look at it another way: the process of R and D continues in all countries, and especially in NATO countries. Some of the results of that R and D will affect people in many other countries, often in major ways. The decision process for the setting of R and D objectives and for the subsequent use of R and D results, as noted earlier, is largely influenced by domestic considerations. Where is the voice in the R and D decision process (or in the election of government officials) of those who also will untimately be affected by the application of the R and D?

Individuals of other countries are often implicitly represented in decisions on immediate issues: foreign offices will usually reflect to some extent likely foreign reactions to specific courses of action. But in R and D, foreign offices have essentially no voice, and even if they did, could hardly represent foreign attitudes toward uncertain future R and D results.

To make this observation is not meant to foreshadow a proposal of an institutional remedy. This relationship would appear to be a fact of life likely to be a characteristic of international relations far into the future, or until some more fundamental changes of the international structure come about. Science and technology will remain national enterprises, whatever their international implications, for the foreseeable future. There may be a few exceptions, however, in which foreign voices do affect R and D goals implicitly, if not explicitly. Nuclear power developments may be the best current example, but it is an exception to the general rule [57] of R and D carried out within nations, little encumbered by the wishes, concerns, or even interests of citizens of other nations.

Trans-national pressure groups

The emergence of interdependence has been accompanied by the development of many informal associations which are as much international as national. Some of these networks are primarily professional, and include people concerned with particular technologies or scientific disciplines. Two examples of well developed networks of this kind are the international agricultural research community, and the nuclear engineering community. Oceanographers and atmospheric scientists also form strong international networks. These professional associations may exercise strong influence within their functional domains, but they are not a particularly strong political force. Nevertheless, they are an important constituency for the international allocation of resources and for the conduct of certain kinds of research, especially in the environmental sciences. The spread of nuclear power and political interest in nuclear energy throughout the world are, in part, the result of the strength and good communications within the nuclear engineering network.

A more recent development is the emergence of networks with a much more consciously political motivation, mostly connected with the environmental movement. These networks share information not only of a technical nature, but also about political tactics for influencing or blocking plans of national governments.

Just as the nuclear engineers emerged as a professional transnational network, so has the anti-nuclear movement emerged as a transnational quasi-political network. Its tactics, well co-ordinated across national boundaries, have caught official bodies by surprise, and have been taken seriously only belatedly by national governments. Indeed, it can be said today that the international nuclear industry is at a virtual standstill largely as the result of the success of the loosely but effectively co-ordinated tactics of anti-nuclear groups throughout the industrialized countries outside the Eastern Bloc. In general the bases of concern expressed by these groups have changed over time, but at any one time are remarkably similar in different countries. We are speaking here of the effectiveness of their tactics, not the merits of their concerns.

The question raised by both professional and quasi-political networks is their probable future role in international politics. Do they represent the beginning of a new constellation of interest groups operating on the international political stage, much as national pressure groups have traditionally operated within democratic nation states? What will be the future relationships of these networks to official international bodies, and what are the channels by which their interests will be exercised? For the present, we can only raise these questions, not answer them. But it is important to identify this new phenomenon as it is likely to become increasingly important, and no discussion of the management of global problems can be complete without recognizing the increasing role that various kinds of international pressure groups will play in the way national governments and international agencies deal with these problems.

18.4. Conclusion

Science and technology have brought the world into a new era, characterized by new issues, or greatly transformed traditional issues. Many of these have taken on dimensions that pose dangers to our own countries, or to all countries; dangers that cannot be precisely defined or easily avoided. The political processes in all countries have been sharply affected, and the institutional and policy framework on the international level is poorly constructed to cope with the changed situation and with unavoidable future needs.

The nations of the NATO alliance have a special responsibility, and a considerable stake, in these issues because of their dominant role in science and technology, as well as their deep involvement in all of the political and not only technical aspects.

For most of the political and technical issues discussed here, or that could have been discussed, there are no 'answers'; only approaches, partial solutions, bypasses, compromises, experiments, analyses, and further R and D. But that is not novel. Rarely do important policy questions have clear-cut answers, since values and interests of affected parties are always different.

That does not mean, however, that events can be allowed to take their course. Leadership in recognizing important issues, even (or especially) those not on the immediate agenda, seeking understanding, and proposing courses of action, can make the critical difference in the nature of an issue and in its international effects.

There is much to be done; the agenda is very long, increasingly difficult, and with steadily larger implications. The nations of NATO have the opportunity and the challenge to respond.

References

[1] For a recent summary of China's economic and technological situation, see Whiting, A.S. and Dernberger, R.F. *China's future: foreign policy and economic development in the post-Mao era,* Council on Foreign Relations 1980s Project. McGraw-Hill, New York. (1977).

[2] Revelle, R. Can the poor countries benefit from the scientific revolution? A paper presented at the meeting of the American Association for the Advancement of Science, December 1967; Brown, L.R. *World without borders,* Random House, New York, Ch. 3. (1972).

[3] Freeman, O.L. Multinational companies and developing countries: a social contract approach. *Int. Develop. Rev.* Vol. 16 (4), 17–19 (1974).

[4] Goulet, D., *The uncertain promise: value conflicts in technology transfer,* p. 83. IDOC/North America, Inc., New York, (1977).

[5] Cooper, R.N., Kaiser, K., and Kosaka, M. *Towards a renovated international system,* p. 26. The Triangle Papers No. 14, The Trilateral Commission, New York, (1977).

[6] *The U.S. and world development: agenda for action.* Overseas Development Council, 1976. Praeger, New York, (1976).

[7] Brooks, H., Policies for technology transfer and international investment, Chapter 12 In *The new Atlantic challenge* (ed. R. Mayne), Chapter 12, Charles Knight and Co. Ltd., London, (1975).

[8] [4], p. 81.

[9] [1], pp. 89–133.

[10] Repetto, R.C., *The relationship of the size distribution of income to fertility and the implications for development policy.* Harvard Center for Population Studies, Research Paper No. 3, March (1974).

[11] *The Public Interest,* No. 11, Spring 1968, pp. 97–8.

[12] NAS–NRC, *World food and nutrition study,* pp. 131, 143. National Academy of Sciences, Washington, D.C. (1977).

[13] Ref. [12] pp. 169, 172.

[14] Brooks, H. Sociotechnical systems: central and distributed goals. In *The National Research Council in 1976: current issues and studies,* pp. 133–49. National Academy of Sciences, Washington, D.C. (1976).

[15] [12], p. 43.

[16] *Bangladesh land water and power studies: final report,* Harvard Center for Population Studies, June 1972.

[17] Moravcsik, J. and Ziman, J.M. Paradisia and dominatia: science and the developing world. *Foreign Affairs,* 53, 699–724, (1975).

[18] Ref. [12], pp. 5–16.

[19] National Academy of Sciences, *Air quality and stationary source emission control.* Prepared for U.S. Senate Committee on Public Works, Serial N1. 94–4, March (1975).

[20] *A growing worry: the consequences of development.* Conservation Foundation letter. January (1978).

[21] Glaser, P.E., Solar power from satellites, *Physics Today,* 30, 30–8, (1977).

[22] Greenwood, T., Feiveson, H.A., and Taylor, T.B. *Nuclear proliferation: motivations, capabilities, and strategies for control,* pp. 125–83. Council on Foreign Relations 1980s project. McGraw-Hill, New York, (1977).

[23] Smith, A.K. *A peril and a hope,* pp. 451–60. A Report on the International Control of Atomic Energy, Department of State Publication, No. 2498, Washington, D.C. USGPO, March 16, (1946).

[24] Ref. [7], pp. 164–7.

[25] Carnesale, A. and Rathjens, G.W. The nuclear fuel cycle and nuclear proliferation. In *International arrangements for the nuclear fuel cycle,* (eds. A. Chayes and B. Lewis), pp. 3–15. Ballinger, Cambridge, Massachussetts. (1977).

[26] *German Tribune,* 823, p. 7, 22 January (1978).

[27] *Information Resources Policy: arenas, players, and stakes, Annual Report 1976-7,* Program on information resources policy, Harvard University, Vol. 1, October, 1977.

[28] Read, W.H. *Foreign policy: the high and low politics of telecommunications.* publication P–76–3, Program on Information Resources Policy, p. A23. Harvard University (February 1976).

[29] Landsberg, H.H., *Natural resources for U.S. growth, resources for the future,* p. 250. The Johns Hopkins Press, Baltimore, M. (1964).

[30] Schumacher, E.F. *Small is beautiful: economics as if people mattered.* Harper and Row, New York, (1973); Lovins, A.B., *Soft energy paths: toward a durable peace.* Ballinger, Cambridge, Massachussetts. (1977).

[31] Brookings Institution, *Economic prospects and policies in the industrial countries,* p. 11. The Brookings Institution, Washington, D.C. (1977).

[32] Heilbroner, R. *Foreign Affairs,* 51(1). 139–53 (1972), especially p. 140.

[33] Leontief, W., *The future of the world economy: A.U.N. study,* New York: Oxford University Press, New York (1977).

[34] An extensive comparison of various energy models has been carried out by the Energy Modeling Resource Group of the NAS–NRC Committee on Nuclear and Alternative Energy Systems, to be published. See also *Energy and the economy, Energy Modeling Forum Institute for Energy Studies Report* 1, Stanford University, (September 1977); Berndt, E.R. and Wood, D., Economic interpretation of the Energy–GNP ratio. In Macrakis M.S. (ed). *Energy: Demand, conservation, and institutional problems,* pp. 21–30. Cambridge, Mass.: M.I.T. Press, Cambridge, Massachussetts, (1974); Jorgenson, D.W. and Hudson, E.A. U.S. Energy Policy and Economic Growth, 1975-2000. *Bell Journal of Economics and Management Science* 5(2) 461–514 (1974), Hoffman, K.C. and Wood, D.O. Energy System Modeling and forecasting. *Annual review of energy,* Vol. 1. Palo Alto: Annual Reviews, Inc., Palo Alto (1976).

[35] Conant, J.B. *My several lives,* Chapter 23. Harper and Row, New York, (1970).

[36] Rice, D.B. (Chairman). *Government and the nation's resources,* Table 9, p. 30. Report of the National Commission on Supplies and Shortages, Washington, D.C.: USGPO (December 1976).

[37] Tilton, J.E., *The future of nonfuel minerals,* pp. 34–40. The Brookings Institution, Washington, D.C. (1977).

[38] Ref. [36], p. 38.

[39] Ref. [37], p. 52.

[40] National Materials Advisory Board, *Contingency plans for chromium utilization,* National Academy of Sciences, Washington, D.C. (1978).

[41] Kemp, G., Scarcity and Strategy. *Foreign Affairs,* 56(2) 396–414, (1978).

[42] Revelle, R., Chairman, *Energy and climate,* report of NAS–NRC Panel on energy and climate. National Academy of Sciences, Washington, D.C. (1977).

[43] Ref. [42].

[44] Stumm, W. (ed.). *Global chemical cycles and their alterations by Man.* Physical and Chemical Sciences Research Report 2, Dahlem Konferenzen (1977); cf. esp. Man and the global nitrogen cycle, Group Report, pp. 253–74.

[45] Tukey, J.W. (Chairman). *Environmental effects of chlorofluoromethane release.* National Academy of Sciences, Washington, D.C. (September 1976).

[46] United States Environmental Protection Agency, *Environmental radiation protection for nuclear power operations.* Supplementary Information (5 January 1976).

[47] Williams, J., Kromer, G., and Weingart, J. (Eds.). *Climate and solar energy conversion.* Proceedings of a Workshop, 8–10 December 1976, Report CP-77-9, December 1977. International Institute for Applied Systems Analysis, Vienna (1977).

[48] Ref. [5], pp. 32–3.

[49] Brooks, H., Expertise and Politics. *Proc. Amer. Phil. Soc.* **119**(4) 257–61, (1975).

[50] Council on Science and Technology for Development: *Objectives, organization, program, budget,* p. 13. Informal council working paper (January 1978).

[51] Ref. [5], p. 38.

[52] Brooks, H., Potential and limitations of societal response to long-term environmental threats, pp. 241–52 in ref. 44.

[53] *Final Communique of the Organizing Conference of the International Fuel Cycle Evaluation.* Doc. 41 (21 October 1977); Deutch, J. M., *Statement on the Office of Energy Research FY1979 Authorization to Sub-committee on Fossil and Nuclear Energy Research Development and Demonstration.* House Committee on Science and Technology. (8 February 1978).

[54] Nye, J.S. Non-proliferation: a long-term strategy. *Foreign Affairs* **56,** *601–23, (1978).*

[55] Blisset, M. (Ed.). Part 2: How the national environmental policy act is implemented, In *Environmental impact assessment,* pp. 25–99. Engineering Foundation (1976); Anderson, F.R., *NEPA in the courts, a legal analysis of the national environmental policy act,* Johns Hopkins University Press, Baltimore, Maryland, (1973).

[56] Ref. [20].

[57] Ref. [5], Appendix D., pp. 59–68.

Commentary
Alexander King

This chapter is exceptionally clear, comprehensive, and well balanced. Indeed its general scope and emphasis is so close to my own thinking that I find it difficult to disagree with its judgements and analysis. This commentary is therefore to be taken as additional remarks and extensions of the discussion of the authors.

Cultural aspects of the international flow of science and technology

While science, which uncovers new knowledge, is universal in its significance and, indeed, forces all cultures to reassess their values, technology which is based on science incorporates the value-systems, life-styles, and goals of the industrialized countries, in which it has been dominantly generated. Thus the impact of Western, materially oriented technology inevitably has a strong impact on the often fragile cultures of many of the less developed countries, and even on those which have long and rich traditions, often based on ancient religious tenets. The European countries still retain some degree of cultural integrity, while they have assimilated the material mores so strongly developed in the United States and indeed have strongly contributed to their arising. This applies to the East Europeans as well as those of the West. The values and appetites of the industrialized societies have also dominated the élite of the less developed countries, who have often enjoyed a Western education, while educational systems in many of the developing countries, established by the former colonial powers or introduced by the aid programmes are often structured on American, British, French, or Russian models with much of the same content as in the metropolitan countries.

It is highly desirable in the next phase of aid and technological transfer that much more attention should be given to the impact on the various cultural systems involved, with a view to maintaining cultural diversity and integrity and thus avoiding abrupt disruption of traditional societies, with the human uncertainty and social unrest which this can produce. There may well be methods of upgrading the subsistence economies of different countries through means which are not copies of the paths by which the Western nations have achieved their present levels of affluence. In this connection it may be useful to look closely at the experience of China.

It may also be advantageous to humanity as a whole to preserve and even encourage a broad cultural diversity on strictly biological grounds, so as to ensure the greatest possible extent of adaptability in face of new problems and situations which are not now foreseen.

The diversity of development levels

It has become customary to classify the nations of the world into three categories — those of the market economy countries, the state economies of Marxism, and those of the Third World at an earlier stage of development. Sometimes the OPEC countries are added as a fourth world. This categorization may be useful politically, but it is somewhat misleading in that it lumps together as the Third World countries that are at very different levels of development and with widely diverse possibilities for it. Rather we must regard the range of development situations in terms of a continuous spectrum of conditions and potentialities, taking account of present levels of economic achievement, minerals and energy possession, and human skills. At one extreme we have countries such as the United States, the Soviet Union, Canada, Australia, and South Africa, rich in raw materials and energy sources, and producers of food beyond their domestic needs, possessing accumulations of capital, highly developed industrial structures, and rich also in scientific and managerial skills. Next come countries such as those of Europe, East and West, as well as Japan, also highly skilled and with capital-intensive economies, but far from self-sufficient in energy and raw materials. Then there are nations such as the oil-producing countries, or mineral-rich countries such as Zambia or Zaire, which are actually or potentially rich in consequence of their possession of natural resources, but lacking the infrastructure of industry and the skills which make affluence possible. A further group, which includes the countries of the Indian sub-continent, are rich in tradition, highly and probably over-populated (and thus possessing enormous work forces, at present insufficiently employed), poor in capital, raw materials, and energy, as well as the capacity to sustain sufficient food production for their rapidly increasing populations. At the extreme, there are countries at a low level of development and with but few advantages of environment and resources, tailing away to those with practically nothing – the poorest of the poor. Thus the simplistic classification into the three worlds hides the diversity of conditions within the so-called Third World; clustering together countries such as Brazil, Iran, Mexico, and Venezuela with aspirations and possibilities of joining the ranks of the so-called developed countries, with those whose possibilities are meagre. It also masks the different problems for the future of the resource-rich and resource-poor industrialized nations, to which we shall refer later.

Science, technology, and development

As Brooks and Skolnikoff have clearly stated, the disparities between the rich and the poor countries are even greater with regard to research and development performance than in economic terms, although it can be argued whether this is a cause or a consequence of development differentials; it is probably both. More than 90 per cent of the world's scientific effort is in the developed nations, while it is stated that only about 2 per cent of the world's effort to generate new technology lies outside the industrialized zone. Given the importance that technological development has had in the advanced economies, it seems obvious that the transfer and diffusion of technology is essential if the economic disparities are to be narrowed. This is extremely important with regard to the pressing need of the less developed countries to grow more food for the expanding populations and, indeed, a good start has been made through the regional institutes for agricultural research. The potential for improvement is great and much more requires to be done – again with more attention to social and cultural prerequisites and consequences. Real increases in levels of prosperity depend, however, on industrialization in the less developed areas, both through import substitution and the manufacture of products for export. Much of the necessary technology exists in the industrialized nations, and much of it is freely available as the patents have long since expired.

It is not surprising, therefore, that the transfer of technology has become a major issue, but it is highly regrettable that the subject has become over-politicized, with stress on the problems of the nature of industrial property and mercantile conditions of transfer, while ignoring the deeper issues of how to achieve effective innovation, relate it to the basic needs of the country, and identify the conditions necessary to ensure that the imported technology will be organically assimilated and effectively exploited.

Development is a complex socio-economic process, within which technological input, either indigenous or imported, is only one of many elements, albeit essential. To consider the problem of technological transfer in isolation from other elements, and especially in political terms, can be dangerous. Unless the ground is well prepared and the technology well selected as appropriate to national objectives, it can be seed cast on stony ground which may germinate initially, but takes root and spreads only with difficulty.

It is extremely important, therefore, that the conditions for successful technological transfer be well understood and taken into account. There is still much uncertainty in developed countries as to the optimum national conditions for technological innovation, so it is not surprising that few of the less developed countries, lacking experience in this field, pay attention to these matters. The main prerequisite for transfer is perhaps the existence in the receiving country of a scientific and industrial competence. This is not achieved easily or quickly, and while the need for such indigenous capacities is generally appreciated, it is seldom given sufficiently serious attention by either the donor or the recipient countries. The mere establishment of a national research institute, a measure adopted by some aid programmes, while useful, is itself insufficient. In several instances, such institutes, after an initial period under a foreign scientific director, have withered away; they may be regarded locally as prestige institutions, but their influence on the national development process is often marginal. For indigenous scientific activity to be fertile, it must be intimately coupled with both the educational and the productive processes, and this is extremely difficult to accomplish in most countries at an early stage of development, when the new universities are struggling to achieve high academic standards, and the new industries have little in the way of general technological competence and see little value in research.

One of the essential needs is, of course, to ensure that the necessary skills are available when new plants and processes are introduced, and this entails the introduction of training schemes both technical and managerial, well before the new plants come into operation. Still more difficult is the question of the selection of the technologies to be introduced. There is no simple answer to this question; it will vary greatly with the size, existing level of development, nature and availability of energy and raw materials, and many other factors. Many of the less developed countries suffer a high degree of unemployment and underemployment, and capital is generally scarce. There is a need, therefore, to create jobs, but, on the other hand, there is also need to provide basic infrastructural capacity, such as steel production, which is capital-intensive and labour-non-intensive. In addition, many countries are planning to include some industrialization of an advanced type — again capital-intensive — which will produce goods for export and thus provide an investment income and foreign currency. There is thus a mix of different types of technology needs which will vary greatly according to national circumstances. While national economic and development plans exist in many countries, these do not always balance the different technological needs with the conflicting development objectives. In some instances, as the chapter indicates, increase in the national GNP has not resulted in social advancement and reduction of gross income disparities. In such cases, there is the suspicion that much of the imported techno-logy has been selected for the benefit of the small elites, whose appetites are much the same as ours in the developed countries. It has had little impact on the con-ditions of the masses who still live in subsistence, practising age-old agricultural and craft activities.

There has been much discussion of late of *appropriate technology*. This is often merely a euphemism for *intermediate technology*, a concept which is indeed im-portant for certain purposes, but by no means a complete solution. It is regarded in pejorative terms by many of the less developed countries as a neo-colonialist device to ensure that they remain permanently backward and non-competitive. It would be useful if the term 'appropriate' were to be used only and generally with the real meaning of the word. Many of the technologies in the developed world are beginning to be regarded as inappropriate in the sense that they may be deteriorating to the environment or reducing work-satisfaction. For the less developed world, the appropriate technology mix will include many types of process – a proportion of advanced industries for infrastructural and export needs and to act as pacemakers of future industrial development, a number of more traditional industries for im-port substitution purposes, a spread of intermediate technology, especially for rural areas, and a service to aid the improvement of traditional tools and methods by the application of well-known scientific principles.

The place of the transnational corporations in the transfer of technology is im-portant. They are needed, but not liked, in the Third World, where many countries are not reconciled to their continuing role. There is, at present, no alternative and it is to be admitted that the state enterprises of the Third World are no less harsh in their terms of transfer, while being much less effective. It is extremely important, nevertheless, that the transnationals improve their image in these countries; the un-wise actions of a few have prejudiced the rest. It is to the long-term interest of the multinationals that they attempt to reconcile the normal profit motivation with future possibilities, and take a deeper interest in the continuing needs of the Third World countries, helping in the construction of the indigenous capacities and being willing to assist, for example in permitting, and even encouraging, a proportion of their trained workers to migrate to other firms in other sectors. It is one of the most bitter complaints of the less-developed countries that technology introduced by the transnationals remains secret, and does not permeate the economy generally. This should not be necessary.

With regard to the problems of the terms of transfer, the claim that the con-ditions set by the transnational are unduly harsh, considering that they are nor-mally spreading existing markets for technology for which the development costs have long since been covered, is exaggerated, as is also the contention that there is willful withholding. Nevertheless, it might be useful to explore whether preferential terms might be given to Third World countries, the balance being paid by inter-national or national aid programmes.

In summary on this point, if science and technology are to make a more effective contribution to the development of the Third World in the future, a number of changes in policy and attitude would seem desirable:

(1) For both donor and receiver countries in the transfer of technology, a longer-term perspective is necessary to enable selection of processes, their modification to meet local materials, work methods, and cultural conditions, and the training of local technical and managerial personnel.

(2) For the receiving countries, an appreciation of the different elements involved in successful technological innovation.

(3) Understanding of the place of technology as part of the socio-political system of development, rather than as an autonomous factor.

(4) Conscious effort to build up an indigenous capacity for research and development in each country, articulated with the educational and the productive processes, in which university research scientists, research institutes, industrial enterprises, and the relevant government agencies collaborate towards the common objective.

With regard to the last of these desiderata, it will be necessary for the Third World countries to have help and guidance from the more industrialized countries. In the past, scientific considerations have played a minor role in many of the national aid agencies, being regarded essentially as a mechanism for the solution of specific problems. If real assistance is to be given in the building up of national capacities, a much broader and fundamental use of science and the scientific method will be necessary. The rather narrow approach of some national aid agencies to science derives from their isolation from the overall science policy considerations of the countries concerned. Some national bodies for science policy, having no responsibility for development problems, seem to resent involvement in such matters as detracting from the attention and resources needed for their internal problems. It would seem desirable that the scientific component of aid should be closely associated with general scientific thinking.

Problems of the global environment

These issues are well described by Brooks and Skolnikoff, who stress the complexities and uncertainties involved. There are many reasons why these problems should be discussed at the political level, and not least because of the uncertainties. The problems are partly within the natural system itself and partly related to human activities. Nature has always provided threats to man and to other organic species – volcanoes, earthquakes, and ice ages – and as yet our understanding of the formidable complexities of the climatic and hydrological systems, of the fluctuations of solar radiation, and many other aspects is inadequate to make other than short-term predictions.

Man's impact on the environment is no new phenomenon. Bad agricultural practices in ancient times, including 'slash and burn' methods which still persist, have helped to create the deserts of the world, and it is calculated that about 60 per cent of the topsoil once available to man has been lost, while much of the rest is deteriorating. What is new in the present situation is the enormous increase in human activity, resulting from the overall increase in world population, together with substantial increases in the consumption *per capita* of materials and energy particularly in the industrialized countries. Whether the increased extent of human intervention constitutes a real and irreversible threat to the global environment not yet certain, but the evidence is sufficiently strong to warrant its being taken seriously. This is especially so because of the long time that must elapse between the introduction of a potentially dangerous practice and the appearance of significant consequences.

But there are further and more specific reasons why there is need for public and political understanding of the global environment issues. One of these relates to the interaction of energy and environmental policies. With the mounting energy consumption which will result from world population increase, and the demands for continuation of reasonably high levels of economic growth in the industrialized countries, there are fears of a serious shortfall in petroleum supply beginning in the last decade of this century, if not before. National energy policies, while relying on some alleviation through once-and-for-all conservation measures, are seeking non traditional means of energy production, such as solar and geothermal. Because of the long lead time of the research and development process, it is unlikely that such approaches will contribute significantly to supply before serious energy shortage occur; significant additions from nuclear fusion are likely to be still later. During the transition period, it is likely that a much greater use of coal, of offshore petroleum, and of nuclear-fission power will be necessary, all of which are likely to have environmental impacts. There are thus many options to be considered and many decisions of long-term significance to be made, in which environmental factors will have to be considered and where governments will have to steer a steady course between counsels of fanaticism and those of complacency. The dilemma is well illustrated in the case of coal, the burning of much greater quantities of which raises the question of the potential dangers of accumulation of carbon dioxide the atmosphere; this is well described by our authors. We are already beginning hear the somewhat ironical statements (from nuclear scientists, of course) that 'coal is too dangerous to use'. The veracity of such a statement is, of course very doubtful, but there is, at least a point of interrogation here.

Finally, one must remark that the question of deterioration of the climate caused naturally or by human intervention, is especially important at this time because of the vast increase of food production required to feed the expanded population.

From the institutional point of view, these problems are extremely difficult to tackle since, although they are potentially of importance to all human beings, they are not and cannot be the responsibility of any particular country, and the existing international machinery is incapable of solving them.

The carbon dioxide issue illustrates well the new type of difficulty which may face decision-makers increasingly in the future. The number of uncertainties is very great, and it is improbable that they will be removed by new research findings time for decisions on the future energy options to be made on the basis of undisputed facts. For example, the role of the oceans in absorbing the increased carbon dioxide content of the atmosphere is obscure. Natural warming up of the oceans through climatic change might, by decreased solubility of the gas in water, aggravate the problem and induce a run-away process. Again, it could become much more serious earlier if the present rapid cutting down of the tropical rain forests is allowed to continue. The influence of this would not only be in the earth's albedo, as Brooks and Skolnikoff state, but also by reducing the assimilation of carbon dioxide by vegetation it would add significantly to the atmospheric carbon dioxide concentration, as recent calculations have shown.

We are likewise ignorant concerning the detailed climatic effects which a few degrees increase in the world mean temperature would induce. Certainly these would not be uniformly distributed over the surface of the planet; the effects would be greater at high latitudes, giving rise to disquiet concerning the possible extent of polar ice melting. Again, if the warming up were to result in a northern displacement of the temperate wheatlands, northern soils, such as those of Canada, would not be able to compensate for the loss of food from the prairies of the United States, which would have become semi-arid. The as yet non-existent art of the management of uncertainty may well be important for our future.

Problems of population increase and demographic structure

In the chapter, problems of the world population explosion are touched upon rather lightly and it may be useful to discuss some aspects, important to the NATO countries and not only to the Third World. World population increase is very rapid, an extra million inhabitants being added about every four and a half days. The overwhelming proportion of this increase is in the less developed countries, while in the more affluent nations, fertility rates are low, and in a few cases even below replacement level. There are indeed some recent signs of a reduction in the birth rate in some Third World countries, and of marginally increased fertility rates in industrialized countries, but in consequence of the very low average age in many parts of the world (which has resulted from recent demographic growth – in many cases 50 per cent of the population is under 15), a doubling of the world population is inevitable in about 35 years, unless natural catastrophe intervenes. It should also be realized that a doubling of the population in many of the developing countries entails a threefold increase in the workforce, mainly in places where there is widespread unemployment and underemployment. These trends present many difficulties to countries at all stages of development and involve many scientific and technical elements. The immediate issue is how the new masses will be fed and provided with the other basic human needs, greatly increasing the importance of speeding up agricultural research and the transfer of its technologies to the Third World. It also stresses the importance of matters of water resource availability and management, especially as we shall see a large growth of the cities of the Third World, due both to the gross population increase and the urban drift. It has taken many centuries to create the infrastructure of the world. Will it be possible to double this in the short span of 30 to 40 years? If so, what will be the consequences in terms of capital provision, demands for materials and energy, and many other factors? If not, what will it mean in terms of human suffering, social unrest, mass migration, and even in military terms?

While it will be necessary to improve agricultural production as far as possible areas where the population increase is greatest, it would be over-optimistic expect local improvement to be sufficiently fast to meet the new imperatives; tempo of agricultural research is inevitably slow, and changes from traditio practices and conventional food habits even slower. Thus, while we are assured the experts that it is technically feasible to feed a population many times that the present, it is highly probable that there will be increasing pressure on the countries still possessing food surpluses or capable of producing more. Amon these, the United States is by far the most important, although contributions c come to a smaller degree from Canada, Australia, New Zealand, and the Argenti For the Americans this could raise grave issues, both internally in terms of chan in price structure, but also with regard to North–South matters and also East–We as well as possible competitive demands from East and South.

It may be useful at this stage to say a few words concerning the consequences the present changes in demographic structure between developed and und developed countries. We have already noted that, while the population of the Th World is exploding, fertility levels in the industrialized world are low, and in so cases, even below replacement level. This trend is particularly striking within Soviet Union, where the fertility rate of European Russia is remarkably low, wh that of the Eastern Republics is very high. The consequence, if this continues that the Soviet Union will be predominantly of Asian stock in a few decades. T North–South demographic disparities mean that by the early years of the ne century, the proportion of the world's population in the presently industrializ countries will have fallen below 20 per cent of the total, and this could dimin further. Furthermore the average person in the developed countries will be midd aged by the beginning of the century, while the average person of some of the m populous of the Third World countries will be not much over sixteen. The thinkable would be the existence of an enclave of rich, elderly countries heav armed with sophisticated weapons and surrounded by the great majority of t world's population, poor, unemployed, hungry, and extremely young. But t shadow of such a possibility should be sufficient to spur us towards a m balanced world order.

The demographic changes may well have an impact on the national scien policies of the industrialized countries, which, with the ageing of the populatic will have a smaller work force, while at the same time the higher proportions of t elderly will make great demands on the health and welfare systems. There will th be a need for the production of greater resources by smaller numbers, a resu which can only be achieved through increased productivity and technological inn vation. In addition, the industrialization of the Third World, with its need f export markets, its increasing demands for materials and energy, with greater co petition for these resources and other factors, will greatly modify the world indu trial and trade patterns. For the developed nations this would suggest grea emphasis on the production of capital goods for export to the Third World, wh the European NATO countries, with their limited natural resources and vuln ability to the withholding of supply, will be encouraged to seek new politi

elationships with less developed, resource-rich regions and to stimulate the creation
of innovative industries producing products of high added value and low materials
and energy content. Such changes would greatly modify the international flows of
science and technology.

Some general considerations

The emergence of the global problems and the issues associated with the diffusion
of science and technology as agents of growth and development raise many doubts
as to the capability of existing institutions of government to face the new conditions; this is potentially even more serious with regard to the international
machinery. The structures, procedures, and attitudes of governments, designed for
earlier, simpler times and expanded, rather than modified, as the scope of governmental intervention widened and the scale of operation increased, have increasing
difficulty in coping with situations of scale, complexity, and uncertainty. Indeed,
some of the new imperatives question the adequacy of the political system itself,
both in free market and Marxist economies. Many of these issues are dealt with by
Brooks and Skolnikoff, but it may be useful to stress a few of them here.

We are increasingly struck by the interactions of the problems: the impact of
energy on the environment, the probable influence of changing demographic
structures of the world on international industrial patterns, the intricate interactions of land, water, and energy availability, the ability of increasing food production to meet the needs of an exploding world population, as well as the political
repercussions of all these situations and, indeed, the impact of political events on
them. In fact, the problems of contemporary society are essentially 'horizontal'
in nature and spread untidily across the whole spectrum of governmental activity
whose structures are essentially 'vertical'. The policies of government are mainly
evolved on a sector-by-sector basis while, despite the activities of interdepartmental
committees, reconciliation of sectoral policies tends to be forced upward to a high
political level where it is difficult to find adequate time and interest to consider the
various conflicts and reinforcements between them. Governments have not, in
general developed the 'staff' function strongly, in contrast to the 'line' function.

A second inadequacy concerns the difficulty of governments in giving serious
consideration to the longer-term problems in the face of political pressure on short-
term issues, although the latter may be much less fundamental. With an electoral
cycle in the NATO countries of 4-5 years, neither administrations nor opposition
parties can ignore those issues which excite the immediate interest of the voters,
and there is little political time to give deep consideration to the basic and long-
term problems that are usually complex and intractable. The consequence is a
tendency to improvise, to tackle symptoms rather than causes, as one crisis succeeds
another. Governments are well aware of this difficulty and many countries have
created prospective units in departments, think tanks, and commissions on the
future to look at the longer-term trends, consider alternative policies and strategies,
foresee difficulties ahead, including the unforeseen consequences of current policies, and, in general, to plan their national policies within the global perspective.

Some of these units are within the government machinery, as in the case of th Swedish government's Secretariat for the Future, which is attached to the Prim Minister's Office, the British Central Review Staff or U.S. Office for Technolog Assessment, which is an organ of Congress rather than of the Executive. Othe are independent or semi-autonomous bodies. The difficulty here is to reconcile th need for such bodies to be in intimate relationship with the points of decisio making and at the same time freely in touch with the swiftly moving currents world change and scientific advance. While many of these new bodies have a us ful function, it would seem that they contribute only marginally to the capaci of governments for the evolution of long-term policies, evaluation of risks, or th analysis of alternative options.

Brooks and Skolnikoff present an excellent analysis of the competence or i competence of the international machinery. In the science and technology fiel the importance of the international organizations is increasing with the growi interdependence of the nations, which technological factors are inducing, and wi the rising uncertainty of the global problems. Here the difficulties are much great than on the national scale. Discussions on issues where scientific elements or i pacts are significant are far more distant from reality than on the national lev where advice is easily obtained from the government's own scientific establishme or from academies of science, etc. Discussions in the UN and its agencies seem to increasingly politicized and dominated by the consortium of the Third Wor countries, having little scientific competence at their disposal and often disinclin to allow advice to be sought from the international non-governmental bodies, whi are felt to be unduly influenced by developed country thinking. This is clear evidenced in the preparations for the United Nations Conference on Science a Technology for Development, to be held in mid-1979. It is true that one of th major inputs to this meeting will be national papers submitted by the memb countries, in which the views of local scientific bodies will presumably be i corporated. However, there will be nearly 150 of these, and it is uncertain wheth they will even be discussed, important as they may be for background informatio The impact of the international non-official bodies is likely to be marginal and it to be feared that in the plenary discussions, matters of substance will be ignored favour of a continuation of the debate between the rich and the poor nations the terms of technological transfer and the fear of the power of the transnation corporations.

The problem of the long- versus the short-term considerations is particularly i portant where technological matters are involved. The tempo of science is great different from the tempo of technology. In many instances, the lead time of search and development followed by production on a significant scale is a matter decades, although this can be greatly reduced at times by crash programmes – at high cost. It is implicit in much economic thinking that new technology is evolv at the right moment by the interaction of economic forces. This may have be largely true in the past, with with increasingly sophisticated technology deriv from systematic research rather than from invention, and with rapid rates of soc and economic change, it cannot be relied upon for the future. There is always th fear that if we wait until crises appear, shortages become apparent, costs of materials soar, or old technologies become unacceptable, new technologies or substitutions will be ready too late. These considerations are particularly relevant with regard to energy, where the number of technical options for the future are large, but for many of them, research and development will be costly and require perhaps more time than we have before traditional energy sources become scarce and costly. Until the recent oil crisis, the low price of petroleum provided no incentives to industry or government to invest heavily in the development of alternatives, even to the extent of pressing on with the production of high calorific gas and oil from coal. The long lead time of research and development makes it impossible for technology to respond rapidly to the forces of the market.

It may be wise for the industrialized countries to prepare contingency plans and to develop a number of technological options, up to the stage of the engineering prototype or the chemical pilot plant, in preparation for the emergencies that forward scanning of economic and political trends may indicate. Such an insurance plan would be costly, but like all insurance, could give a high degree of protection. The costs could be greatly lowered if, through NATO or perhaps more appropriately OECD, such work could be undertaken in co-operation between government and industry by the cost-sharing device of international co-operation between government and industry by the cost-sharing device of international co-operation. Much experience has been gained in recent decades concerning modes of international research co-operation, with a few successes and many disappointments, and a basis exists for more effective working together in the future. This might well take the form of planning developments in common and sharing out elements of each project to those countries interested and willing to participate in each case, the work being undertaken in the most appropriate laboratories — academic, government, or industrial. The case for a new look at the need for international scientific *and* technological co-operation, its modalities and problems, is very strong.

19

Defence planning and the politics of European security

Johan Jørgen Holst

19.1. Introduction

The present chapter does not purport to share the results of the author's consultations with his crystal-ball. Nor does it aim at presenting any kind of a planner'
manual. It is very simply the attempt of one individual to outline certain basi
considerations of importance to long-term defence planning. I do not aspire t
provide precise and definitive answers. The task at hand has been one of identi
fying and formulating certain key issues and questions. It is not an exercise i
prediction as much as an attempt to circumvent some of the inherent limitation
of prediction concerning long-term developments in the external environment.

The approach is deliberately eclectic. I remain relatively unimpressed wit
attempts at writing scenarios of alternative futures, where the limits appear to b
fixed more by individual ingenuity and imagination than by any requirement fo
verification. Consequently, this chapter attempts to deal with the future in term
of certain broad parameters rather than integrated scenario descriptions.

I have spent many years engaged in policy analysis and a few years only in
policy-making role with political responsibility. The approach and concerns in thi
chapter probably reflect to some extent my personal history. I am well aware o
the fact that my observations and judgements are likely to reflect to a considerabl
degree my own values, prejudices, and experiences. Approximations to objectiv
insight will emerge only as a result of interaction with experts with different ou
looks, concerns and backgrounds.

The chapter deals mainly with the defence aspects of security, not because th
social and societal dimensions are not important, but in order to delimit the area o
discussion. The discussion will focus, however, on some of the novel sources o
potential conflict which reflect the emerging social complexities in the internationa
environment. The overriding objective of the present era is that of assuring peacefu
change.

In this chapter I shall consider the following themes:

the structure of the security order in Europe;
the role of nuclear weapons in the maintenance of the security order;
the spectrum of potential conflicts; and
the impact of technology.

19.2. The structure of the security order in Europe

The structure of the security order in Europe is determined primarily by the fact of Soviet military power and position of political dominance in Eastern Europe. That position is buttressed and legitimized by an ideological outlook which is predicated on the notion of permanent struggle. Countervailing power has been supplied by the United States. The credibility of the American commitment to maintain a balance of power in Europe is in part inherent in the positioning of American troops (more than 310 000 in 1977) and weapons in Western Europe in peacetime.

It is true, of course, that the aggregate GNP of NATO is almost three times that of the Warsaw Pact countries ($3.4 trillion[1] versus $1.2 trillion), and that NATO has a population advantage of about 50 per cent over the Warsaw Pact countries (540 million versus 360 million). Hence, in the abstract it would seem unnecessary for the American weight to be thrown permanently into the European scales in order to contain Soviet power in Europe. However, the salient fact in this connection is the fragmented political structure of Western Europe – the European Community notwithstanding – the narrow geography involved, and the history of national diversification and strife in modern Europe. American protection has provided an important political and psychological precondition for the process of reconciliation and nascent integration in Western Europe. It continues to provide assurance against external aggression and internal rivalry. That condition is likely to constitute a stable factor in the security calculus in Europe.

The security order in Europe differs from the traditional balance of power systems in its structural rigidity compared to the flexible rules of adjustment of previous periods. The alliances have become institutionalized mechanisms for political concurrence and defence organization. They are *not* symmetrical in structure or process, but share a quality of permanence. The structural rigidity of the present balance-of-power arrangements in Europe is determined in part by the ideological determination of power alignments, but principally and fundamentally it is a reflection of developments in weapons technology and, most importantly, of the arrival of nuclear weapons.

Nuclear weapons carry the potential of instant large-scale destruction. Manned bombers and ballistic missiles extend their range beyond immediate territorial proximity. Distance can no longer provide protection for anybody. The calculation of real distance in terms of military access has become rather complex. The condition of permanent vulnerability is not radically new to the states of Europe. To the states of North America it constitutes a fundamental change. Apart from her favourable geographical location, the security of the U.S. had since the Civil War rested on the superior war potential which could be mobilized within her own frontiers and on the balance of power in Europe. At the end of the Second World War the European balance had suffered a complete collapse and the invulnerability of the American industrial capacity promised to be but a temporary asset, since it was generally accepted that the Russians would sooner or later acquire the atomic bomb.

[1] Here a trillion is taken as 1 million million or 10^{12}.

As the chilly winds of Soviet–American controversy came to prevail in the international arena the United States attempted to re-erect the twin-pillar security of the past. A restoration of the European balance of power was first attempted by the buildup of indigenous European strength to withstand Soviet pressure (the Marshall Plan and the Mutual Defense Assistance Program) and, when that proved inadequate, by the substitution of American for European power (the Truman Doctrine, NATO). The policies which aimed at a reconstruction of the second pillar of American security included the abortive plan for international control of atomic energy and the development of a doctrine to cope with the potential contingency of bilateral nuclear war.

The emergence of a system of mutual deterrence based on large forces in being has hardened the pattern of security arrangements while softening the protective shells of the nation states. The maintenance of stable deterrence has become an overriding policy-goal for the two principal system-wide powers of the present era. That priority tends to reduce the flexibility of interstate relations.

A dissolution of the alliances in Europe would not only introduce uncertainties with respect to the relative positions of the superpowers within Europe and outside. The prospects and problems of nuclear proliferation would almost inevitably undergo a qualitative change. The fact of Soviet regional dominance would persist and latent regional rivalries and uncertainties would presumably exert novel pressures on national decisions and priorities. Most of the developed states in Europe have the capacity to make nuclear weapons. A nuclear scramble in Europe would inevitably affect perceptions with respect to the utility of nuclear weapons and prospects of further proliferation outside Europe. Postulating the existence of NATO and the Warsaw Pact as long-term parameters in the security order in Europe does not, of course, preclude internal changes with respect to the division of labour, responsibility, and influence within the alliances. Indeed such changes are more than likely within NATO in the years to come, although the fact of American leadership is likely to persist.

The more dynamic perspectives relate to the possible evolution of inter-alliance relations. Here we can identify three distinct levels of potential transactions: the rules of the game; the exchange of goods and services, and the limitation of military forces. The quality of intra-European relations may change within the broad parameters of the existing structures. The Conference on Security and Co-operation in Europe embodies a codification of some basic ground rules which are predicated on the *de facto* existence of present territorial realities. The inviolability of frontiers and the abrogation of force as a means of changing the territorial realities constitute the primary ground rules which were first suggested by the bilateral eastern agreements of the Federal Republic. Ambiguities persist with respect to the applicability of general norms in specific contexts, most notably with respect to the Soviet hegemonial system in Eastern Europe. Such issues will not be negotiated around the bargaining table, but rather through a process of political linkage and containment.

Certain confidence-building measures (CBMs) have been agreed upon for purposes of giving some operational precision to the general commitment to refrain 'from the threat or use of force'. They constitute small building blocks in a process of reducing the role of military force in the political interplay in Europe. Confidence-building measures have the advantage that they may be made precise and specific and hence lend some focused substance to the process of political reconstruction in Europe. They are in some sense prior to arms control and disarmament. They are not directed at the level of military effort and competition, but form elements, rather, in a framework for the indirect alleviation and reduction of the incentives for competition which derive from uncertainty and possible misunderstanding. From a psychological point of view they introduce novel sources of transparency.

The current system of CBMs is not very extensive, but it constitutes a beginning. The obligation to preannounce major manoeuvres and the recognition that reciprocal observation constitutes a desirable objective is but a beginning. The Belgrade meeting demonstrated how difficult it is to move beyond the arduous beginning. However, practice has become more liberal and reciprocal coincidental with the convocation of the review conference at Belgrade. A policy of mutual example and stimulation of bandwagon-effects is perhaps more likely to result in further amplification, reinforcement, and expansion of the present CBMs than formal negotiations. [1]

Greater transparency and interpenetration between the states of the two alliance groupings in Europe may be attained also on the level of commercial transaction and cultural exchange. However, the causal connection between trade patterns and the outbreak of war is at best very tenuous. Transfer of technology may influence internal priorities and allocations, but there can be no assurances that this will in fact prove beneficial from the point of view of peace. Interpenetration may exacerbate latent tensions between state and society, particularly in those states where the social order is based on a considerable degree of repression with respect to human rights. The promulgation of human rights as a focus of foreign policy has served the important function of realigning the external purposes of Western states with the values upon which their social order is founded. It highlights a concern and provides a modicum of protection to courageous defenders of the right of the individual as against the state. However, credibility may suffer as results fall short of expectations and compromise intervenes in the sphere of categorical imperatives.

There is no automatic leverage in expanded exchange, but it does provide arguments and vested interests in favour of continued co-operation as countervailing influences to those associated with *abgrenzung* and confrontation. The possibility of another 'Budapest' or 'Prague' cannot be eliminated, but the thresholds may be raised by increasing the opportunity costs involved. [2]

The military confrontation in Europe is maintained at a very high level of effort and forces in being. The dangers of accidental conflagration may not be very high, as the lines are clearly drawn and the military organizations maintained under fairly rigid control. However, the spiral of reciprocal fears and incentives to exploit or prevent initiatives in a tense crisis may create rather acute problems of pre-emptive instabilities. The military efforts entail economic and social costs which, however, probably impact in a differential way as between the states in the two alliances.

Negotiations aiming at mutual and balanced force reductions are predicated on the notion that it should be possible to establish a condition of undiminished security for either side with a lower level of effort. The results have been singularly unspectacular. The currency of negotiations may very well be that of force levels, but the real and symbolic objectives concern the longer-term parameters for the security arrangements in Europe.

The Western states have insisted on negotiating towards a common ceiling on the level of troops in the reduction area. The Eastern states have tended to favour the imposition of national sub-ceilings for each of the states located within the reduction area. This dispute is important, of course, to the integrity of the integrated defence arrangements in NATO as well as to options for future political evolution in Western Europe. Furthermore, the regime proposed by the Eastern states, not surprisingly, would accentuate the power differential between the Soviet Union and the states within the reduction area, as the latter would have absolute limits imposed on their total military effect while the Soviet Union, situated outside the reduction area, would not be subjected to similar constraints. This structural asymmetry is at the core of the Western insistence on Warsaw Pact reductions down to parity levels in the reduction area, while the Eastern states have argued in favour of equal percentage reductions which would perpetuate a numerical distribution in their favour. The Eastern and Western participants differ in their estimates of the force levels in the Warsaw Pact portion of the reduction area. (The discrepancy is on the order of 150 000 men.) The data discussion has become a surrogate for bargaining over the reduction method. A resolution of the data dispute is a political and practical necessity as long as the method of reduction involves percentage reductions and common residual ceilings. Alternative approaches involving absolute and asymmetrical reductions might provide a way out. Verification procedures could then be tailored to checking net flows.

Any complex multilateral arms control negotiations will tend to become encased by self-contained logic which becomes increasingly removed from the original condition giving rise to the need to negotiate constraints. Even marginal changes in the negotiating position will loom large from the point of view of those who manage consensus collectively agreed at. Internal stalemate will then tend to stimulate political circumvention by the major powers through direct initiatives, with potentially serious costs to alliance cohesion. Political vulnerabilities in this respect are much stronger in the West than in the East. Negotiations may indeed be pursued with the contingent aim of exploiting much structural asymmetries.

Negotiations may become increasingly removed from the evolution of the security problem. Hence, the permanent MBFR 'congress' in Vienna is no longer needed as a means of dissuading the Congress in Washington from voting in favour of unilateral American troop withdrawals. Secondly, the manpower level approach may be losing relevance as the Soviet ability to initiate standing-start, rapid reinforcement attacks appears to be growing.

The Western states have taken the view that any reduction agreement should encompass certain associated measures designed to stabilize such an agreement. We can distinguish broadly between stabilization measures, verification measures, and non-circumvention provisions. Measures of stabilization and verification would involve commitments to implement mutual observation through the exchange of observers during manoeuvres, to establish fixed observation posts at key transportation nodes, as well as to observe certain codes of conduct with respect to troop manoeuvres, movements, and redeployments. Non-circumvention provisions would be designed, for example, to protect against the transfer of military pressure to other areas in Europe. As they would apply to areas outside the reduction area as well, they could serve the function of preserving the coherence and cohesion of the security order in Europe. Hence, we are once more confronted with the need to consider arms control arrangements from the point of view of their impact on the structure of the political order. Arms control remains a political currency.

The basic goals of arms control include:

(1) Reducing the probability of war;
(2) Reducing damage and suffering if war should occur;
(3) Reducing obstacles to a quick and equitable end to a war;
(4) Reducing the costs and burdens of the arms competition;
(5) Reducing the role of military force in international relations.

This normative enumeration does not, of course, imply that arms control policy and defence policy are or should be at variance. Arms control objectives should be integral to defence policy. Conflicts may, of course, arise. The arms control objectives in and of themselves may not suggest an internally consistent logic. The policy problem is one of optimizing complex outcomes rather than simply maximizing single goals.

As suggested in the above discussion, particular arms control measures for Europe cannot be considered as politically neutral technical arrangements for enhancing stability; they are instruments for building and managing the security order in Europe. Arms control proposals have been advanced for constraining political change undesirable from a particular interest perspective. They have also been deployed as a means of exploiting favourable asymmetries and minimizing the impact of unfavourable ones, blocking undesirable departures and promoting political advantages. The ostensible commitment not to strive for unilateral advantage can eliminate neither political competition nor incentives and opportunities for exploiting ambiguities and spin-off effects. However, the very concept of arms control is predicated on the notion of certain shared interests transcending the competition of the moment. International politics is not a zero-sum game, and the shared danger of nuclear holocaust has forced states to some extent to regard security as a shared value — an idea derived from notions of community and a regulated system of power balance, based on interdependence and reciprocal restraint [3]. But such ideas are not equally strong in the political traditions and outlooks of all the states in Europe. There are also competing traditions, emphasizing hegemonic ambitions and a commitment to struggle.

19.3. The role of nuclear weapons in the maintenance of the security order

The two superpowers will remain the major custodians of the nuclear peace for the forseeable future. Management of the central balance will continue to constitute a primary area of attention in defence planning. Its stability cannot be taken for granted. The balance of deterrence has in fact proved to be more robust and less delicate than some analysts have feared. However, technological development and the propensity to compete for relative advantage may introduce elements of instability into the strategic equation. Stability in this connection has two major dimensions:

Crisis stability prevails when the configuration and size of the opposing forces do not promise substantial advantages with respect to the outcome of a war to the side which takes the initiative in a crisis. The situation would be unstable if the elasticity of the outcome with respect to who strikes first is high. Hence, stability inheres in the absence of strong incentives to take the strategic initiative in an intense crisis in which the fears of imminent large-scale violence would dominate the atmosphere.

Arms race instability prevails whenever the configuration and size of the opposing forces are such that the search for crisis stability or unilateral advantage will cause the forces of one side to expand in ways which will compel the other side to augment its own efforts as a response in order to preserve the balance of mutual deterrence. It should be interjected here that we understand very poorly how the decision processes of the two superpowers interact. The simple model of an action-reaction process simply does not stand up to analysis. [4]

The existence of nuclear weapons has necessitated the elaboration of a theory of nuclear deterrence. The concepts are rough and the theory inadequate as a basis for policy prescription. Deterrence is essentially a psychological variable. What constitutes deterrence will inevitably depend on perceptions with respect to expected outcomes of alternative courses of action. The calculus will never be precise. Certain distinctions are, however, of considerable importance. It is now generally understood that the distinction between first strike and second strike postures is important, particularly from the point of view of crisis stability. It is the forces capable of surviving a first disarming strike which will be available for retaliatory purposes. Hence, the relative vulnerability of offensive forces to preclusive attacks constitutes an important parameter in defence planning with respect to the configuration of the forces. The distinction between counterforce and countervalue attacks is likely to remain important from the point of view of escalation control. However, with improvements in accuracy more forces will contain options for employment against different classes of targets. The latter distinction is different from that between first and second strike attacks. The sizing of the strategic forces will depend on targeting doctrine, the potential target structure, mission assignments, the size of the opposing force, etc. Here, there is considerable room for judgement. The concepts of sufficiency are vague. The notion of essential equivalence does not imply a requirement for symmetrical force matching. It suggests rather that any advantage in force characteristics enjoyed by one side be offset by equivalent advantages of the other side. Perceptions may vary, of course, with respect to assessments of relative advantages. [5]

In the U.S. an important criterion for the sizing and configuration of the strategic forces has been the notion of assured destruction. However, the extent of the destruction which has to be assured seems to be based more on marginal productivity calculations than on any real analysis of how much destructive potential is necessary in order to deter. And deter what action? Historically, the concept of assured destruction emerged in the context of a major discussion about a possible modification of the system of deterrence by the introduction of active ballistic missile defence systems. The structural issues involved in that debate transcended technical estimates of relative system efficiency. The *de facto* condition of mutual vulnerability was by many presented as a normative goal irrespective of available alternative postures with greater defensive emphases. Deterrence in its specific Soviet–American configuration had become institutionalized. [6]

That factual condition was subsequently codified in the ABM-treaty which was negotiated during SALT-I. It remains questionable, however, whether that treaty should be taken as a normative affirmation of the American 'theory' of mutual assured destruction. The ABM treaty probably reflected a low valuation of the then existing state of the ABM-art, particularly in the Soviet Union, rather than a doctrinal alignment between the two superpowers. Since the ratification of the ABM treaty the U.S. BMD effort has declined. The SAFEGUARD system has been deactivated except for the Perimeter Acquisition Radar (PAR). Funding for BMD has dropped substantially and the research programme has been reoriented. The Soviet BMD activity has not similarly declined. It is now acknowledged that the 'lead enjoyed by the United States in BMD at the time (the US) entered into the ABM treaty has greatly diminished [7]. The Russians 'are developing a BMD system which appears to be rapidly deployable and they have an active interceptor flight test program' [8]. The introduction of long-range cruise missiles into the operational inventories of the superpowers may provide new incentives for the upgrading of air defence systems and the deployment of surface-to-air missiles for purposes of impeding the penetration of the relatively slow-moving cruise missiles to their targets. However, the small radar cross-sections of cruise missiles will make detection and interception difficult to accomplish. Such efforts at damage limitation could alter the doctrinal framework with respect to the configuration of strategic forces in general.

Third country forces could conceivably alter the calculus. The initial American decision in favour of the SENTINEL ABM-system was explicitly argued in terms of containment of Chinese missile threats. The Chinese missile capabilities have, however, matured much more slowly than many expected in the late–1960s–early-1970s. The operational capabilities are limited to liquid-fuelled MRBMs and IRBMs as well as some 80 TU-16 medium bombers. Hence, China constitutes primarily a strategic strike problem for the Soviet Union and the peripheral states in Asia. The Soviet incentives to exploit BMD technology for purposes of countering Chinese offensive capabilities are likely to remain stronger than those of the United States. The diplomatic *rapprochement* between Peking and Washington has done more to change perceptions with respect to the role of China within the international power geometry than the retardation of the Chinese ICBM programme, (the two may be related, of course). French and British forces constitute in effect adjuncts, triggers, and residual insurance with respect to the US strategic deterrent.

Widespread proliferation of nuclear weapons could also cause a breakdown of the SALT system. Here, predictions with respect to the future environment have tended to overstate the propensity of states to translate potentiality into actual capability and thus to downgrade the intervening political calculus. The notion of a general process of dominoes had contributed to the depoliticization of the analysis of the problem of proliferation. To some extent the NPT is predicated on the notion of a general process of nuclear proliferation. However, in terms of political management attention has to be paid to regional configurations of incentives and disincentives. [9]

The relationship between the policies of the nuclear weapon states and the would-be nuclear weapon aspirants is particularly complex and ambiguous. To some extent a nuclear emphasis posture on behalf of the existing nuclear weapon states will set a standard with respect to the currency of power for the potential hegemonic regional powers aspiring to exploit and reinforce the transformation of the international system into a more flexible and fractionalized order. However, a number of the potential nuclear-weapon states enjoy the protection from extended deterrence. A reduced role for nuclear weapons could be seen as a weakening of the nuclear guarantee, particularly as power relations have changed within the central balance of power between the two principal nuclear-weapon states.

The convertibility of nuclear power into political power has in fact proved to be extremely limited. It is to an important extent unusable power except in the context of its own negation within the balance of deterrence. The coat-tails of nuclear deterrence probably do not extend too far beyond the system of East–West relations. Uncertainties prevail, however, with respect to the boundaries of that system. In some sense nuclear weapons constitute a Great Power burden. Their existence has contributed to the partial divorce of power from authority and influence in international society. That divorce has in turn provided flexibility and mobility to international relations. It is unlikely, however, that world order has been enhanced in the process.

The issue of nuclear proliferation will inevitably be perceived in the context of the dominant perspectives in international society. The globalization of so many issues of distributive justice will affect the framework of perceptions also with respect to nuclear weapons. Any anti-proliferation policy which is championed by the nuclear haves will collide with demands for equity from the have-nots. The days of mechanical policies are gone.

Major policy issues are emerging with respect to the relationship between nuclear energy technology and the availability of near-to-weapon grade fissile materials. A *de facto* contraction of the lead times involved in the acquisition of nuclear weapons, without actually deciding, could lower the threshold for eventual decisions. An important debate is emerging about the need to close the fuel cycle by separating plutonium and unburned uranium from the spent reactor fuel for purposes of recycling through the reactor. The economics do not look too promising in the short term. Interest in recycling is tied, however, to the availability of reasonably-priced supplies of nuclear fuel. The breeder reactor may provide an economic justification for closing the fuel cycle. A pattern of international co-operation cross-cutting the traditional East-West and North-South barriers may make the need to control nuclear proliferation an issue of considerable impact on the structure and process of international society in the long term. The 15-member (London) nuclear suppliers club and the 40 + member International Nuclear Fuel Cycle Evaluation (INFCE) constitute interesting institutional building blocks. [10]

The Soviet Union has become the most active competitor within the central balance. It has been estimated that over the period 1967-77 the level of Soviet strategic force efforts (RDT and E excluded) was almost two and a half times that of the United States measured in dollars. In 1977 the Soviet level was approximately three times that of the United States [11]. That disparity in effort is unlikely to persist for very much longer.

Constraints may be imposed by agreement in SALT or by an acceleration of the American effort drawing upon a superior GNP and a technological base of much higher productivity. The SALT process is likely to continue and transform and mute the competitive aspects of the Soviet-American nuclear confrontation. Competition and rivalry will not cease but it is likely to be constrained by mutual agreement. Such agreements will include aggregate ceilings, sub-ceilings on specific capabilities, constraints on modernization, guidelines for future developments and negotiations, postponement of non-equivalent force deployments, data exchange, etc. Most of these ingredients form part of the SALT-II package. As the process develops and the constraints become more specific the need for an authoritative and verifiable data exchange will grow. We have seen that fact reflected in the SALT-II as well as MBFR negotiations. Hence, the very process of negotiation is likely to generate imperatives and requirements which may break down some of the traditional barriers to greater transparency.

The negotiating process will affect areas outside the narrow focus of negotiations. Proposals and positions will be designed with a view to maximizing spin-off effects. In the Soviet armoury of proposals those relating to forward-based system (FBS), non-circumvention, and non-transfer have clearly been made with a view to the potential impact on U.S.-European relations. All such proposals carry the potential of unexpected boomerang effects. There is also the fundamental issue of technological circumvention or undermining of the framework and conceptual categories of the ongoing negotiations. Cruse missiles, the Soviet SS-20 missiles, and the BACKFIRE bomber illustrate the difficulty of establishing viable boundaries for long-term negotiations about broad areas of activity with a very volatile technology.

The most critical issues in the near-to-medium term with respect to strategic forces will be that of first strike instabilities due to the increasing vulnerability of silo-housed ICBMs. That issue has, of course, not been brought forward by SALT. It is a function rather of significant improvements in the accuracy with which weapons can be delivered on fixed targets, even over very long distances. There will, of course, remain very large uncertainties with respect to the possibility of executing disarming strikes against the silo-housed missiles of the adversary. Fratricide, reliability, operational degradation, launch on warning responses, etc. constitute sources of uncertainty. However, it is the official American view that a more survivable ICBM-basing mode will be needed 'or a considerably more capable silo-based missile to maximize the retaliatory effectiveness of the small percentage of missiles expected to survive an all-out Soviet attack on the Minuteman Force in the mid-to-late-1980s [12]. The emergence of a fifth generation of Soviet ICBMs as successor systems to the SS–17, –18, and –19 force, the deployment of which proceeding at a rate of about 125 a year, could exacerbate the American dilemma and make the deployment of a semi-mobile MX inevitable. The MX would be an effective silo-killer and some observers have cautioned against its impact on crisis stability. It should be remembered, however, that the Russians already have deployed the systems which constitute Minuteman-killers and create the problems in the first instance. An MX deployment could be designed to require a greater expenditure of payload in the attack than would be destroyed by the attack. The appropriate basing mode (hardened trenches, dispersed shelters, etc.) is still being studied. The issue of pre-emptive instability is likely to provide a basic *leitmotiv* to the SALT–III negotiations linking the latter to the negotiations which took place in the abortive 1958 surprise attack conference.

The offensive force postures of the two superpowers are likely to maintain basic triadic structure in spite of the emerging vulnerability of silo-based ICBMs. Each of the three force components: air breathers, SLBMs, and ICBMs have different system vulnerabilities and confront the adversary with distinct and independent uncertainties. In the absence of severe constraints on accuracy-yield combinations producing hard-target kill capabilities some kind of mobile ICBM-basing mode seems likely. Constraints on mobility could in principle be envisaged as part of SALT–III regime, particularly since verification would become much more difficult with Soviet style land-mobile ICBM deployments. Long-range cruise missile (CM) forces constitute a novel air breathing option. They could be deployed in air launched, sea-launched, or ground-launched modes. A full U.S. initial operational CM capability will not occur till 1982–7, i.e. after the expiration of a three-year SALT–II protocol entered into in mid-1979. Cruise missiles are ambiguous from the point of view of arms control. They pose serious verification problems. However because they are slow-moving, they would be ill-suited for first strikes and well suited for second and third strikes. That characteristic may look particularly attractive in the context of the increasing vulnerability of silo-housed ICBMs. They would be relatively invulnerable to pre-emptive strikes, but their in-flight vulnerability would be higher than for ballistic missiles.

The SLBM force will remain the most survivable element in the current TRIAD. It has the ability to maintain a multitude of potential threat azimuths and would seem to be particularly conducive to crisis stability. However, strategic force arms limitation regimes based upon SLBMs as the principal residual component could entail rather extensive geopolitical consequences, particularly if the constraints include such schemes as sanctuaries and zones of preferential access.

Concern about the intrinsic stability of the central balance will inevitably increase the political centrality of the central balance. Such centrality will tend to bring the conflictual aspects of the East–West competition to the fore. The impact on extended deterrence is, however, ambiguous and will depend on the totality of political relations in Europe.

The Soviet buildup and momentum in the strategic force sector constitute a long-term challenge to the arms race stability of the present situation. SALT has from the American point of view become a vehicle for curtailing Soviet developments. Ambiguities persist with respect to Soviet objectives. To the extent that the latter can be extrapolated from the emerging capabilities the evidence is still ambiguous but disturbing: 'Much of what they are doing both offensively and defensively coincides with the actions that would support a damage limiting strategy. And it is within the realm of possibility that they are attempting to acquire what have been called "war winning" capabilities'. [13]

It will in general remain obscure what margins of qualitative of quantitative superiority will be 'meaningful'. To the extent that they be understood to produce significantly different war outcomes for the two superpowers they are likely to affect *expectations* as to which side would have to back down in intense crises for less than vital stakes. Such expectations are likely to affect the calculations determining policy choices in the normal flow of interstate relations. It has been observed also that 'a perceived nuclear superiority of a more limited, even purely quantitative kind, could become politically significant if combined with conventional superiority at selected fronts'. [14]

The continuity of deterrence inheres in the linkages between strategic forces, theatre nuclear forces (TNF), and general purpose forces. The structure of theatre nuclear forces is likely to change somewhat in the years to come. Conventional options will probably be given priority, particularly in the West. Expanded flexibility and social acceptability point in that direction. However, conventional defence and theatre nuclear defence will not be considered as alternatives but rather as mutual supplements. Attention is likely to focus on the synergisms of combined arms options. Increased attention will be devoted to the survivability of theatre nuclear forces so as to reduce the pressure they would exert on decision-making in a crisis. Improved SAS (special ammunition site) security both with respect to resistance to weapons impact and unauthorized access will be another area of concern in the West. The structuring and sizing of the forces will reflect a re-examination of the balance obtaining among long-, medium-, and short-range systems. The structural force-balance will be influenced *inter alia* by developments with respect to command, control, and communications (C^3) and target acquisition. A subset will include the balance between onshore and offshore longer-range systems. There is likely to be an evolutionary upward adjustment of NATO's long-range TNF forces, in order to circumvent the pre-emptive threat to NATO's quick reaction alert (QRA) air forces posed by the mobile and MIRV'ed SS-20 and for purposes of confronting the Soviet leaders with a continuous spectrum of armaments under conditions of essential equivalence at the level of intercontinental systems. Extended range PERSHING-II and Ground Launched Cruise Missiles (GLCM) seem the more promising candidate systems. The process of modernization has to be considered in a political context as well. The following three principles are likely to be salient: (1) no NATO country should be excluded from participation; (2) no NATO country should be asked to carry exclusive risks or burdens as a result of deployment; (3) no NATO country which has adhered to long-standing policies

excluding deployment of nuclear weapons should be asked to change the groun
rules which have become elements in the established order in Europe; and (4
modernization should be implemented in a fashion which is consistent with th
continued pursuit of arms control.

Nuclear air defence and ADMs contribute but marginally to the decisiveness c
the NATO response. They would also require very early decisions. Hence, their ro.
is likely to diminish. Battlefield systems designed for deployment close to the fo
ward edge of the battle area (FEBA) will be given greater survivability, highe
accuracy, and reduced collateral damage effects. The nuclear warheads in gener:
may be provided with selectable yields and tailored weapon effects. However, th
distinctiveness of nuclear weapons will be preserved and the nuclear threshold kep
high. The need for flexibility and escalation control will require selective an
limited nuclear options as well as the option of theatre force employment in th
context of general nuclear response. Hence, the interlocking of theatre- an
strategic-nuclear forces into a seamless web of deterrence will continue to const
tute a primary goal of defence planning. However, the absolute size of the NAT
TNF force is likely to diminish as the Western alliance move more decisively in th
direction of a combined arms options posture. Such a development is likely also i
view of public attitudes and the need to make defence policy credible in terms c
public opinion as well as the perceptions of the adversary.

The dispute about reduced blast/enhanced radiation (RB/ER) weapons const
tutes a rather vivid illustration of the kinds of irrationalities and distortions whic
are likely to accompany the modernization of the TNF posture in the absence o
careful attention to presentation and diplomatic management (this is not to sa
that there are no serious and rational arguments against RB/ER weapons – there ar
particularly in relation to arms control, but those arguments did not dominate th
public debate). [15]

In future, arms limitation negotiations are likely to comprise TNF. The bounda
ries in SALT are becoming less tenable in view of the evolving technology. Th
Russians have raised the issue of Western forward-based systems (FBS) and in th
West the exclusion of the Soviet MRBM/IRBM force – particularly with the phase
in of the SS–20 – is causing increasing concern. The definitional categories of SAL
could in the future process carry the implication of defining Western Europe withi
the Soviet system of strategic interest. However, the modalities for including TNI
in arms limitation negotiations are politically very significant. Separate talks shoul
be avoided. Inclusion in SALT would probably provide the greatest assuranc
against decoupling from the central balance. Hence, we are back to the primacy o
the political framework for management of the nuclear balance.

19.4. The spectrum of potential conflicts

The design contingency for the NATO force posture in Europe has been a majo
Warsaw Pact move to achieve mastery in Europe. That contingency will continu
to define the minimum requirements. The ability to control escalation in the centr
of Europe constitutes a *conditio sine qua non* for the ability of NATO to cope wit

more limited and off-design contingencies on the periphery in and off Europe. In the event that the balance in Central Europe should shift decisively in favour of the Warsaw Pact, the local preponderance of Soviet power on the flanks would loom darker and heavier on the horizon. Equivalence of power on the central front will remain the key to the balance of power in Europe.

Soviet military power in Europe has not remained a static quantity [16]. In recent years the Soviet posture has improved considerably. Moscow has very clearly assumed the initiative in terms of force modernization and restructuring. Soviet ground forces have been strengthened by large numbers of self-propelled artillery, modern tanks, infantry combat vehicles, anti-tank guided missiles, attack helicopters, bridge-crossing equipment, and organic air defences. The range and weapon-load of Soviet aircraft has been markedly increased. The MIG–23s and 27s (Flogger B and D), SU–17s (Fitter), and SU–19s (Fencer) have provided a capability which was previously lacking to conduct deep air superiority and interdiction missions. Over the last decade the number of Soviet main battle tanks have increased by 31 per cent, artillery pieces by 38 per cent, armoured personnel carriers (APC) by 79 per cent and fixed wing tactical aircraft by 20 per cent [17]. About 1000 men have been added to each of the tank divisions and approximately 1500 to each of the motorized rifle divisions; 150 000 men have been added to the Soviet forces in Eastern Europe over the last decade, including 70 000 men and five divisions which were deployed in Czechoslovakia in 1968. However, most of the Soviet force buildup has taken place in the Far East where the 1965 force level of 20 divisions and 200 fighter/attack aircraft has been expanded to around 40 divisions and 1000 fighter/attack aircraft.

The maturing Soviet threat has increased the range of Soviet options for major offensive operations. The option of attacking with the forward deployed forces has become more real. The offensive accent has been strengthened as the ability to attack without, or with, but minor reinforcements has been strengthened. The ability to bring the forces of the western military districts rapidly into Eastern Europe adds to the surprise attack problem in Central Europe. NATOs major defence planning problem in the years ahead will consist of ways of blocking Soviet offensive options across the central front. [18]

A long-term defence programme has been agreed in NATO. In preparation thereof studies were carried out in ten priority areas by special task forces on readiness, reinforcements, mobilization of reserves, maritime strategy, air defence, communications, command and control, electronic warfare, rationalization (including interoperability and standardization), consumer logistics, and the modernization of theatre nuclear forces. A consolidated report was approved by the NATO Summit meeting in Washington in Spring 1978. It established the foundation for long-term defence planning in NATO and Western capitals. To complement this effort an immediate programme of short-term measures was also approved. The Soviet challenge necessitated a co-ordinated, focused response based on explicit priorities.

The spectrum of potential contingencies include:

(1) Central war (limited and general);
(2) Major attack across the central front in Europe;

    (3) Limited attacks on the NATO flanks;
    (4) Soviet suppression of East European countries;
    (5) External involvement in intra-Western disputes;
    (6) Hostile exploitation of socio-political turmoil in Hostile Europe;
    (7) Hostile interception of major sources and lines of supply to the West;
    (8) Hostile intervention in extra-European areas of vital importance to the West
    (9) Demonstrations of force (implicit or explicit) for purposes of inducing adaptive behaviour;
  (10) Conflict over ocean resources and jurisdiction.

Soviet local superiority on the flanks could conceivably be exploited for pur poses of local aggression in situations in which the balance in Central Europe should falter or the coupling between periphery and centre in the NATO system of guaran tees should weaken. Numerical preponderance at the level of strategic forces could affect perceptions about the credible extension of deterrence. However, the concen tration of large central war assets on the Kola peninsula on the Northern flank would presumably constrain local aggression for limited territorial gains as the consequences of escalation could be very far-reaching. Fear of such consequences could, of course, deter intervention from the major Western Powers, but the resi dual uncertainties seem more likely to constrain limited aggression in the first instance. However, due to the geostrategic significance of North Norway and the Turkish Straits those areas are likely to be involved in the early phases of a major war in Europe [19]. The ability of the major powers in NATO to project rapid reinforcements, including naval and air power, to the flanks of the alliance, will constitute a primary source of deterrence to Soviet attempts at limited aggression in those areas.

Eastern Europe constitutes an area of great ethnic and cultural diversity. Nation alism has been a powerful determinant of the history in this turbulent part of Europe. It has to a large extent been directed against hegemonial power exerted from outside the region (the old Ottoman, Habsburg, and Russian empires, Nazi Germany and the Soviet Union). The Soviet Union views Eastern Europe as a defensive glacis and legitimation of the ideological basis and socio-political struc ture of the Soviet state. The Soviet army is a major instrument for the exercise of Soviet hegemonic power; defending the territorial integrity of the 'socialist' repub lics of Eastern Europe, assuring order, and preserving the *status quo*.

The potentials for future conflict embrace a broad spectrum of possible contingencies:

    (1) Popular protests against the established regimes;
    (2) Socio-political reforms transcending the threshold of Soviet tolerance;
    (3) National resistance to Soviet domination;
    (4) Nationality conflicts within a single state;
    (5) Nationality conflicts involving neighbouring states.

Soviet military intervention remains a latent possibility. Conflicts arising from such sources could pose dangers of spill-overs for the neutral states and NATO in

Europe. Forces should be structured for purposes of containing such spill-over effects. Flexible, conventional options will be required.

NATO will not adopt a posture of deliberate counter-intervention. However, the issue of how much residual ambiguity ought to prevail with respect to such options may become a matter for consideration from the point of view of deterrence. In any event NATO will have to consider in what ways non-interventionist responses could entail deterrence against Soviet military suppression in Eastern Europe. The pan-European perspectives which have been stimulated by *détente* and the CSCE-process pose issues with respect to the general credibility of the Western commitment to the fundamental values of human rights. The dilemmas posed by the competing rationalities of *raison d'état* perspectives and the protection of individual liberties, could exacerbate tensions between state and society in the West.

The countries of Eastern Europe are likely to remain in a state of uneasy equilibrium between internal and external pressures. Uncertainties with respect to the outcome of a post-Brezhnew succession struggle in the Soviet Union could have serious repercussions upon Eastern Europe via transnational factional alignments. The Soviet leaders are likely to be sensitive to the dangers of contamination from centrifugal nationalism in Eastern Europe, as the multinational character of the Soviet state constitutes a source of potential vulnerability. By the year 2000 the Russians which now make up 53 per cent of the Soviet population will account for only 45 per cent, and the growing population of Central Asia will result in some 25 per cent of the Soviet population being of Moslem origin. Such concerns may constrain Soviet propensities to intervene in a post-Tito struggle between the major South Slav nationalities of Serbs, Croats, Slovenes, and Macedonians. The dangers to the European peace from external intervention into a situation of communal conflict in Yugoslavia can hardly be overrated. However, the incentives to avoid such a situation are likely to prove very strong indeed in the post-Tito phase of collective bargaining and management in the Yugoslav state. The need to establish international rules and commitments to non-intervention on the basis of the CSCE Final Act may have to be faced by the states of Europe and North America.

Should the pressure of Soviet hegemonic power ease in Eastern Europe 'traditional' conflicts over ethnic territories may re-emerge (Yugoslavia–Bulgaria over Macedonia, Hungary–Romania over Transylvania, etc.). The Soviet Union may exert pressure in order to improve its strategic position, by creating a corridor to Bulgaria through Romanian Dobruja or by establishing Adriatic naval facilities in Yugoslavia or Albania. The strength of such incentives will depend on the general direction of Soviet policy and the framework and content of East–West relations.

Intra-Western disputes may provide openings for the Soviet Union to exert pressure and influence via the political processes of the countries involved or by the demonstration or actual projection of military power. The calibration of moves and instrumentalities will, however, be extremely demanding as the volatility of such situations as the Greco-Turkish dispute over Cyprus and the continental shelf in the Aegean do not lend themselves easily to external manipulation. However, the objective of Soviet policy might focus on the disruption of NATO-links rather than on the establishment of new ones with Moscow.

A certain Third-Worldization of the politics of Mediterranean Europe is introducing an important element of instability into the security order of Europe in the wake of the diminished saliency of the East–West cold war. Southern Europe exhibits strong tensions and wide gaps between society and state, between the aspirations with respect to development and distribution of values in society and the policies pursued by governments. Conditions in Southern Europe do not reflect any single unifying trend, however. National diversities constitute effective limits and barriers to domino effects. Each country tends to respond more to internal factors than to the examples of neighbours.

Worst-case predictions based on the notion of an inexorable Marxist advance through the manipulation and power of Euro-Communist parties, have not been vindicated by actual events. In fact, the political reconstruction and integration of the Iberian countries of Portugal and Spain into the twentieth-century European reality has so far been a remarkable success. France has been rather resilient towards the option of a popular front control of the Fifth Republic. In Italy the situation remains tenuous, partly because of the incomplete renewal of the leadership élite and the absence of a social democratic alternative. But there is no single Euro-Communist challenge either. There is a variety of Euro-Communist parties trying to adapt to the social conditions in their respective countries and in the process becoming useless or unavailable as Soviet instrumentalities of power, and even dangerous in terms of their impact on Soviet orthodoxy and ideological legitimacy in Eastern Europe. Some countries, most notably Turkey, are in the midst of a serious identity crisis. The general tenor of the situation is such as to give primacy to internal politics as compared to external military threats. In the context of a significant Soviet military buildup that confluence may produce conditions in which the shadow of military force may unexpectedly come to weigh rather heavily on the volatile politics in Southern Europe. A tendency to denigrate the military aspects of the European balance and to ignore the objective changes taking place in the military situation is likely to persist.

In such a context NATO will need to develop credible means to block attempts at external military intervention – explicitly and implicitly – into the socio-political processes of the states in Southern Europe, to protect so to speak the internal preoccupations from some of the unpleasant trends in the external environment. It is important to strike a new balance between the logic of domestic political development on the one hand and the structure of the Western common defence on the other.

Access to raw materials has seldom if ever been a sufficient condition for war, nor has it been a necessary one. The scenario of the Soviet Union coalescing with a raw material suppliers' cartel to stop or curtail deliveries to the West is very hard to envisage. Real shortages are unlikely to emerge. Even assuming no substitution as a result of rising prices, no raw materials will be in a significantly shorter natural supply condition by 1990 than they are today. The only commodity which would seem to offer the objective opportunities for cartelization is bauxite where two-thirds of Western consumption is provided by Jamaica, Australia, Surinam, and Guyana – a group which is unlikely to become manipulable by Soviet fiat.

Oil is a special case with very low elasticities and substantial macro-economic effects. The supply of oil could become a potent weapon indeed if coupled to Soviet military power via political alignments concluded in the aftermath of radical *coups* in Iran and/or Saudi Arabia. That extreme scenario seems more likely than a Soviet attempt to disrupt the lines of supply by the application of naval power as an isolated, limited contingency. However, a Soviet forward policy with respect to radical regimes in Iran and Saudi Arabia would inevitably raise the danger of a general confrontation with the United States due to the extreme dependency of the West on Middle East oil. A good deterrent to such policies would seem to inhere in American determination to maintain essential equivalence in all the key areas of military power. In the context of a major war in Europe, efforts aimed at protection and interception of the lines of oil supply could constitute possible extensions of the conflict.

The Soviet Union is acquiring increased options for force projections into distant areas. Soviet capabilities will, however, for many years be inferior to those of the US. The propensity to exercise the option may not, however, parallel the distribution of capabilities. Intervention may, of course, be counter-productive, distort priorities, and tax will and resources. It is not necessarily encountered in the most efficient manner by counter-intervention — directly or indirectly. Pressure can be exerted through diplomatic means. Successful and unopposed intervention is likely, however, to affect perceptions with respect to the relative power and roles of the major powers in international society. The steady modernization of the Soviet navy into a potent force of global reach constitutes, perhaps the most visible demonstration of Soviet capabilities for force projection.[20]

In the contextual fluidity of the post-confrontation period of East–West relations, the contingency of indirect use of force as an accompaniment to diplomatic activity is likely to assume greater saliency. The shadow of force rather than its substance may come to weigh heavily on the peripheral areas of Western Europe where Soviet local power preponderance constitutes a permanent background factor. The availability of credible Western reinforcements will constitute a major requirement for purposes of containing Soviet indirect pressure. Attention has to be paid to demonstrations of countervailing power in peacetime and not be confined by cost-effectiveness considerations of how to carry out a given mission in wartime. This observation is particularly pertinent to the planning of Western naval force structures and operations.

The emerging new law of the sea is likely to contain substantial sources for potential conflicts over the distribution of resources, competing uses of the oceans, and jurisdictional delimitations. The most dramatic development is the international acceptance of the concept of the 200-mile exclusive economic zone, EEZ. Within such zones the coastal states will have sovereign rights 'for the purpose of exploring and exploiting, conserving, and managing the natural resources'. They will have the right to determine the allowable catch of fish and to grant other states access to the surplus. The EEZ is designed to protect the interests of coastal populations who will be given preferential rights with respect to access to their traditional sources of livelihood.

The concept of the EEZ is, however, problematical from the point of view of global equity. Its institution will amount to a virtual appropriation to national jurisdiction of over one-third of the last commons of mankind, a surface area roughly equivalent to the total landmass of the earth. Nearly half of the ocean which will be enclosed by EEZs will go to the high-income countries which have less than a quarter of the world's population. But the poor countries will be better protected by the institution of EEZs than by a prolongation of the past regimes of *laissez-faire*.

Maritime boundaries between adjacent and neighbouring states will require many years to settle, and quarrels over resource allocation in particular areas may be anticipated. In the absence of a global treaty, access to the minerals of the deep sea bed could also become an important source of international conflict. Disputes can arise between the maritime powers attempting to preserve freedom of navigation and coastal states interested in its regulation. Military activity and research on the continental shelf could be another area where the creeping jurisdiction of the littoral states may collide with the interests of the major powers. [21]

Obviously, it is possible to envisage combinations and escalations across the spectrum of potential future contingencies. The policy-planning problem is one of generating *options* rather than constructing alliance consensus about design *contingencies*. Many limited contingencies would affect the various members of NATO in different ways. They would be reluctant to lock themselves in by prior commitment to a certain response in a given set of less than extreme circumstances. The issues should be addressed parametrically, that is, based on an examination of things that may happen, alternative ways of dealing with them, and the consequences of dealing or not dealing with them in certain ways. The purpose is to identify, to the extent possible, generic capabilities for action that are *not* tied to specific contingencies.

The differential impact of various contingencies on the members of the alliance may, of course, strengthen their interest in the option of quitting at times and places of their own choosing. The extent of that 'danger' is probably inversely correlated with the availability of less than suicidal responses to aggression. Too many options may, of course, increase collective indecision. Deterrence considerations would seem to require that the members of the alliance that are most directly affected by a given challenge possess the means for effective response. Increased capability to act on their own may increase the chances that the smaller allies may be on their own in the event of a limited challenge. Hence, it is important also that the options be such as to establish boundaries to the conflict without deterring other members at the alliance from lending their support. The options must provide for flexibility in the availability and orchestration of individual and collective responses.

In the current context of increasing tension between state and society in industrial countries, which has brought into question the legitimacy of established institutions and the governability of democracies, the need for plausible responses to external dangers may become an important prerequisite for credible government. This is particularly the case in a situation of reduced acceptability of force as an instrument of policy in Western democracies. Reliance on an application of force that is clearly incommensurate with the challenge at hand, might cause alliance cohesion to disintegrate as the alliance could be viewed as irrelevant or even as a threat to national security during emergencies in Europe. Technology may become an important parameter in this connection.

19.5. The impact of technology

The roles within the arms competition of the two superpowers have changed fundamentally over the last decade. The Soviet Union has become the driving force behind the momentum of the competition. The level of Soviet expenditure on defence has reached about 14 per cent of GNP while the U.S. is spending slightly more than 5 per cent. In 1977 the estimated dollar costs of Soviet defence activities were roughly 40 per cent more than the U.S. outlay. During the 1970s Soviet defence spending in roubles has grown at a rate of 3-5 per cent per year. U.S. spending in real terms declined until 1977. Measured in dollar costs estimated Soviet military investments (procurement, construction, and RDT and E) are now about 75 per cent greater than those of the US.

In response to that challenge NATO has established the target of a 3 per cent annual increase in expenditures on defence measured in real terms. However, the current economic crisis may cause postponements and slippage. In the absence of agreements to curb the defence effort on both sides, the level of activity will rise in the years ahead.

In competition, NATO's major advantage inheres in its superior industrial and technological base. The industrial culture of the West, its flexibility, versatility, and innovative capacity is far superior to that of the Soviet Union which lacks a strong civil science and technology base.

In contrast, the structural and procedural weaknesses of the Soviet system has been described in the following terms: 'There is little flexibility in the centrally administered system; the supply system is unreliable; managers are confronted with a complex series of regulatory constraints and disincentives to innovate; and compartmentalization and discontinuities among the research, development and production phases create interface problems which inflate Soviet weapon costs, delay production and tend to reduce innovation, technical sophistication and performance'[22].

There are also some important differences in style between the United States and the Soviet Union with respect to testing, evaluation, and deployment which may inflate Western perceptions of the volume of the Soviet effort. Since the Russians proceed incrementally with the evolution of new systems from older ones there is a great deal of commonality of components. They typically field new systems in small quantities despite apparent deficiencies, and subsequently introduce follow-on modifications on newer models and upgrade the older ones. The United States attempts in contrast to eliminate most deficiencies during the R and D process and to introduce modifications before operational deployment. Thus the Russians tend to deploy earlier in the process than the United States.

In military technology the two superpowers are much closer than in civil technology, but the U.S. probably maintains an overall lead. The Russians cannot reap the synergistic effects which are possible in a richer and broader technological environment. What is more important is the fact of a clear American lead in the technologies which are particularly relevant to the development and production of precision guided munitions (PGM). PGMs involve the three separate technologies of target sensors, precision guidance, and warheads. They rely on miniaturization, large-scale integration, advanced sensors, digital computers and displays, and they draw on Western expertise in systems engineering.

PGM technology include sensors that are capable of 'seeing' targets on the battlefield day and night (FLIR-forward looking infra-red systems) and in all-weather (tactical SLAR-side looking airborne radar) conditions. PGMs can be delivered by artillery, missiles, aircraft, etc. New delivery systems like cruise missiles or RPVs (remotely piloted vehicles) involve technologies where the U.S. enjoys substantial comparative advantages at present such as light-weight guidance and propulsion technologies. The U.S. also maintains a significant lead in jet engines and avionics technologies.

It is perhaps too early to announce a revolution in the effectiveness of new weapons technologies (NWT), but we may very well be on the threshold of a quantum jump in effectiveness. What needs to be assessed, however, is the tactical, operational, and strategic implications of NWT. Here we are still moving in the dark. It is important also to remember the cost obstacles to rapid and broad-scale acquisition of PGM systems [23]. However, there can be no doubt of the force effectiveness multiplier resulting from the increased accuracy with which PGMs may be delivered on targets. PGMs may also permit conventional substitution for nuclear weapons in many missions due to their increased accuracy. The signalling effects of nuclear employment are however on a different level since they derive from the perceptions of danger involved in crossing the nuclear threshold. There are many conceptual and practical issues involved in signalling determination by the demonstrative use of nuclear weapons.

It is obviously too early to advance firm conclusions about the operational implications of PGMs. Certain tentative indications may, however, be useful as a baseline for long-range planning. In summary form they would include:

PGMs will result in an increased pace of war;
Conflicts will tend to become wars of equipment attrition rather than man-power attrition;
Local actors will become more dependent on outside resupplies;
The increased pace of war will require additional emphasis on warning time and readiness, tactical intelligence, reconnaissance and target acquisition, command and control, equipment repair capability, and rapid reinforcements;
PGMs permit decentralization of decision authority, but the pace of war may require centralized control over the process of escalation;
PGMs may permit, rapid *fait accompli* actions;
Prepositioned and mobile PGMs deployed across potential routes of enemy advance may increase the defensive capabilities even of smaller powers against superior invaders;

There will be strong incentives to disperse rather than concentrate military value in one position or vehicle;

Concealment, hardening, and deception will constitute important countermeasures (ATWs and MANPADs need protection);

There will be incentives to rely on small mobile units;

Countermeasures may focus on interfering with seeker and guidance systems;

PGMs could be effective terrorist weapons but require substantial training;

PGMs will reduce collateral damage in a war;

PGMs will raise the nuclear threshold and provide new conventional options.

It should be emphasized that these are but tentative hypotheses which need further analysis and verification. The options provided by NWT have to be examined from the point of view of cost-effectiveness. Fewer but more costly units for a given mission may produce a higher degree of effectiveness for a given cost than more numerous and less costly units. But there is, particularly for the small countries, a quantity/quality calculus to be kept in mind. The territory to be defended will pose certain minimum requirements with respect to numbers. Certain threshold numbers have to be reached in order to perform the mission in question at all and to permit a 'graceful degradation' in war. Here the issue of absolute costs is essential. Increased standardization and interoperability may reduce maintenance costs and improve overall efficiency. But the escalation of costs will pose more dramatic dilemmas for the smaller powers. They will not be able indefinitely to maintain balanced national defence structures. Certain missions will have to be dropped, based on low-performance systems, or incorporated into a multinational division of labour which would permit the co-operating countries to reap the economies of scale and specialization. Such developments would, however, imply a rather dramatic break with the symbols and rules of the system of sovereign states.

Table 19.1 Costs of post-war generations of fighters in the Royal Norwegian Air Force

Type	Vampire	F-86G Thunderjet	F-86F Sabrejet	F-104G Star-fighter	F-5A Freedom-fighter	F-16
Period	1948-55	1952-61	1957-66	1963-	1965-82	1980-
Peacetime operating costs (1978 kroner pr flight hour)	1450	1970	2060	6060	3630	8800*
Procurement cost (ten million kroner)	0.5	1.8	2.5	7.5	5.1	40.0
Procurement cost (1978 million kroner)	2.6	7.4	8.7	22.0	13.6	44.0
Relative procurement cost (fixed prices)	1	2.9	3.4	8.4	5.3	16.9

*Estimate.
Source: Ref. [24].

Table 19.1 depicts the costs of the post-war generations of fighters in the Royal Norwegian Air Force. The trends contained in Table 19.1 illustrate the general point above.

Thus we see that technology and its costs may impose a logic on decision-makers that competes with that embodied in the historical traditions of the modern European state system. We do not know how statesmen will resolve the emerging dilemmas, but they cannot be ignored or postponed indefinitely.

References

[1] For a more thorough discussion of the confidence-building process see Holst, J.J. and Melander, K.A. European security and confidence-building measures. *Survival* **19** (4) 146–54 (1977).

[2] For a fuller discussion of various *détente* strategies see Andrén, N. and Birnhaum, K.E. (Eds). *Beyond détente: prospects for East–West cooperation and security in Europe.* A.W. Sijthoff, Leyden (1976).

[3] The best books on arms control are still Bull, H. *The control of the arms race.* Fredrick A. Praeger, New York (1961, 1965); and Schelling, T.C. and Halperin, M.H. *Strategy and arms control.* The Twentieth Century Fund, New York (1961). Subsequent studies have brought few if any conceptual insights. There is a plethora of technical studies of specific issues of arms control. The best most recent general study is Barton, J.H. and Weiler, L.D. *International arms control: issues and agreements.* Stanford University Press, Stanford, California (1976).

[4] For a stimulating discussion see Wohlstetter, A. Racing forward or ambling back? *Survey* 3/4 163–217 (1976). See also Holst, J.J. What is really going on? *Foreign Policy* **19**, 155–62 (1975); and Gray, C.S. *The Soviet-American arms race.* D.C. Health, Lexington, Massachussetts (1976).

[5] The basic inquiries into the theory of nuclear deterrence include: Brodie, B. *Strategy in the missile age.* Princeton University Press, Princeton, New Jersey (1959); Kahn, H. *On thermonuclear war.* Princeton University Press, Princeton, New Jersey (1961); Schelling, T.C. *The strategy of conflict.* Harvard University Press, Cambridge, Massachussetts (1960); Schelling, T.C. *Arms and influence.* Yale University Press, New Haven, Connecticut (1961); Snyder, G.H. *Deterrence and defense.* Princeton University Press, Princeton, New Jersey (1961). The seminal systems studies were Wohlstetter, A.J., Hoffman, F.S., Lutz, R.J., and Rowen, H.S. *Selection and use of strategic air bases.* RAND, R–266 (April 1954) and Wohlstetter, A.J., Hoffman, F.S., and Rowen, H.S. *Protecting US power to strike back in the 1950's and 1960's.* RAND, R–290 (September 1956). A useful summary is found in Morgan, P.M. *Deterrence: a conceptual analysis.* Sage Publications, London (1977).

[6] For a discussion of the issues involved see Holst, J.J. and Schneider, W., Jr. *Why ABM? Policy issues in the missile defense controversy.* Pergamon Press, Elmsford, New York (1969).

[7] Brown, H. (Secretary of Defense). *Department of Defense annual report fiscal year 1979,* p. 124. (2 February 1978).

[8] Perry, W.J. (Under Secretary of Defense, Research, and Engineering). *The FY 1979 Department of Defense program for research, development, and acquisition,* p. V–30. (16 Febraury 1978).

[9] Disincentives to exercising the nuclear option are studied in Holst, J.J. (Ed.). *Security, order, and the bomb.* Universitetsforlaget, Oslo (1972).

[10] The most important recent studies of the nuclear proliferation problem are: *Report of the Nuclear Energy Study Group: nuclear issues and choices.* Sponsored by the Ford Foundation, administered by the MITRE Corporation. Ballinger, Cambridge, Massachussetts (1977). Greenwood T., Feiveson, H.A., and Taylor, T.B. *Nuclear proliferation, Motivations, capabilities, and strategies for control.* McGraw-Hill, New York (1977). Wohlstetter, A., Brown, T.A., Jones, G., McGarvey, D., Rowen, H., Taylor, V., and Wohlstetter, R. *Moving toward life in a nuclear armed crowd?* Pan Heuristics, Los Angeles, California (1976).

[11] *A dollar cost comparison of Soviet and US defense activities 1967-77*, p. 6. Central Intelligence Agency. National Foreign Assessment Centre, SR78-10002 (January 1978).

[12] Ref. [12], p. 62.

[13] Ref. [12], p. 62.

[14] Azrael, J.R., Löwenthal, R., and Nakogawa, T. *An overview of East–West relations*, p. 39. The Triangle Papers: 15. The Trilateral Commission, New York (1978).

[15] A moderate and well-argued defence of the RB/ER option is found in Carr, Bob. The neutron bomb: proceed, but with caution. *NATO's fifteen nations*, pp. 50–4. (December 1977/January 1978).

[16] For a good overview see Wolfe, T.W. *Soviet power and Europe 1945–1970.* The Johns Hopkins Press, Baltimore, Maryland (1970).

[17] *Statement on the defence estimates 1978*, p. 7 Cmnd 7099. Her Majesty's Stationery Office, London (1978).

[18] For some recent analyses of the maturing Soviet threat see Blechman, B.M., Berman, R.P., Binkin, M., Johnson, S.E., Weinland, R.G., and Young, F.W. *The Soviet military buildup and U.S. defense spending.* Brookings Institution, Washington, D.C. (1977); Record, J. *Sizing up the Soviet army.* Brookings Institution, Washington, D.C. (1975); Berman, R.P. *Soviet air power in transition.* Brookings Institution, Washington, D.C. (1978); Karber, P.A. *The tactical revolution in Soviet military doctrine.* The BDM Corporation (2 March 1977).

[19] Holst, J.J. *Var Forsvarspolitikk: Vurderinger og Utsyn*, Chapter A.7. Tiden Norsk Forlag, Oslo (1978).

[20] It is interesting to note that the Soviet presentation of the rationale for its maritime efforts reads like a modern version of Alfred Mahan. See Gorshkov, S.G. *Morskaya Moshch Gosudarstva.* Voenizdat, Moscow (1976). For a discussion of the competitive situation in the North Atlantic Ocean see Bertram, C. and Holst, J.J. (Eds). *New strategic factors in the North Atlantic.* Universitetsforlaget/IPC Science and Technology Press, Oslo/London (1977).

[21] For a general discussion of the issues see: Oda, S., Johnsten, D.M., Holst, J.J., Hollick, A.L., and Hardy, M. *A new regime for the oceans.* The Triange Papers: 9. The Trilateral Commission, New York (1976).

[22] Ref. [7], pp. 11–15.

[23] For further analyses of the emerging capabilities and their implications see Holst, J.J. and Nerlich, U. (Eds.). *Beyond nuclear deterrence: new aims, new arms.* Crane, Russak and Company, Inc., New York (1977); Digby, J.F. Precision guided weapons. *Adelphi Papers* No. 118 (1975); Burt, R. New weapons technologies: debate and directions. *Adelphi Papers* No. 126 (1976).

[24] *Forsvarskommisjonen av 1974*, NOU 1978: 9, p. 175. Universitetsforlaget, Oslo, 1978.

Commentary
Karl Kaiser

The following critique is in substantial agreement with the major points made by Johan Holst. It therefore enlarges on some of the points that he has made, and comments on some others. The analysis will proceed in three steps, first, by looking briefly at the central trends and issues of security and defence; second, by looking at some of the problems of research and technology in specific problem areas; and then, finally, by drawing a few conclusions for science policy and institutions.

Central trends and issues in security and defence policy?

In his analysis, Johan Holst has mentioned and discussed a number of such central trends. Viewed from a long-term perspective, the most important seems to be a process of expansion in the scope of security. What we mean by security today is quite different from what we meant some ten or twenty years ago. It has developed into a much greater problem than in the past. This process of expansion raises the important question whether the rules, institutions, and organization of security policy that were invented in the past are still adequate.

This re-interpretation and expansion of security started, to some extent, with the 'Harmel exercise' in the Alliance [1], when it decided that defence against aggression was its one great purpose, but that the attenuation of the roots of the conflict that could lead to aggression was the other major purpose of the Alliance. *Détente* and defence were regarded as the two sides of the same coin and ever since then, the Alliance has gone through a process of expanding its purpose. NATO is no longer a mere defence organization, but a security organization.

We witness an expansion of security issues into many areas. They contain realms that were not perceived as relevant to security ten or twenty years ago. The ten contingencies for future conflict, elaborated by Johan Holst, contain seven 'non-classical' cases, such as resource problems, supply lines, ocean issues, etc. The analysis by Brooks and Skolnikoff in Chapter 18 produces another range of issues where we can see that any narrow conception of security is obsolete and that some of these issues are actually of a global nature. Consequently, the institutions that we have created in the past are, of necessity, too narrow to deal with some of these questions. Whether they relate to subversion or economic security, disputes on the oceans or environmental dangers, they go beyond the traditional scope of the institutions that the Alliance built up in the past. Unlike the narrower area of defence, solutions in these fields require co-operation between very divergent countries, including even non-members of the Alliance. Inevitably, such a task is inherently more difficult than defence co-operation among allies united by certain common goals. In fact, security policy in some of the new areas, such as security of supply of important raw materials, sometimes requires co-operation with countries that are adversaries in the other areas. [2]

Consequently, this expansion of the scope of security raises different and often difficult problems of recognizing what the issues are and of accommodating the new issues, either within the existing framework of security organization, or of finding in both politics and in research the appropriate new institutional response or the right kind of research organization. Often a type of co-operation between institutions and countries may have to be established that differs substantially from what exists in the Alliance.

Research and technology in specific problem areas

Johan Holst has identified a multitude of areas that are of growing importance for research, and to which we have to address ourselves. I cannot deal with all of them, but I would like to select a few that seem to be particularly important.

In the area of defence technology, he mentions precision guided weapons as one important area for future research and work. What appears striking in this respect — and I find myself in agreement with the various propositions he makes — is that this field seems to be typical of many contemporary problems, for it is not the technology itself which poses the most important problem for scientific work, but the assessment of its impact. With these weapons, the old lines of division between the tactical and the strategic, the conventional and the nuclear, become blurred. It becomes difficult to assess what they really mean for the armed forces structure, for the nature of war, or for political stability. The same is true for some of the other areas that he mentioned, notably three of them: the expansion of nuclear energy and its impact on future proliferation; the change in the law of the sea, leading to a different ocean regime which might potentially become a major source of conflict in the twenty-first century; and the confidence-building measures, where East and West have made some modest steps to introduce more calculability and transparency of behaviour between the two Alliance systems.

Now all of these areas, though rather different, have one thing in common. If one wants to deal with them, one has to bridge the gap between very different realms, both institutionally in the existing political organization, and from the point of view of disciplines which, in the field of science and research, deal with them. If, for example, in the field of nuclear energy one wants to discuss meaningfully the question of proliferation, types of expertise have to be brought together which are normally not united in a person or in an institute. These are challenges for which we are badly prepared, but to which we have to address ourselves.

In connection with the expansion of the scope of security we have to ask ourselves where the 'defence of freedom', to use a classical term, stands today. Traditionally we have meant by that notion defence against aggression, from outside the Alliance, but as time went by, the defence of freedom increasingly assumed the additional meaning of relating to the survival of functioning societies with freedom of choice, justice, and freedom from fear. If we were to conduct a poll among the contributors to this volume asking for the probability of possible conflicts today, in all likelihood a majority would suggest that it does not lie in direct military aggression, but that it is the defence of freedom in the sense of a survival of a functioning free society where the threat and the probability of conflict is greatest.

If this proposition is reasonable, it means that the West has to deal with an additional set of problems that cannot be handled and solved by soldiers or military technology, but by a whole range of actors such as social reformers or economic advisers, and by effective measures of security of supply, or successful management of interdependence. In this sphere of security policy, therefore, we have to bring together a multitude of rather different specialities, and probably pursue such an approach in a somewhat changed institutional framework and with different research organizations. The countries of North America and Western Europe have an increasing array of social and societal issues as the major problem of the Alliance.

Finally, the Harmel exercise of the Alliance has, as mentioned already, resulted in the important conclusion that in addition to dealing with outside aggression through adequate defence, the Alliance should, as its major second purpose, deal with the roots of potential aggression by addressing the origins of the conflict. By creating new mutual dependencies between the two Alliances, by stabilizing the conflict, by introducing calculability and transparency in political behaviour, both sides can attempt to lower tension and contribute to what is generally called *détente*. If human history is a guide to the future, it can be argued that if NATO remained a mere defence organization confining its activities to military measures the end result on both sides is likely to be a catastrophy.

It is, therefore, for good reasons that the Alliance has decided to pursue defence and *détente* jointly. But such an attempt is inherently difficult, since it constantly has to reconcile conflict and co-operation with the same adversary, who also is a partner. There is no precedent in history for this kind of undertaking on such a scale. Politicians, the public, or the media are badly prepared for such a complex and difficult exercise, which attempts to reduce conflict between systems that have profound and deeply rooted differences, which cannot be reduced rapidly and which will take many years to achieve substantial progress in bridging the gap. It is in this area where the social sciences have a major and a particularly difficult task in helping the Alliance to make sure that its double purpose of pursuing defence and *détente* at the same time can be effectively fulfilled.

Conclusions

The limits of space permit only a few conclusions for science policy and organization. If the preceding analysis is correct, then the first and main conclusion for a NATO science policy that is devoted to security should be to shift the emphasis of support from specific research areas to links and bridges between them. This is not to say that that has not been the case in the past. Indeed, some of the contributions presented in this volume are evidence for this. I am suggesting only a shift of emphasis from specific research to joint research, from specialized work to creating bridges and links between different areas. For example, in the case of non-proliferation, research can only produce meaningful and applicable results if the social sciences and the nuclear sciences are brought together.

A second conclusion with regard to research on new technologies appears to suggest itself. The emphasis of research should be put more than in the past on the problems of application. This requires a greater involvement of non-military and of researchers outside the scientific areas traditionally associated with the defence community. In the past, natural scientists and engineering specialists have conducted this kind of research and rarely expanded it, as seems to be desirable, to the social sciences. Now, not all social scientists are interested only in Marx and Freud, as some critics suggest. If challenged to an investigation of important issues, most are likely to respond. If the defence and foreign policy community is willing to accept their sometimes uncomfortable results, more can come forth from the social scientists if they are actually asked and involved in research on security. Governments can influence this process, as can NATO as a whole, by shaping priorities accordingly and by expanding the scope of research to these fields.

My third and final conclusion is coloured by my background as a social scientist. Given the expansion of the security problem to new areas which will be decisive for the survival of our free societies, there is a good case for involving social scientists in research on security much more than in the past. This should be done as much as possible by joining projects and by co-operating with other sciences relevant to the issues under consideration. These thoughts lead me to conclude that the main issue is probably the transfer problem. The allocation of resources for research on science policy should be less concerned with pushing specific scientific fields to great depth and further sophistication, because we are not doing so badly, particularly if we compare out state of affairs with that of the Soviet Union. Instead, science policy should shift some of these resources to the problems of application to practical issues, to research pursued jointly in different disciplines, and to work on problems which cut across the traditional lines of division of our research and development organization.

References

[1] The Harmel exercise was a political study undertaken by NATO and published as *Report of the Council on future tasks for the Alliance.* NATO, Brussels (1967).
[2] Some of these problems were examined in greater detail by a Study Group of the Research Institute of the German Society for Foreign Affairs, published in Kaiser, K. and Kreis, K.M. (Eds.) *Sicherheitspolitik vor neuen Aufgaben.* Alfred Metzner Verlag, Frankfurt (1976).

20

The third phase of NATO

Harlan Cleveland

20.1. Introduction

Among the more than a thousand standing international committees in which the United States Government is represented, the NATO Science Committee has long been an outstanding example, both of courage and of self-starting initiative, as it develops the brains of the Alliance – without (surprisingly to most people outside NATO) insisting that their research fields be closely tied to the NATO Defence Programme.

I think it is fair to think of NATO as having three phases, of which we are at the beginning of the third. The first was getting defence and deterrence organized and establishing an astonishing stability with essentially unusable weapons, despite the technological dynamism of the military inventors. A second phase was the beginnings of a caucus on how to make peace with the Russians – SALT, MBFR, Helsinki, and the rest. At one point in the negotiations about the Harmel Report I tried to introduce into the text my own favourite definition of *détente* as 'the continuation of tension by other means'! I still think that is a properly hard headed way to think about *détente*.

And now we come to what Karl Kaiser in his commentary in Chapter 19 called a time when 'security problems are more than anything else the product of how we govern ourselves' – and how we govern ourselves in accordance with what Minister Judd called 'a cause' [1]. We needed this golden moment, when all the nations of the Alliance are under democratic rule. The leaders of the Alliance, in this room and elsewhere, will have to go where the peoples of the Atlantic nations want to go.

If I were my former colleague and Dean, Ambassador André De Staerke, I would probably start with a quotation from St. Thomas Aquinas. As it is, we will have to make do with Mao Tse-Tung. In one of his more trenchant thoughts in that little Red Book, Mao Tse-Tung explains about leaders following the followers. 'Our comrades' he says, 'must not assume that the masses have no understanding of what they themselves do not yet understand. It often happens that the masses outstrip us and are eager to advance a step and that nevertheless our comrades failed to act as leaders.' Then, translated into American, Mao goes on to say to the cadre (or leaders), 'Look, you cadres, don't you get the idea that you are making the policy. It is the people at large who make the policy. Your task is to sniff around

uess at the sense of direction at which things are already moving, and *then* do your
hing, which is quantify, programme, and organize to reach the goals to which the
eople in their inchoate way are already pointing. After that', he says, 'you can act,
ut always test the action as it goes along by going back to the people to get their
eaction.' This cyclical process, says Mao, 'over and over again in an endless spiral
vith the ideas becoming more vital and richer each time, is the correct theory of
nowledge.'

On the basis of some experience in China, I very much doubt that this describes
ow decisions are made there these days. But it is not a bad description of how
hanges in policy get made, in 1978, in the nations of the North Atlantic Alliance.
o let's think not about where the experts are pointing. Let's consider the direction
a which the NATO masses are already heading.

Because time does not permit I will skip, with some reluctance, a discussion of
ne North and South issues – even though North–South issues, by and large, are
ow driving East–West relations and Atlantic and Pacific relations. We are not
aving our confrontations with the Soviets these days in Berlin; we are having them
a Angola and the Horn of Africa. Our most difficult issues across the Atlantic
robably have to do with oil from the Middle East. But I shall skip the South be-
ause I want to focus on the governance of the science-based 'technology societies'
f the Atlantic Community.

0.2. Crisis in governance

he central problem is clear enough. It is not 'limits to growth', it is limits to
overnment. Let's not fudge the facts: in the industrial democracies, being deve-
•ped has come to mean a chronic crisis of governance – our leaders baffled by
rban dirt, danger, and disaffection; our young people educated for non-existent
•bs; our middle classes squeezed by inflation and harassed by bureaucracy; our
urms and factories hosting an enormous migratory proletariat; our economies
ritically reliant on Middle Eastern oil for their swollen energy needs; our govern-
tents revolving in endless and ineffective coalitions.

The Marxist industrial societies are no better off. They, too, face a chronic crisis
f governance. They feel unable to plan for the succession of power. They depend
n military force for internal stability and international clout. They cannot main-
ain momentum for their kind of revolution in the developing world. And they are
nable, in the case of the Soviet Union, even to feed their own people without
aning on American farmers. It is not only in the poor countries that things are
tanaged poorly in our interdependent world.

And yet the yeast is rising. In every industrial nation a large number of people,
ften beginning with young people, have started to do some rethinking of growth
atterns. In the United States, for example, the size of families has declined to
out the population replacement level. The historic trek from urban to rural
abitats has slowed down, levelled off, and then gone into reverse. The ecological
hic in its many manifestations has started to make its influence felt in the market-
lace and in politics. A revolution has begun in the roles and status of ethnic

minorities and the female half of the population. Local communities insist on gaining more control over their own growth.

What is going on here? My ancestors, like yours, were among those 'curious apes', so I have tried to define the nature of the transition we already seem to be in.

Where we are moving *from* is easy. We are moving from an ethic of indiscriminate growth as the central organizing principle – growth unfairly distributed, growth wasteful of our resources and damaging to our surroundings, growth that neglects needs, growth preoccupied with the supply not the requirements side of the equation, growth symbolized by a GNP that includes weapons, traffic jams and drug abuse, but leaves out housework, week-end work, or environmental improvement.

We know what we are coming *through*. We have just come through a dark tunnel, a tunnel of love in which we flirted, briefly, with 'no growth'. This is very familiar ground to this audience, and I will not beat it to death. I will remind you only of the paragraph in the *Limits to growth* study that was the most neglected and the least read, the paragraph that started with the words 'the final, most elusive and most important information we need deals with human values'.

I do not, then, have to argue the first two steps of this Hegelian dialectic. The question I want to address is this: if indiscriminate growth is for the birds, and if 'no growth' is unpopular and unnecessary, what do we do next? 'No growth' did you will agree, turn out to be both unpopular and unnecessary. Professor Ashby (Chapter 2) has told us that, what with substitutions and industrial flexibility, we are not unduly dependent on any one raw material (except, at the moment, on petroleum). And, as if to test the popularity of 'no growth', we even tried it. We had a real-life empirical experiment called the Great Recession of the mid-1970s. Nobody seemed to like it – not the developing countries which saw their export markets dry up and their raw material prices swing low, not the industrial countries which saw inflation and unemployment fuse together at the same end of a new kind of business cycle.

The last time I studied Keynesian economics was at Oxford with a young don named Harold Wilson. He taught me something that I remember very well: he taught me that business economics is cyclical, but that one thing you can depend on is that the recession will be at one end of the cycle and the inflation will be at the other end. That did not prove to be a particularly helpful piece of information either to Harold or to me.

If neither 'growth' nor 'no growth' is acceptable as an organizing principle for modern industrial society, then we will have to move – indeed, we are already moving – toward some kind of selective growth, purposeful growth, growth (in Schumacher's phrase) 'as if people mattered'. For convenience in discussing this thought at length, my co-author and I baptized this emerging ethic with a made-up name 'humangrowth' [2]. I take that to be something like what Professor Danzin in Chapter 14 means when, in describing European destiny, he spoke of a 'nouvelle croissance' – a 'newgrowth'.

20.3. A new growth ethic focused on human development

What are the elements of this emerging ethic?

The nature of the new growth ethic that focuses on human development can only be described from scattered evidence spotted during our present transition. But that evidence strongly suggests:

a new scepticism of science and technology;

a new emphasis on ecological causes and effects;

a new concept of information as a basic resource;

a new willingness to think in global perspective and longer time-frames;

a new surge in life-long 'continuing education';

a new insistence that jobs be available and that work be interesting;

a new turn from quantitative to qualitative criteria as measures of personal well-being; and

new styles in governance for societies with less power at the 'top' and more kinds of people involved in planning and doing.

The *first* part of the transition is a new scepticism of science and technology. Science is now regarded as too important to be left to the scientists, and the technologies produced from their insights too morally ambiguous to entrust to the engineers.

The tale of the Manhattan project is still a warning in living memory. That secret laboratory that produced the first atomic bombs did not have on its staff a single person whose assignment was to think about the implications of the project if it were to succeed. (I checked this point with Professor Rabi who corrects me only to the point of saying that he and Dr. Oppenheimer used to discuss the subject at lunch.) Systematic study of the 'software' — that is, nuclear arms control — began only after the hardware was manufactured. We have been playing catch-up ever since.

The ethic of indiscriminate growth was associated with the illusion that science would always think up something to offset whatever damage was done. Then, when it appeared that a 'technological fix' wasn't available, and the social fall-out of science was sometimes severe, critics arose to damn technology as the main reason for the human predicament.

What seems to be emerging now is a more balanced reaction. It assumes an indispensable role for science and technology in purposeful growth, but does not look to scientists and technologists to decide for themselves what the social purposes and applications of their discoveries should be.

We are educating ourselves to consider the options, not the inevitabilities, of science. Until recently the popular assumption was that if we could invent something we had better manufacture it. But now the agreements with the Soviets not to build ABM systems, the decision (despite the Concorde) not to build an American supersonic transport, the popular reaction against nuclear power, and the hold up on development of the B-1 and the neutron artillery shell, are all straws in a new wind. And something analogous is happening in basic scientific research: the cautious approach to recombinant DNA and weather modification bear witness.

'Il faut vouloir les consequences de ce qu'on veut.' We have learned that we must not only want what we want, we must also want what it leads to.

I was witness to the sceptical mood recently when an expert on genetic manipulation – that was his word, not mine – told us that the hazards were small and warned us against 'irrational hysteria'. My colleague Ambassador DeRose, leaned over and whispered, 'Are you thoroughly reassured?' 'Well,' I whispered back, 'I feel just the way I would if I heard a political scientist say that the hazards of political manipulation are small.'

It is now taken for granted, I would argue, that institutions should be built to contain, channel, and control new technologies *while they are being developed*, that the human purposes should be clarified and the social constraints should mature *in parallel* with the hardware, not be spliced in later on.

A *second* aspect of the transition we are in is the development of an ethic of ecology – a new emphasis on ecological cause and effect.

The modern environmental movement earned its first spurs by colliding head on with economic growth. But that was before growth meant more than production – and before 'environment' meant more than the protection of natural systems.

Three unexpected changes in public perception, three elements of the transition we are in, may now bring growth and environment into converging paths – in that wonderful Italian phrase of former Prime Minister Moro, the convergence of parallel lines, *paralleli convergenti*. One of these is the mutation to a more purposeful growth ethic, partly under the pressure of the organized environmentalists.

Another mutation is occurring in the very idea of stewardship: it started with the aim of protecting nature from people and is now broadened to encompass the protection of people from themselves. The Declaration of Stockholm, in 1972, called it the 'human environment' – not only the natural environment or biosphere, but also the man-made environment or technosphere. Yet another change, slower in coming, is the shift within the environmental movement from the no-saying ethic of protecting nature toward a doctrine of enhancing the whole human environment by prudent affirmative action – not only what Peccei and Mesarovic in Chapter 13 called 'earth-keeping' but something more than that.

The emerging ethic of ecology can thus be seen as an interlocked system of socially determined limits. They are not 'limits to growth'. They are limits to environmental damage, to wasteful use of resources, to thoughtless behaviour, to social injustice and to destructive conflict.

Third, there is a new concept of information as a basic resource. We have dealt at some length here, during the past two days, with what Professor Danzin in Chapter 14 calls 'a civilization of knowledge'.

Information is increasingly recognized as a new kind of resource. And thinking of information as a basic resource blows away many of the earlier assumptions about the very nature of economic growth. It places the determination of the purposes of growth directly in our hands, not in 'invisible hands' or 'the market' or 'the proletariat' or any other formulation that avoids the individual's moral responsibility for the social effects of his or her actions. But in consequence an information society must depend more heavily on educated people, many of them educated in new ways for functions never before performed.

The *fourth* postulate is a new willingness to think in global perspective and longer time-frames. The essential paradox of modernization is clear enough. Some years ago, Sir Isaiah Berlin in one of his conversations with Henry Brandon described it in five luminous sentences.

As knowledge becomes more and more specialised, the fewer are the persons who know enough about everything to be wholly in charge. One of the paradoxical consequences is therefore the dependence of a large number of human beings upon a collection of ill-coordinated experts each of whom, sooner or later, becomes oppressed or irritated by being unable to step out of his box and survey the relationship of his particular activity to the whole. The experts cannot know enough. The coordinators always did move in the dark but now they are aware of it, and the more honest and intelligent ones are rightly frightened by the fact that their responsibility increases in direct ratio to their ignorance of an ever-expanding field.

I heard a scientist saying, 'Because of the limitations of the human mind, we cannot take everything into account.' What is this we are hearing? Is this a statement of scientific fact? Or of scientific timidity? As John Maddox in Chapter 1 said, 'Would we have stopped the Portuguese explorations?' It was only a few years ago that a television camera aboard *Apollo 8* provided visual confirmation for hundreds of millions of us of what we previously grasped only intellectually, the planet Earth was small, lonely, unitary, finite, vulnerable. Every night we look at the television and see that synoptic view of the world in the weather pictures. The lengthening agenda of planetary problems, as masterfully summarized by Brooks and Skolnikoff in Chapter 18 will fit into no other perspective than the now-familiar notion that the world is round.

Nor will they fit into short time-frames; this, too, is becoming better understood, and has been another theme of these volumes. There is a well advertised reluctance on the part of elected officials to take unpopular positions in the interest of future benefits; yet the Joint Economic Committee of the United States Congress recently completed a two-year study of long-term economic growth prospects, and after commissioning 41 research papers, examining more than 100 other studies, and holding extended hearings, its final staff report stresses the imperative need for decision-making on the basis of long-term goals. Most of the chapters in these volumes do likewise.

A *fifth* part of the transition we are in is the new surge in life-long continuing education – again a theme that has been dominant in these volumes. If in 'a civilization of knowledge' there is likely to be a growing demand for people who can 'get it all together', then the most functional forms of education may turn out to be not the vertical patterns of specialization, but the creation over a lifetime of greater sensitivity to the interconnectedness of things. When I was managing a university system, I noticed that we had a number of interdisciplinary courses in the catalogue. When I looked more closely, I found that most of them were 'team taught' – that is the academic euphemism for an arrangement by which each professor teaches his own discipline and the students are expected to be interdisciplinary. I even complained about this in a meeting of Deans one day, and the Dean of our medical school said, 'Harlan, don't take it so hard. It is the same all over. In a modern urban hospital, the only generalist left is the patient.'

Sixth, there is a new insistence that jobs be available and also that work be interesting. In our own country we have been declaring one-fourteenth of our work-force surplus to our national requirements for several years now. Both the employers and the employee organizations have dropped the ball – and bounced it to the government.

I am, I must confess, not impressed with the likelihood that creating enough jobs and interesting work is going to be accomplished through bilateral discussions between management and trade unions – leaving out all the other organizations and interests that also have a stake in the outcome. I know there is a lot of enthusiasm in Europe for worker participation in boards of directors, but as a consumer I cannot get enthusiastic about the development of business–labour cabals dedicated to raising prices for the rest of us to pay.

There are probably a lot more unemployed than any of our labour-statistics people are willing to admit. One of our best pollsters in the United States, Daniel Yankelovich, now claims that if you count all the people who would take a job if it became available as our 'true unemployment' then the figure in the United States is more like 24 to 27 per cent, rather than 6 or 7 per cent of the labour force as calculated in the traditional way.

Something can be done about this. The staff of one committee of the US Congress has been suggesting that if everybody took one year off in seven it would help and it is true that the job opportunities thus opened up would almost exactly match the number of Americans that were looking for jobs in the early winter of 1977. That would not solve the problem: most of the jobless would not qualify for the jobs that opened up. But it would be a start on creating an economy – which should surely be our aim – in which people are scarcer than work.

A *seventh* transition is the new turn from quantitative to qualitative criteria as measures of personal well-being. The effect of putting aside the old view of growth is to liberate mind and spirit from the more strictly economic cost-benefit analysis as the controlling criterion for public policy, from the 'bottom line' as the overriding measure of entrepreneurial success, and generally from a primarily economic view of the needs and wants of mankind.

The evidence of this shift is accumulating fast. The search is on, for example, for new ways to express personal and organizational objectives and new ways to measure progress toward them, in terms that place human beings, not averages or institutions, in the centre of the picture. That is the meaning of the shift away from GNP toward more complex social indicators than quality-of-life yardsticks; away from average consumption *per capita* toward individual and group entitlements; away from measuring the health of the health delivery system toward measuring the health of people; away from educational standards expressed in years of schooling, toward standards of relevant educational accomplishment.

20.4. New styles of governance needed to implement the new growth ethic

We can now come closer to seeing, as a whole, the transition we are in. We are moving toward a scientifically responsible, ecologically aware, information-rich, option-maximizing, pluralistic society whose citizens work as a matter of right and also as a matter of right, gain life-long learning in a global and future-oriented prospective.

Can we adapt to such a society? Or do we have what Peccei and Mesarovic (Chapter 13) fear may be an inherent incapacity to adapt? I think we can adapt — indeed, that in ways I have suggested we are already adapting. How much of our brain are we already using? Not very much yet.

But man's capacity to adapt *does* require constant stimulation — and in stimulating us all, Peccei and Mesarovic have of course performed a signal service. The computerized simulations are impressive, but the stimulation cannot always be by the numbers. IBM's chief scientist, Lewis Branscomb, in his commentary to Danzin's contribution defined common sense as 'the models we carry in our heads'. We have to learn the numbers but not act by them; we have to learn to take them or leave them, we have to learn to see them as only a pale reflection of the complex images those models in our heads already contain.

An *eighth* part of the transition we are in — and the last one I will mention here — is the active search for new styles of governance — new ways to go, as John Maddox put it, from A to B. That is the missing chapter in the futurist literature these days. You read these books about how bad it's all going to be and toward the end the author says we simply must do this, we have to do that, and you turn the page to find who 'we' are: instead of addressing the 'how to get there' question, the author has gratefully written 'The End'.

Who *is* the 'we' and what do 'we' do? We've had a discussion here of the tools that 'we' can use — the computers and their limitations, the adversary process, and even the Science Court which I thought had been buried and was again buried by Brooks and Skolnikoff in their contribution (Chapter 18).

At a small meeting in New York recently, Henry Kissinger was ruminating on his experience in government. He commented that the only questions that get to the President are disagreements. And these disagreements tend to be settled, he said, by an adversary proceeding which leaves no room for the possibility that both sides of the argument are wrong and the probability that a compromise between them — a quite likely outcome, for the President naturally values the morale of his subordinates — will be the worst outcome of all.

In sum: a tidal wave of change of values is well under way, and the main obstacle to converting these new values into policies and institutions is not the limits to physical resources or the limits to physical resources but the limits to government.

In one sense these limits have been the central issue of political theory ever since there has been political theory, but the issue has usually been objections to governing authorities that were viewed as having too much control over people and events. The condition we now face, I think, is the reverse — the apparent incapacity of existing political institutions, whatever their ideological colour, to manage unprecedented complexity.

Political leaders keep up a brave front, but their incapacity for decision-making is becoming more and more visible to the rest of us. Central economic planning, popularized around the world partly by industrial democracies who won't touch it with a ten-foot pole themselves, is nearly everywhere in disarray. The new migratory proletariat streams across national frontiers whether national immigration laws permit it or not. Ethnic and religious rivalries and subnational separatists threaten the integrity of long-established nations: South Africa, Nigeria, Ethiopia, Jordan, Lebanon, the United Kingdom, and Canada are only the most current examples. Power is leaking out of national governments in three directions: to local communities seeking more discretion, to non-government enterprise that can do things so much faster and more flexibly than governments can, and to international agencies which must attempt somehow to manage new technologies that transcend national jurisdictions.

The institutions of government, in short, are left over from the era for which they were designed — the era of undifferentiated growth in which the many different kinds of growth did not have to relate to each other. Watching my own government and other governments, I have come to a disturbing conclusion: when it comes to governance there is one thing worse than doing bad things on purpose, and that is doing good things but not relating them to each other.

20.5. Conclusions

In conclusion, two quick suggestions for and about the NATO Science Committee, which, like the rest of us, will have to adapt to the transition we are in.

If our greatest shortage is of situation-as-a-whole people, why don't we use NATO fellowships to 'bribe' more integrators and not bribe only the best of the specialists. I mean 'integrators', not 'generalists'. I mean people who are graduating from quality work in a specialty to face the ambiguities and puzzlements of getting it all together. Somebody, in speaking of the hundred thousand people or so who have passed through the NATO Science Programme, spoke of the need for 're-cycling the alumni'. Maybe we should have a programme focused on integrative thinking, futures research and risk assessment, designed for people who have already passed through the NATO Science Programme once, have become first-rate specialists, but are now graduating to general management, or broader policy planning. The NATO Science Programme is, as Dr Rabi has said, a kind of university. Like the other universities in our societies, it is going to have to move more rapidly to offer opportunities for lifelong education.

My other suggestion is this: if the industrial democracies are in trouble because they are not yet wrapping humanizing institutions around runaway technologies, why don't the NATO Council and the Secretary General bring the NATO Science Committee and the people it can attract to its work (of which the conference that preceded this book has been an impressive example) into the mainstream of thinking about the Alliance — about R and D strategies, for example. The Alliance continues to depend on how we deploy our guns, but will also be more and more dependent on how we deploy our brains. More policy-planning input from the

NATO Science Committee – as has been advocated by the leader of the American delegation, Frank Press. [3]

I have argued that we are already in transition to a new ethic of purposeful growth. This new ethic is not yet splashing into the headlines. It may even have a hard time elbowing its way onto the agenda of the North Atlantic Council. But great ideas are never noisy on arrival. They slip into our minds unannounced. Remember Albert Camus: 'Great ideas', he said, in a lecture in Rome in 1954, 'come into the world on doves' feet. If we listen closely we will distinguish amidst the empires and nations, the gentle whisper of life and hope.'

References

[1] Judd, F. In *Statements and declarations: the NATO Science Committee Twentieth Anniversary Commemoration Conference,* p. 21. NATO, Brussels (1978).

[2] Cleveland, H. and Wilson, T.W. *Humangrowth.* Aspen Institute for Humanistic Studies, Princeton, N.J. (1978).

[3] Press, F. In *Statements and declarations: the NATO Science Committee Twentieth Anniversary Commemoration Conference,* p. 25. NATO, Brussels (1978).

21

Problèmes socio-politiques et culturels d'importance majeure pour les pays occidentaux posés par les progrès scientifiques et technologiques

François de Rose

Si j'ai finalement accepté, après une longue hésitation, de contribuer à cette investigation ce n'est pas que j'aie la prétention de traiter le sujet qui m'est imparti. Tous les documents qui ont été produits pour ce livre, si nombreux et si compétents qu'ils soient, ne peuvent avoir l'ambition d'être complets et exhaustifs encore moins sans doute d'apporter des réponses à toutes les interrogations qu'ils provoquent.

Ce qui m'a déterminé c'est le souvenir d'une réflexion de Robert Oppenheimer qui, venant de cet esprit supérieur, m'a toujours paru de nature à donner du courage au commun des mortels: «What we don't understand,» disait-il, «we explain to one another» («Ce que nous ne comprenons pas, nous nous l'expliquons les uns aux autres»).

Admirable confidence d'un homme de science qui éclaire la démarche du progrès scientifique en posant pour base l'humble aveu de l'ignorance et cherche à en réduire l'étendue par la confrontation des idées, des interrogations, des intuitions, des hypothèses et naturellement des erreurs.

C'est donc dans cet état d'esprit que j'aborderai aujourd'hui le problème des interactions entre science, technologie et société en m'efforçant de le voir sous son aspect politique au sens large du terme. Conscient du risque que je prends de naviguer entre l'enigme et le lieu commun, si j'ose prendre la parole c'est dans l'espoir que sur un tel sujet il n'y ait que des degrés dans la non-compréhension ou mieux la non-appréhension des problèmes.

Un autre sentiment s'impose à moi et, me semble-t-il, peut être partagé par d'autres. C'est celui de la fierté mais aussi du sens de la responsabilité qui est celle des Occidentaux et, pour être plus précis, des Européens pour avoir donné au monde la science moderne.

Pourquoi la science est-elle née en Europe? La réponse est difficile. Mais, personnellement, j'ai toujours trouvé une certaine force à l'idée que cette naissance ne pouvait intervenir que dans un milieu qui, sur le plan spirituel, rejetait la prédestination. Dès lors que dans les religions judéo-chrétiennes la destinée de l'homme dans l'au-delà n'était pas décidée de toute éternité mais qu'il en était le maître, il se trouvait dans un système qui donnait un motif et une signification aux efforts qu'il pouvait faire pour s'élever au-dessus de lui-même et dominer sa condition, un système fondé sur l'espoir.

Mais la science, contrairement aux autres composantes de la culture, s'est trouvée parfaitement transférable aux hommes issus d'autres civilisations. Si l'art, dans la plupart de ses formes, est lié à certaines zones géographiques dont il est originaire, il n'en va pas de même de la science qui peut être assimilée en tous les points de la planète et qui bénéficie des apports de chercheurs venus des quatre coins de l'horizon. En présence des problèmes que nous pose l'impact de la science et de la technologie sur nos sociétés, cette universalité de la communication scientifique peut être l'un des atouts qui nous aideront à les résoudre.

Pour clarifier le débat, je me propose de mentionner quelques-uns des problèmes politiques, sociaux et culturels résultant des développements scientifiques et technologiques et cela à deux niveaux différents: le niveau mondial et celui de nos sociétés occidentales. N'ayant pas compétence pour décrire ou analyser ces développements scientifiques ou technologiques, je les prendrai tels qu'ils apparaissent dans leurs conséquences à qui n'a ni les connaissances de l'expert ni les vues du philosophe, mais se sent concerné pour lui-même et pour ceux qui viendront après lui.

21.1. Problèmes planétaires

Peut-être la tentative de prospective à vingt ans d'échéance est-elle trop courte à l'égard du phénomène qui transcende tous les autres. Je veux dire l'explosion démographique, conséquence s'il en est des développements scientifique et technique qui ont apporté l'hygiène, la médecine, la chute de la mortalité infantile, la lutte contre les épidémies, l'allongement de la durée de la vie. S'il est vrai que dans quarante ou cinquante ans la terre peut compter huit ou dix milliards d'habitants, nous sommes dès maintenant tenus par la nécessité de répondre à deux questions: comment les nourrir et comment leur donner du travail?

Quarante ans! Le temps qui s'est écoulé depuis le début de la dernière guerre. Nombre d'entre nous seront encore de ce monde, nos enfants à coup sûr et nos petits-enfants seront, dans tous les aspects de leur existence, concernés par ce problème.

Pour nourrir ces hommes, il ne paraît guère concevable que ce soit encore la production du continent Nord américain qui supplée aux insuffisances des deux tiers de la planète. Les aptitudes locales, tant sur le plan géographique que sur le plan humain, devront être pleinement exploitées. Un rapport présenté par M. Colombo à la dernière réunion de la Commission Trilatérale, tenue à Bonn en octobre 1977, a conclu à la possibilité de doubler la production de riz en Asie du Sud et du Sud-Est en quinze ans. La réalisation reposerait essentiellement sur des travaux d'irrigation et coûterait environ 50 milliards de dollars. La science pourra peut-être proposer des solutions plus radicales et plus révolutionnaires : élargissement des bases de la chaîne alimentaire en «cultivant la mer» par la production des algues, du plancton et la suite; entreprises de modification des climats pour transformer les zones désertiques ou semi-désertiques? Mais n'est-ce pas une utopie dans le laps de temps très restreint qui nous reste et compte tenu du fait, comme le souligne le rapport de MM. Peccei et Mesarovic, que la désertification, bien loin de reculer, progresse? Et si cela s'avère possible, quelles seraient les conséquences sur d'autres régions? D'éventuelles manipulations des climats ne créeraient-elles pas des menaces de catastrophes écologiques et des tensions internationales encore plus graves que celles connues jusqu'à présent?

Quant au problème du travail de ces milliards d'hommes, il est en grande partie lié à celui de l'énergie qui pourra leur être fournie. La violente réaction que provoquent les plans de développement de l'énergie nucléaire et notamment les surgénérateurs est-elle justifiée? Est-ce raison ou folie de vouloir cette énergie ou de la refuser? Ni les partisans, ni les adversaires ne se laisseront convaincre. Mais quelques observations devraient pourtant aider à dépassionner ce débat. Tout d'abord, que la principale difficulté que nous rencontrons ici et que nous recontrerons presque à chaque pas de cette exploration est de prévoir et de vouloir les conséquences de ce que nous voulons. Certains qui réclament l'élévation du niveau de vie, n'acceptent ni la pollution, ni le développement de l'urbanisation. D'autres qui disent vouloir préserver à tout prix la nature ne sont peut-être pas près à renoncer à l'avion ou l'automobile au bénéfice de la bicyclette ou la marche à pied.

Nous concéderons donc qu'il n'est pas fatal de développer, suivant les plans élaborés, la production d'énergie nucléaire. Mais il faut en voir les conséquences. Et celles-ci sont une modification profonde de nos économies et de notre genre de vie et très certainement une montée importante du chômage chez les pays qui sont dépourvus d'autres ressources énergétiques. Il nous faudrait vivre différemment. Le débat entre ceux qui redoutent le risque nucléaire et ceux qui l'acceptent est mal posé. Il devrait porter sur le refus ou l'acceptation du genre de vie qui nous serait imposé si nous décidons de nous passer de cette énergie. Si dans vingt ans l'Europe devait compter par dizaines de millions les sans travail permanents, structurels, serions-nous plus ou moins heureux, malheureux, maîtres ou esclaves de la situation?

A une échéance moins immédiate, sommes-nous autorisés à espérer que l'énergie à bon marché sera à notre portée? Le document soumis à notre examen par MM. Häfele et Sassin montre que si des solutions sont possibles, elles nécessiteront des efforts d'ordre institutionnel à l'échelle internationale qui seront véritablement révolutionnaires. Une circonstance favorable est que les zones où se trouveront les plus grandes concentrations humaines seront aussi celles où l'énergie solaire, si elle est maîtrisée, sera la plus abondante. Alors, espérons, mais en sachant que, faute d'une solution, les problèmes que nous connaissons depuis cinq ans seront tenus pour jeux d'enfants et notre situation pour l'âge d'or.

Parmi les nombreuses conséquences de cette expansion démographique, je n'en relèverai qu'une mais qui nous concerne au premier chef. Je veux parler de l'énorme déséquilibre des nombres entre les pays européens, y compris l'URSS et l'Amérique du Nord d'une part, les habitants du reste de la planète de l'autre. Dans son rapport datant de 1970, le Conseil Mondial de la Population prévoit que dans un siècle (et c'est l'hypothèse optimiste) l'Europe aura progressé de 10% et le Tiers Monde aura triplé. Dans une telle situation, les différences de niveau de vie seront de moins en moins supportables et supportées, tandis que le manque de ressources énergétiques et de matières premières aggravera encore la vulnérabilité de l'Europe, du Japon et, à moindre degré, des États-Unis' Les données de base des relations internationales ne peuvent manquer d'en être sérieusement altérées.

Mais, dès maintenant, dans le domaine des relations internationales deux phénomenes paraissent transcender tous les autres.

Le premier est l'émergence de ce que l'on est convenu d'appeler les superpuissances. Certes, il y a toujours eu des puissances grandes, moyennes ou petites. Mais elles différaient par des degrés dans la possession de moyens analogues. Aujourd'hui, la combinaison du nombre des hommes, de l'étendue du territoire, de l'abondance des ressources naturelles, d'une compétence scientifique et de la maîtrise technologique crée, dans l'ordre militaire, une disparité ou plutôt une discontinuité qui est presque de nature entre les superpuissances et les autres.

Il s'agit bien de l'ordre militaire car dans les autres branches d'activités, nous voyons que les pays européens, si peu organisés qu'ils soient à l'échelle de leur continent, surclassent complètement l'Union soviétique sur les plans du développement économique; du volume des échanges extérieurs, du progrès du niveau de vie, de la contribution au progrès des sciences et de la technologie (non liées aux programmes militaires) et des capacités d'aide au Tiers Monde.

Mais, seules les superpuissances peuvent consacrer une part très substantielle de leur produit national à l'éditication de leurs forces armées, seules elles peuvent entretenir des forces terrestres navales et aériennes à la fois nombreuses et utilisant les techniques les plus modernes, seules, en un mot, elles disposent des moyens d'une politique globale. D'ou l'établissement de rapports spécifiques entre elles dont le point central est évidemment la négociation sur les armes stratégiques. Mais du fait de leur compétition scientifique et technologique dans tout ce qui touche de près ou de loin à la puissance militaire, du fait qu'elles incarnent des systèmes rivaux en matière politique économique et sociale, du fait qu'elles sont les seules à avoir les moyens d'une politique globale, des relations particulières s'établissent entre elles qui tendent à passer au-dessus et en dehors des rapports de cohésion à l'intérieur des systèmes dont elles sont chef de file.

Et sans doute devons-nous ici constater en passant que l'Europe, quelle que soit sa richesse intellectuelle et quel que soit le degré d'organisation auquel elle parviendra restera handicapée par le manque de deux des attributs des superpuissances: l'espace et les ressources.

L'autre phénomène est encore plus important. C'est celui de l'impact des applications scientifiques aux techniques de guerre.

Je ne dirai rien de la prolifération des armes nucléaires présente à tous les esprits. Nous sommes tous maintenant plus ou moins au courant des effets à attendre des armes atomiques, des fusées, de la propulsion nucléaire, des observations par satellites, des applications de l'électronique à l'acquisition et au traitement des données, etc. Nous savons que les dangers de la guerre atomique, bactériologique et chimique sont tels que l'humanité est aujourd'hui capable de se détruire et de rendre méconnaissable la planète qui l'a enfantée.

Aussi et pour la première fois est-il patent que les profits à espérer d'une victoire ne sont plus comparables aux risques de destruction qu'entraîneraient des hostilités et la notion de dissuasion est entrée dans les moeurs internationales. Il est vain d'imaginer un déroulement de l'histoire différent de celui qui a pris place, mais il n'est pas interdit de penser que le monde a fait, au cours des trente dernières années, l'économie d'une guerre entre les deux superpuissances laquelle, sans l'existence de la menace nucléaire, paraissait pour le moins probable. Qui peut dire également si en d'autres temps, les bouleversements économiques causés par le quadruplement du prix du pétrole n'auraient pas entraîné une réaction de force et la prise des champs pétrolifères sous contrôle des consommateurs?

Aussi longtemps que les règles subtiles de la dissuasion continueront de s'appliquer, un conflit entre les grandes puissances paraît moins vraisemblable qu'à aucune autre époque du passé.

Ce bienfait sans précédent comporte pourtant une contrepartie qui nous intéresse ici directement. La guerre quand elle n'a pas été seulement un jeu de princes, a servi de porte de sortie en présence de situations ou de contradictions devenues intolérables. Le marxisme la tenait pour inévitable en milieu capitaliste et lorsque les Chinois nous avertissent de la fatalité d'un conflit il est bien évident que ce n'est pas à une guerre entre puissances libérales qu'ils songent.

Mais, si la guerre est aujourd'hui reconnue absurde ou simplement irrationnelle peu importe le qualificatif, cela signifie que la formule fameuse de Clausewitz est périmée, ce qui serait certainment la mutation la plus importante jamais intervenue dans les relations internationales. Mais cela signifie aussi que le monde devra vivre avec les problèmes qu'il résolvait ou oubliait avec la guerre, qu'il devra les résoudre ou faire de la scepticémie économique ou sociale. Cela signifie, entre autres, que le cycle classique : crise, conflit, destruction, reconstructions, crise . . . ne s'applique plus et qu'il nous faut trouver de nouveaux rythmes d'expansion qui nous feront certainment regretter celui que nous avons connu jusqu'en 1973. A moins que nous n'adoptions la solution de cauchemar prévue par George Orwell celle d'un monde réduit à trois puissances se faisant une guerre perpétuelle où le plus fort s'opposerait aux deux autres, mais sans jamais rechercher la victoire et sans jamais employer l'arme nucléaire, le but des hostilités étant seulement d'assurer la destruction des armements, condition nécessaire à leur production et au maintien des masses populaires au travail mais en esclavage.

Autres bouleversements des rapports internationaux : la fin, grâce aux satellites d'observations, du secret sur l'état réel des préparatifs militaires des différents pays. A ce jour, seuls les États-Unis et l'URSS sont en mesure de jouir de ce privilège sans lequel les premiers accords SALT n'eussent même pas été concevables. Il s'en faut que ces progrès aient fait disparaître la suspicion dans les rapports entre ces deux pays. Mais celle-ci porte sans doute davantage sur les intentions ou les arrières-pensées que sur les réalisations matérielles. On comprend que le Gouvernement français ait proposé la création d'un système international d'observation qui rendrait accessible à la communauté internationale le contrôle d'engagement de limitation ou réduction des armements.

Mais les utilisations de satellites ne sont pas seulement liées à la prévention ou la conduite de la guerre. Leur application la plus spectaculaire se place dans le domaine des communications. Nous sommes déjà habitués à assister chez nous au premier pas d'un homme sur la lune, aux jeux olympiques qui se déroulent à Tokyo ou Montréal et l'on sait le rôle que la projection en direct des scènes de la guerre du Vietnam a joué dans la politique américaine.

Dans peu de temps, il sera possible de recevoir, en direct, les émissions venues d'autres pays. Les hommes politiques ne sont pas les seuls à réfléchir au puissant moyen de propagande que constituera cette possibilité d'action sur les opinions publiques des pays étrangers et sur le désavantage qui frappera ceux qui n'en disposeront pas. Les sociologues ne peuvent pas sous-estimer le choc sur les cultures et même les civilisations d'une telle pénétration d'images venues de l'autre bout du monde et qui ne seront plus filtrées au choix sur les réseaux de télévision. Ce que le cinéma a fait pour le rayonnement de l' «American way of life» et la propagation du 'blue-jeans' se répétera multiplié au centuple dans tous les domaines et pour toutes les idéologies.

Faut-il insister sur la pollution et son antidote l'écologie?

Les journaux, les mass media nous ont épouvantés, à juste titre, par le spectacle des rivières charriant tous les rejets de nos industries, de plages rendues inaccessibles par l'accumulation des immondices, de vastes sections de littoral frappées de mort parce qu'un pétrolier s'est éventré au large. On a avancé l'hypothèse que la grande sécheresse du Sahel pouvait résulter de l'affaiblissement de l'évaporation due à la mince pellicule de pétrole répandue sur l'océan. La diminution de la couche d'ozone qui entoure notre planète supprimerait l'écran qui nous protège contre des rayonnements dangereux. La fonte des glaces polaires, par réchauffement de la température due à l'augmentation du gaz carbonique dans l'atmosphère, pourrait noyer la plupart des grandes villes du monde. Les fléaux que sont la drogue et la délinquance juvénile sont sans doute liés à la déshumanisation de la vie dans les grands ensembles.

Une des caractéristiques de notre époque est que les forces dont nous nous sommes rendus maîtres sont si puissantes que la nature n'a plus le temps de corriger les excès qui modifient son cours.

La prise de conscience de la portée de ces désordres a été brutale, mais profonde et bénéfique. Pourtant le danger de cette grande peur est de faire tenir pour acquis ce qui n'est qu'hypothèse. On ne sait pas si l'augmentation du CO_2 entraînerait un réchauffement ou refroidissement de l'atmosphère. L'économie pastorale est sans doute un des agents de désertification les plus puissants conduits par l'homme avec l'exploitation sauvage du manteau forestier. La nature elle-même n'est pas exclusivement bonne qui connaît les tremblements de terre, les typhons, les sécheresses, les inondations. . .

La prophétie est toujours un exercice risqué et Antigone n'a pas toujours raison contre Créon. Cette grande peur reste utile à condition de ne pas préconiser des remèdes contre nature, c'est-à-dire contre l'instinct de progrès ou irréalisables tels que : croissance zéro, appel à des énergies qui ne sont pas maîtrisées, retour à une économie dont la fonction serait seulement de satisfaire les besoins de survie de l'homme.

Les hippies ne sont pas seuls à nourrir ces idées. La littérature proposant des modèles de construction idyllique de la société non polluante est aussi nombreuse que la littérature utopique du siècle dernier lorsque les sociologues ont commencé à prendre conscience des problèmes créés par la première révolution industrielle.

Le paradoxe est que ce péril de la pollution dans l'immédiat s'accompagne de celui d'une pénurie dans un avenir variable suivant les produits, mais dès maintenant assez proche. Pénurie du pétrole et de certaines matières premières encore que pour celles-ci, il semble bien que, dans la plupart des cas, le problème de leur extraction soit essentiellement fonction du prix de l'énergie. Il est certain que l'on voit mal, à échéance d'un quart de siècle, la solution à une équation dont les facteurs sont l'augmentation de la population, l'élévation du niveau de vie et la production d'énergie à bas prix et le financement du développement des pays du Tiers Monde.

Les hommes de science et les techniciens ont, au même titre que les politiques et les sociologues, le devoir d'être attentifs à l'angoisse qui étreint tant de nos contemporains, en particulier les jeunes si nous voulons éviter qu'à la confiance béate dans l'avènement prochain de l'âge d'or en l'honneur il y a cinquante ou cent ans ne succède un rejet systématique de la notion de progrès.

Mais un excès n'en corrige pas un autre et l'homme ne vit pas que de fromage de chèvre. Ce que l'écologie doit apporter c'est la recherche (et si possible la formule) d'un équilibre, non pas entre la nature et l'homme individuel, mais entre la nature et la société humaine avec ses caractères irréversibles et ses lois de développement et de progrès. L'équilibre qui nous sauvera ne sera pas la rêverie bucolique de Jean-Jacques Rousseau mais un équilibre nouveau adapté aux réalités de notre siècle.

21.2. Problèmes pour les nations de l'Alliance

Le titre qui nous est proposé pour ces réflexions met l'accent sur les problèmes qui se posent aux pays occidentaux.

Je crois que tout ce qui a été dit jusqu'à présent concerne ces pays au même titre que les autres. Alors convient-il maintenant de se demander si certains questions se posent à nous d'une manière plus spécifique en notre qualité de pays différents les uns des autres à bien des égards, mais unis par le même souci de protéger notre sécurité et par le sentiment de la communauté de notre civilisation.

Pour l'observateur de l'extérieur, l'Occident présente une entité distincte dans le monde d'aujourd'hui, certes caractérisé par la liberté de ses institutions, mais plus encore sans doute par les niveaux de vie les plus hauts du monde.

Toutefois, ce niveau de vie a été construit en partie sur une division internationale du travail héritée du XIXème siècle et de l'âge colonial. Aussi un fait politique majeur est-il qu'avec la perte du contrôle des sources d'énergie et des principales matières, nous sommes maintenant vulnérables et menacés?

Vulnérables, parce que l'augmentation des coûts des produits importés peut à tout moment bouleverser notre équilibre économique comme on l'a vu avec le pétrole, menacés parce que les pays à bas niveaux de salaires s'équipent de plus en plus (en général avec notre concours, notamment par les ventes d'usines clefs en main) pour concurrencer nos propres produits dont la vente nous sert à financer nos importations. C'est le cas pour le textile, l'acier, la construction navale. Il n'y a pas de raison pour que les choses en restent là. L'automobile déjà est atteinte. Qu'arrivera-t-il lorsque la Chine pourra exporter les surplus d'un marché intérieur d'un milliard de consommateurs? Comment l'Occident, avec ses hauts salaires, sa législation sociale avancée, ses débouchés plus étroits, pourra-t-il demeurer competitif? Sans doute les pays du Tiers Monde absorbent–ils 35% des exporations de la communauté européenne, c'est a dire plus que les État Unis. Mais tant va la cruche à l'eau qu'à la fin elle se casse . . . Le jour où les grandes masses de travailleurs se sentiront frappées par la chômage, du fait d'importations qu'elles jugeront contraires aux lois normales de la concurrence, n'exigeront elles pas une protection quelles que soient les objections des économistes.

Une contre-mesure moins dommageable sera sans doute que l'Occident se consacre de plus en plus aux activités qui incorporent le plus de matière grise et la plus haute technicité. C'est peut-être l'amorce de la révolution que prévoient MM. Danzin et Branscombe à la suite de laquelle la matière première essentielle ne serait plus l'énergie mais la connaissance. Heureusement, nous disposons d'excellents atouts. Car, s'il est possible de faire tourner une usine livrée avec la méthode pour s'en servir, le passage du laboratoire à l'usine exige de longs délais qui tiennent au niveau de développement des hommes, tant les scientifiques que les administrateurs, les techniciens, les ouvriers. Ainsi constate-t-on que, partant de données de base qui ne sont pas secrètes, certains pays seulement sont capables de les faire passer dans la technologie et l'exploitation industrielle. En d'autres termes, s'il est possible aux pays en voie de développement d'acquérir nos technologies, c'est en Occident à peu près exclusivement qu'elles sont élaborées et mises au point. Devrons-nous envisager un non-transfert de nos avantages si celui des ressources qui nous sont nécessaires nous était refusé?

Dans la compétition impitoyable à laquelle nous sommes exposés, nous sommes tenus de rester en tête sur les plans scientifique et technologique, ou à ne peser que du poids de notre nombre et de nos ressources. Mais il y a un double prix à payer pour cette aptitude au recyclage et à la reconversion. L'un est celui de l'enseignement et de la formation qui devront être conçus pour permettre ces adaptations, d'autant plus délicates qu'il s'agira de productions plus sophistiquées. Ce n'est qu'un des aspects, même s'il est important, de l'ensemble des problèmes d'éducation dont le document de Lord Vaizey fait le tableau. L'autre peut présenter une difficulté ou plutôt un obstacle d'ordre économique. Comment concevoir, en effet, l'amortissement des investissements industriels pour des entreprises qui devraient faire preuve d'une telle souplesse d'adaptation dans un marché mondial en évolution constante. Enfin, rien ne prouve que, même si tous ces problèmes étaient résolus, les nouvelles activités où nous aurons un avantage suffiraient à absorber les travailleurs d'industries de masse, comme la sidérurgie ou le textile.

Un autre aspect du même problème est traité dans le rapport de MM. Brooks et Skolnikoff. C'est celui des transferts au profit des pays socialistes.

Certains pensent que nos rivalités pour vendre cette technologie libèrent pour des productions d'armement le potentiel encore limité de ces pays et illustrent la prophétie de Lénine annonçant que, lorsqu'il s'agirait de pendre les derniers capitalistes, ceux-ci se feraient la guerre pour vendre la corde aux pays socialistes.

Si l'aspect mercantil de l'opération était seul en cause, sans doute le jeu n'en vaudrait-il pas la chandelle. Aussi la justification qui en est donnée est-elle avant tout d'ordre politique. Ces transferts tenus pour un élément essentiel de la détente, se fondent sur l'espoir que l'élévation du niveau de vie qui devrait normalement en découler contribuerait à promouvoir la liberté dans les pays marxistes et à atténuer les tensions.

Mais la question peut-être la plus grave qui se pose est celle de l'unité de l'Occident. nous en parlons comme d'un tout parce que nous nous sentons effectivement liés par d'innombrables racines qui plongent dans notre passé comme par un commun attachement aux mêmes valeurs et par des aspirations partagées pour le développement d'une civilisation de liberté et de progrès. Mais ces points forts qui se placent surtout au plan moral et politique ne sont-ils pas menacés par des développements en face desquels nos dispositions ou même nos volontés d'adaptation ne sont pas les mêmes.

En d'autres termes, la question est la suivante : face aux développements scientifiques et technologiques, les Européens et les Américains restent-ils à niveau? Jean Fourastié a analysé cette question à propos de la France, mais il estime que d'une manière générale ses conclusions sont valables pour les Latins plus que pour les Anglo-saxons et pour les Européens plus que pour les Américains.

«En ce début du troisième quart du XXème siècle», écrit-il, «la pensée dominante chez les intellectuels est largement tributaire du passé et n'est guère influencée par les résultats de la science expérimentale... Rien n'est moins expérimental que la pensée des hommes en ce siècle où l'économie, la société, la culture et le milieu de vie ont été transformés par la science expérimentale... Le comportement quotidien des intellectuels [ce mot étant entendu comme englobant tous les hommes dont l'activité professionnelle est mentale par prépondérance] leurs conceptions politiques, morales et philosophiques n'en sont pas modifiées : ils restent ceux du siècle dernier.»

Que ce soit cause ou effet, il en résulte pour le problème qui nous occupe dans l'approche des sciences humaines, morales, économiques, sociales, etc. une préférence pour la rationalité sur la science expérimentale, pour la construction logique sur l'adaptation au réel et, dans l'ordre politique pour l'utopie sur le concret. Lorsqu'on vient de vivre une période électorale particulièrement longue et animée l'on est très sensibilisé sur ce point.

La science n'est pas admise au même titre dans nos différents pays comme une composante de l'humanisme et même plus simplement de la culture dont elle est pourtant devenue le plus puissant agent d'évolution, de progrès ou éventuellement de régression. Quant à la technologie, elle est purement et simplement tenue pour une activité d'ordre secondaire et indigne d'un esprit supérieur. Le cas d'un professeur quittant sa chaire pour diriger les recherches d'une grande entreprise industrielle ou mieux encore pour fonder sa propre entreprise est à peu près inconnu dans certains de nos pays dont le mien. Il est monnaie courante aux États-Unis. De même, est-il plus rare de ce côté de l'Océan de voir un universitaire, un professeur exercer pendant un temps des fonctions gouvernementales ou administratives ou servir de consultant sur tel ou tel sujet normalement objet de ses travaux. Il résulte pourtant de cette pratique une interpénétration beaucoup plus nourrie entre le monde des intellectuels et celui des politiques souvent au niveau des plus hautes responsabilités.

Il est vrai qu'en contrepartie, cette prépondérance donnée à la méthode scientifique expérimentale risque de s'opposer à l'élargissement des réflexions sur la finalité de la société, sur le sens de la vie, sur la philosophie et disons ce mot sur la spiritualité. Sans doute sur ce point l'Europe garde-t-elle dans la force de ses traditions un spectre de préoccupation plus ouvert.

La synthèse est difficile, mais la complémentarité des propensions est une source de richesse pour l'ensemble.

J'ai dit que le problème était avant tout une affaire de volonté. En effet, si les États-Unis font preuve d'une plus grande aptitude à assimiler et utiliser les développements scientifiques et techniques, ce n'est pas par une quelconque supériorité congénitale. Deux facteurs sont essentiels.

Le premier est celui de l'éducation. Trop longtemps en Europe l'enseignement a été conçu comme la transmission aux jeunes générations d'un héritage culturel à peu près achevé, au lieu de les associer aux problèmes de la recherche présente et à venir. Sur ce point l'enseignement aux États-Unis paraît supérieur au nôtre, ce qui entraîne nécessairement une plus grande disposition, au cours de la vie, à accueillir les innovations et les progrès.

Il est à craindre qu'un décalage profond et permanent dans les attitudes face à ce problème ait des conséquences désastreuses pour la cohésion des héritiers que nous sommes et que seront les générations à venir, de la civilisation occidentale. Pourtant les Européens sont capables de grands efforts et la mise sur pied de la Communauté Européenne quelques années seulement après la fin de leur dernière guerre civile est un acte qui a exigé un renversement complet des politiques, des traditions économiques et naturellement des habitudes de pensée et des réflexes populaires sand exemple jusque-là. Mais il serait vain de nier que le mouvement paraît se ralentir, soit que les difficultés pour passer aux stades ultérieurs paraissent insurmontables, soit que la tentation soit trop forte de faire la pause et de profiter des résultats acquis comme si une telle entreprise pouvait s'accommoder d'une progression par paliers successifs.

L'autre atout dont dispose l'Amérique est à la fois plus structurel et pourtant plus facile à équilibrer. Il s'agit de la puissance des moyens dont disposent les chercheurs.

Je voudrais ici donner un exemple qui illustrera ce que j'ai à l'esprit.

Après la guerre, il est apparu que les progrès en physique nucléaire exigeaient des appareils de recherche de plus en plus importants, de plus en plus coûteux exigeant un nombre de scientifiques, d'ingénieurs et techniciens dépassant les capacités de chaque nation européenne prise individuellement.

Ce qui était en jeu c'était de savoir si l'Europe, qui avait donné à la physique nucléaire ses plus grands noms, allait s'éclipser devant la montée des deux très grands pays, seuls en mesure de fournir à leurs savants les moyens nécessaires à de nouveaux progrès. Il ne s'agissait de rien moins que de l'influence que, dans un domaine essentiel, le Vieux Continent serait en mesure de conserver sur le développement d'une civilisation qu'il avait, puisque les connaissances de base sont à la source de tout développement technologique.

Bon nombre d'hommes de sciences européens en étaient conscients. Mais bon nombre d'autres craignaient qu'un effort entrepris sur une base internationale ne vint tarir les ressources disponibles pour les centres de recherche nationaux.

C'est l'honneur d'hommes, comme le Professeur Rabi, qui savaient ce que les prodigieuses réalisations américaines devaient aux découvertes de base, aux enseignements et aux travaux accomplis en Europe, d'avoir compris qu'une décadence des nations européennes en ce domaine créerait une situation dommageable pour l'Amérique elle-même. Ils apportèrent donc les encouragements et l'appui moral de la communauté scientifique des États-Unis.

C'est ainsi que naquit l'organisation européenne de recherche nucléaire, le CERN, qui, en rétablissant les conditions nécessaires à une compétition à armes égales, a permis à la physique européenne de se maintenir en position de pointe avec ses soeurs américaines et soviétiques et qui, bien loin de nuire aux travaux dans chaque pays a, au contraire, joué le rôle d'une puissante locomotive entraînant derrière elle la plupart des laboratoires d'université et centres de recherche du Vieux Continent.

Si je me suis arrêté sur ce point c'est parce qu'il contient une leçon valable pour le sujet qui nous occupe. Et c'est aussi, malheureusement, parce que cet exemple n'a pas eu beaucoup de repliques dans d'autres branches du savoir, à l'exception de l'organisation européenne pour les recherches astronomiques dans l'hémisphère australe, de l'organisation européenne pour la recherche spatiale (encore que sur ce dernier point il s'en faille de beaucoup pour que l'ampleur des réalisations européennes soient au niveau de celles des États-Unis et de l'Union soviétique) et quelques autres concernant l'exploitation scientifique de la Méditerranée ou encore les télécommunications. Et que dire des discussions et indécisions des gouvernements européens qui ont différé de plusieurs années le choix d'un site pour l'édification d'un centre de recherche sur la fusion thermonucléaire contrôlée? Quand on songe au problème de l'énergie qui nous guette, ces délais sont plus qu'incompréhensibles.

Ce n'est par sortir de ce problème de la parité entre l'Europe et les États-Unis que de dire encore quelques mots de l'impact de la science et la technologie dans le domaine militaire.

Sans revenir sur ce qui a été dit plus haut de l'impact des armements nucléaires sur le problème de la guerre ou la paix, il nous faut être conscients des conséquences d'ordre politique des perfectionnements introduits dans la technologie des armes classiques. Ces développements ont, en particulier, pour effet de permettre la détection et l'acquisition des objectifs, le traitement instantané des données relatives à leur identification et à la détermination de leurs emplacements et mouvements et enfin leur attaque et leur destruction avec un coefficient de réussite amélioré par des facteurs de cent ou mille par rapport aux opérations du dernier conflit. Au plan opérationnel, il en résulte un avantage pour la défense, à condition qu'il y ait un équilibre minimum dans les nombres d'armes des systèmes militaires qui s'affrontent, sous peine de voir le plus faible rapidement désarmé. Dans l'ordre stratégique cela signifie que la recette traditionnelle des démocraties pour gagner les conflits, c'est-à-dire recourir à la guerre d'usure et céder du terrain pour gagner du temps, est totalement périmée. La nécessité de présenter constamment à l'adversaire un dispositif de forces qui le contraignent, pour atteindre ses objectifs politiques, à des opérations militaires à un niveau de violence qu'il ne veut pas lui-même risquer est pour les Occidentaux une nécessité absolue.

Or, nous constatons qu'en raison de leur puissance industrielle, de leurs moyens de recherches et de l'ampleur de leur besoin, les États-Unis possèdent dans le domaine des armes classiques des capacités de modernisation souvent supérieures à celles des pays d'Europe. Ce qui s'explique dans le domaine nucléaire est en passe de s'établir dans le domaine classique. Déjà, les missiles de croisière sont pratiquement des monopoles americains, lesquels ont aussi ouvert la voie aux munitions guidees avec precision et aux vecteurs pilotés à distance. Or, ces engins nouveaux sont appelés à jouer un rôle croissant sur l'équilibre des forces et par conséquent sur l'influence de ceux qui les possèdent dans le maniement des crises.

On sait l'attention avec laquelle plusieurs gouvernements européens suivent dans la négociation SALT le sort qui sera réservé aux missiles de croisière tant sur le point des limitations de leur rayon d'action que sur celui de la liberté que les États-Unis conserveront de transférer à leurs alliés la technologie les concernant. Pourtant ce transfert de technologie, pour important qu'il soit, ne doit pas constituer pour les Européens la seule ou même la principale réponse à leurs problèmes. C'est dans la voie d'une coopération dans laquelle ils commencent à s'engager qu'ils pourront non seulement trouver à la fois les marchés, les moyens financiers et les ressources technologiques nécessaires, mais aussi acquérir une autorité et une influence sur l'élaboration des concepts de défense les concernant en rapport avec les responsabilités qu'ils pourraient être un jour appelés à axercer.

Mais le domaine dans lequel le décalage entre les pays de part et d'autre de l'Atlantique pose les problèmes les plus graves est probablement celui des ordinateurs.

Lorsque l'on lit dans le chapitre du Professeur Danzin que 60% des capacités d'informatique appartiennent à une seule firme américaine, que déjà des problèmes se posant en Europe doivent être traités aux États-Unis, lorsque l'on admet que l'informatique est le plus prodigieux agent de transformation de nos conditions de travail et bientôt d'existence, depuis l'usine jusqu'à la recherche biologique ou médicale, bouleversant les techniques de gestion des entreprises et des services, agissant sur le niveau de l'emploi, susceptible de conditionner nos libertés, l'on reconnaît que nous sommes en présence d'un aspect nouveau de notre civilisation. Or, ces développements se poursuivent à des rythmes différents en Amérique et en Europe et les conséquences de cet état de chose, s'il devait se prolonger, seraient incalculables. Elles entraîneraient une infériorité de l'Europe par rapport aux États-Unis et du Japon á l'égard des outils au service du travail intellectuel et établiraient une dépendance d'ordre non seulement matériel mais culturel.

Pour M. Danzin cette modification dans «les rapports de forces retentit directement sur les situations d'indépendance, d'interdépendance ou de domination des États Nations.» On peut parler, estime-t-il, d'une «mutation au sens où les biologistes l'entendent pour décrire l'évolution des êtres vivants. L'homme n'est pas muté génétiquement mais il est muté dans son comportement de groupe et ses réactions sociales deviennent radicalement différentes selon qu'il appartient déjà à la civilisation de l'informatique ou qu'il reste attardé aux âges pré-industriels ou industriels. Ce phénomène, conclut-il, doit être regardé comme beaucoup plus important que la supériorité militaire conférée par la disposition d'armes perfectionnées, car il s'agit d'un clivage entre une adaptation culturelle à progresser dans les activités de la connaissance et une inaptitude du système de valeurs intellectuelles à faire un bon usage des moyens technologiques nouveaux.»

Une vision commune de l'intérêt général pourra-t-elle s'établir, tant en Amérique qu'en Europe pour rétablir un meilleur équilibre dans ce domaine? Les intérêts économiques considérables qui sont en jeu ne s'y prêtent guère et pourtant de tous les sujets dont nous avons débattu en ce livre, c'est là sans doute celui qui est appelé à exercer le plus rapidement la plus grande influence sur l'avenir.

Le problème est bien compris en France et l'un des collaborateurs du Président de la République écrivait récemment dans un grand journal français qu'à la fin du siècle «très peu de pays seront capables de maîtriser le processus de l'informatisation de nos sociétés à ses trois niveaux : la mémorisation, le traitement, le transport de l'information.» Les pays qui seront défaillants sur l'un ou l'autre de ces créneaux «seront frappés d'amnésie dans le premier cas, d'idiotie dans le second, de paralysie dans le troisième.»

Ne se plaçant du seul point de vue politique, cette situation appelle une action prioritaire à l'échelle de l'Europe. L'on se prend à rêver que les hommes qui nous gouvernent adoptent à ce propos la formule du Président Roosevelt annonçant la politique du Prêt Bail au peuple américain «It must be done. It can be done. It will be done.»

Soyons conscients aussi du décalage qui se produit dans le domaine de la production industrielle. Les États-Unis dominent complètement les marchés de l'informatique, des techniques spatiales, des télécommunications. Grâce au développement des sociétés multinationales, au régime monétaire international plus soumis aux aléas de la conjoncture économique intérieure américaine qu'aux exigences d'une saine économie mondiale, grâce à un régime douanier qui désavantage les productions européennes sur le marché américain et avantage les productions américaines sur les marchés du Vieux Continent, ce décalage s'accentue.

Je n'ai ni la compétence ni l'intention d'analyser ces problèmes sur lesquels les hommes les plus qualifiés hésitent et trébuchent et je ne veux faire ni le procès de l'expansionnisme américain ni celui d'une certaine passivité européenne. Mais du point de vue qui nous occupe ici, celui de la cohésion de l'Occident, je veux répéter que nous sommes sur une mauvaise pente et que nous serons tous victimes des conséquences de cet affaiblissement du poids spécifique et relatif de l'Europe au plan intellectuel et au plan matériel.

Reconnaissons que nos gouvernements et nos opinions publiques n'en paraissent pas pleinement conscients.

Et pourtant c'est bien au niveau de l'homme politique entendu au sens large du terme, c'est-à-dire le législateur, le gardien de la cité, que finalement toutes les questions vont aboutir. Et la cité dont il s'agit n'est pas seulement la nation dont l'homme politique doit gérer la souveraineté mais la communauté internationale.

C'est le sujet du chapitre «Science, technologie et relations internationales» de MM. Brooks et Skolnikoff déjà cités.

En présence du déferlement des problèmes de complexité croissante, du renouvellement constant des paramètres les faisant évoluer et de leurs interactions de plus en plus nombreuses, est-il possible d'intégrer dans une vision cohérente les éléments d'une décision utile?

Mais si on ne peut laisser seulement jouer les lois du marché, c'est-à-dire admettre que le monde évolue à l'aveuglette, au hasard des forces qui le dominent, faut-il tendre vers une planification dirigée?

Le Chapitre du Professeur Eckaus souligne combien, quelle que soit l'urgence des difficultés immédiates, une vue des problèmes à long terme et des moyens d'y parer est induspensable sous peine d'en rendre la solution encore plus complexe. Or, il ne croit pas que l'économie de marché en vigueur parmi les nations de l'Alliance Atlantique puisse automatiquement traiter les problèmes à long terme. Cela signifierait que «les fluctuations immédiates de l'activité économique requièrent une action gouvernementale intensive et une coopération internationale mais que les difficultés à longue échéance se résoudront d'elles-mêmes.»

La conclusion qui s'impose à la lecture de ce rapport est qu'une action continue et internationale, sinon pour résoudre, du moins pour traiter les problèmes de l'avenir, est indispensable. Et l'on n'empêchera pas le profane de penser que nos pays n'auront pas sérieusement donné la preuve d'en être conscients et de garder leur foi dans leur propre système aussi longtemps qu'ils n'auront pas rétabli, à titre de première mesure, un régime monétaire international fiable permettant des actions à long terme.

Ces constatations ne sont naturellement pas pour sous-estimer l'extrême difficulté d'une planification.

Et ce ne sont pas les résultats obtenus dans les pays qui font du dirigisme le credo de leur vie politique et sociale qui nous inciteront à les imiter.

Sur la base des expériences dont il nous offre l'exemple, le dirigisme n'apporte pas la solution puisque là où il est installé la réussite économique est, pour dire le moins, médiocre. Bas niveau de vie, particulièrement frappant là où existent des possibilités de comparaison comme la Tchécoslovaquie, autrefois l'un des pays les plus développés de l'Europe, aujourd'hui ramené par le jeu de la planification socialiste à un niveau qui ne supporterait la comparaison avec aucun pays occidental. Avec des territoires et des ressources immenses, l'ensemble des économies du Pacte de Varsovie n'a contribué en 1975 que pour 4,10% aux exportations mondiales et 2,90% aux importations.

Sur le plan politique, le marxisme qui se veut scientifique a accouché des sociétés les plus aliénantes qui soient et sa version du socialisme produit la dictature comme le pommier produit des pommes. Un dirigisme erroné est peut-être plus vicieux encore que la croissance dans le désordre des forces en présence et des réactions qu'elles suscitent.

La seule chose qu'il soit possible de dire c'est que les bouleversements allant s'accélérant et les interactions étant de plus en plus complexes entre les conséquences de ces bouleversements, les systèmes fondés sur une analyse prétendue scientifique de l'histoire, sur l'infaillibilité des hommes en des partis qui font cette analyse et sur une planification rigide des conséquences à en tirer cent plus de chances de se tromper que ceux qui, doués d'une plus grande souplesse ou d'une plus grande disponibilité intellectuelle, conservent leur faculté d'adaptation.

Entre le dirigisme rigide et le libre jeu des forces qui ressemble à une sorte de collin-maillard au bord de précipices, il s'agit de trouver les voies d'une planifcation élargissant ses paramètres à l'échelle du monde et concernée autant par les problèmes de l'homme, de son environnement naturel que de son travail et de sa prospérité. La science et la technologie posent des problèmes d'ordre politique, économique et social, mais aussi d'ordre humain qui ne relèvent pas de la mathématique.

Aussi nous devons être reconnaissants au Club de Rome d'avoir sonné l'alarme en montrant qu'il n'était pas possible que les transformations dont le monde est le théâtre se poursuivent longtemps au même rythme et s'étendent à tous les habitants de la Terre et aussi d'avoir lancé un appel de détresse au nom de l'homme, de la nature et de la planète elle-même.

Le chapitre de MM. Aurelio Peccei et Mihajlo Mesarovic nous propose un choix de nouvelles valeurs lorsqu'il déclare «Dans les années ou décennies à venir le progrès humain ou la qualité de la vie et même notre survie vont dépendre beaucoup plus des inventions ou innovations d'ordre social, politique, institutionnel ou éducatif que de n'importe quelle avance de la science stricto sensu.»

Et le même rapport se termine par la définition d'une révolution humaine qui ne propose rien moins qu'un changement radical dans les grandes orientations de la civilisation issue du développement de la science et la technologie et cela non seulement dans nos pays mais à l'échelle planétaire. Il s'agit pour les auteurs de «renouveler et même renverser des principes et des normes aujourd'hui tenus pour immuables et de promouvoir l'émergence de nouveaux ensembles de valeurs et de motivations spirituelles, philosophiques, éthiques, sociales, esthétiques, artistiques, en accord avec les impératifs de ce temps.»

De son côté, le Professeur Malinvaud a préconisé une «transformation de la morale sociale collective» qui lui paraît moins difficile que la transformation de l'homme lui-même.

Peut-être ai-je donné l'impression, au cours de ces remarques d'être plus sensible aux dangers que font courir à notre planète et à notre espèce les développements scientifiques et techniques qu'aux bienfaits dont nous leur sommes redevables.

Cela tient en partie à ce que le sujet proposé me demandait de parler des 'problems' qui se posent à nos sociétés. C'est aussi que ce qui retient l'attention c'est l'accident d'avion ou de chemin de fer et que l'on ne parle jamais des milliards de voyageurs-kilomètres sans incident. Comme disait je crois William Hearst : si un chien mord un homme c'est indifférent. Mais si un homme mord un chien «That's news.»

Rien n'est plus loin pourtant de mes idées que de faire un procès à la science ou de céder à une nouvelle grande peur comme celle qui étreignait nos ancêtres à l'approche de l'an 1000.

C'est l'heure de nous souvenir que «tout ce qui est excessif est sans importance», comme disait Talleyrand. A la confiance béate du siècle dernier qui n'attendait pas seulement de la science qu'elle fasse le bonheur de l'homme mais encore qu'elle réponde à ses interrogations philosophiques ne doivent pas succéder condamnation et désespoir.

Nous avons devant nous, du moins dans nos sociétés occidentales, sur le plan matériel et culturel, le spectacle des prodigieuses améliorations intervenues dans la condition humaine, allongement de la vie, émancipation de la femme, élévation du niveau de vie, éducation, politique de santé et sécurité sociale, loisirs, activités culturelles. Les satellites ont donné aux facilités de communication un essor que n'avait même pas prévu Jules Verne. Ils font de la météorologie une science moins hasardeuse. Ils vont nous permettre l'inventaire des ressources de la Terre, la surveillance de la pollution des océans, la prévision des récoltes plusieurs mois à l'avance.

Pour ne prendre qu'un exemple, la télévision si souvent accusée de nuire à la vie familiale a entraîné le décloisonnement des groupes ou individus isolés. La population tout entière d'un pays est informée au même instant des mêmes nouvelles et assiste aux mêmes programmes.

Grâce aux mass media chacun peut avoir une notion des oeuvres de Shakespeare, entendre chez lui les grandes heures des festivals de Salzbourg, Aix-en-Provence ou Bayreuth, assister aux ballets de l'école soviétique ou à une chorégraphie de Maurice Béjart.

Sans doute cela entraîne-t-il une modification de la transmission de la culture ou en tout cas d'une certaine appréhension de ces manifestations. Ce qui résultait de la lecture et du travail individuel devient en partie accessible par l'audition et la vision. La connaissance s'élargit et atteint maintenant les masses. Et surtout, elle atteint les jeunes mis beaucoup plus tôt au contact des problèmes des adultes et sensibilisés dès leur adolescence aux grandes questions politiques et sociales du monde moderne. Etat de chose qui a motivé, dans presque tous nos pays, l'abaissement à 18 ou 19 ans de l'âge de la majorité.

Cette possibilité d'une accession plus large et peut-être plus facile aux oeuvres des grandes civilisations, peut concourir à la solution de cet autre problème que l'on appelle couramment celui du Troisième Age. Jusqu'à une époque récente, la longueur de la vie a été sensiblement celle de l'utilité du couple pour procréer et conduire les enfants à l'âge adulte. Aujourd'hui, la vie de l'homme et la femme excède de plus du tiers le temps de leurs fonctions indispensable pour le maintien de l'espèce et l'éducation des jeunes. Est-il sage, dans ces conditions, d'abaisser constamment l'âge de la retraite, en tout cas pour les métiers non pénibles? Ne serait-il pas plus conforme à notre nature de diminuer le nombre d'heures consacrées chaque semaine au labeur pour laisser plus longtemps en activité les hommes et les femmes qui s'en sentent la force et le goût?

Mais il est peu probable que l'on renverse la tendance actuelle, même si elle fait peser sur les actifs une charge de plus en plus lourde. Et l'allongement de la vie du couple fait apparaître dans nos sociétés une nouvelle classe d'hommes et de femmes valides et disposant de grands loisirs. Ici et là, on voit éclore des universités du

Troisième Age, des clubs de rencontre, des associations de voyages qui montrent certainement une voie prometteuse par un développement d'activités culturelles que la pression du travail n'avait pas permis antérieurement.

Faut-il enfin, en contrepoint de ces perspectives d'épanouissement, faire ici référence aux possibilités d'action sur le code génétique de l'homme ou ses circuits cérébraux. Certes les recherches en ce domaine visent, comme l'a souligné le Professeur de Duve, à la prévention de maladies telle la leucémie. Mais qui sait si, lorsque cette possibilité, qui fait déjà partie du domaine expérimental chez les animaux de laboratoire, sera transposable à notre espèce, des régimes décidés à mouler l'homme dans un cadre préconçu ne seront pas tentés par cette puissance diabolique. Nous avons vu l'espoir d'une telle mutation avec les théories de Lyssenko qui pensait que les caractéristiques acquises par l'homme vivant dans une société socialiste se transmettraient par hérédité. Nous avons connu une médecine pervertie dans les camps d'extermination nazis. Aujourd'hui, nous savons les traitements psychiatriques auxquels sont soumis dans certains pays des hommes et des femmes qui ont l'ambition et le courage de penser par eux-mêmes.

Qui pourrait douter que de pareilles entreprises se terminent par une catastrophe sans rémission?

Au terme de ces réflexions, je reviendrai sur trois idées qui me paraissent essentielles. La première est celle de la responsabilité particulière de l'Occident dans la situation présente et, par conséquent, dans la recherche des réponses aux questions posées puisque c'est lui qui a dérobé le feu sacré de la connaissance et a donné au Monde les prodigieux instruments de transformation qui ébranlent toutes les civilisations et qui intéressent aujourd'hui jusqu'aux relations entre l'homme et son milieu naturel.

La second est la conviction que la connaissance de la nature est l'aventure la plus enrichissante et la plus exaltante qui s'offre à nous, à condition de nous souvenir qu'on ne lui commande qu'en obéissant à ses lois, c'est-à-dire, en la respectant sous toutes ses formes et en priorité la vie et l'environnement qui la rend possible.

La troisième est la constatation que «nous ne sommes pas embarqués sur un lac mais sur un fleuve» comme le dit Pascal, c'est-à-dire que sous peine de trahir l'aventure humaine il nous faut concilier à la fois notre instinct de conservation et la poursuite de notre marche en avant.

Même si le mot science employé dans la Bible avait une signification différente de celle que nous lui donnons ici, il est troublant que l'un des plus vieux textes sacrés de l'humanité parle de l'Arbre de la Science, du Bien et du Mal. Mais, il nous enseigne aussi que notre lointain ancêtre avait le libre choix de sa destinée.

Nous sommes en présence du même choix.

22

The implications for the future science programme

Eduard Pestel

During the final years of the 1960s there arose among the public in our Western nations the call that science should turn its attention more to the 'real' needs of society, and I remember that this was the time when I first heard the demand for the *relevance* of science. In those years the first pressure groups for the preservation of the ecology appeared, and the United States' appeal to NATO eventually led to the formation of the Committee on the Challenges of Modern Society (CCMS). This led to a discussion in the Science Committee on whether it should take over the tasks with which the CCMS was to be entrusted, but eventually it was decided to let CCMS come instead with questions concerning fundamental research, e.g. in the field of environmental problems, where the 'hard' sciences, represented within the Science Committee, should try to resolve 'clear-cut' problems that could be neatly categorized within the various scientific disciplines.

I still remember Professor Nierenberg explaining to us at the time of our ignorance about the quantitative increase and the climatological effects of CO_2 owing to the ever-increasing combustion of fossil fuels. And as another personal souvenir of ten years ago I cherish a long conversation with Professor Heisenberg on a lovely evening on the island of Ischia that centred on the topic that in the future it might become more important for the scientist as well as for the engineer to help save our ecological environment, including its social aspects, than to improve the efficiency of machines, say of turbines, by another one-quarter of one per cent. At that time obviously, neither of us had any qualms that thereby Science and Technology would step out into the realm of politics or, as Holst put it in Chapter 19, into the twilight zone of scientific analysis and political debate; we did not even touch upon this issue, probably because of our unspoken naive belief that there could be no disagreement among scientists or between them and politicians, when — outside of internal and external power politics — the true welfare of the people was at stake.

I learned better in the ensuing years, even long before I entered the domain of practical politics, and in the course of the past 10 years I have come to be firmly convinced, as most of you will be too, that human irrationality is probably more part of reality than rationality. And it is here that the whole spectrum of social sciences — from economics to political science to psychology — enters when we try to cope with the 'real' problems of our time in a comprehensive manner.

Now, in these two volumes, and in particular the second one, we have read again and again the plea that the Science Committee should shift the emphasis of its attention more to long-term problems, such as those listed in some of the commentaries and, in particular, by the two eminent people preceding me, and thereby to include in our programme the social sciences more than had hitherto been done. It was argued – and this argument was not new to the Science Committee, having been voiced there repeatedly during the past – that the security of the Alliance (security being viewed as a much wider concept than just military security) depended in the long run on the successful resolution of such long-term problems; be they internal to the NATO nations, such as the present economic and social problems, be they external as with the North–South gap as well as the East–West polarization. I believe none of my colleagues in the Science Committee would refute the validity of these arguments concerning our security.

But what can science and technology really do in coping with such long-term problems, where action is presently required lest we lose many of our options for their solution? Will giving more support to the social sciences really help matters? Did we not also read that the lack of a universally accepted fundamental theory in the social sciences was at the bottom of the fact that so far political decision-makers were receiving mostly contradictory advice from different social scientists? And what could even good solid advice do when the political decision-makers would just have their eyes glued on the next election term, being aware that they would not stand a chance to resolve these long-term problems if they were unsuccessful in alleviating the short-term problems as perceived by their electorate? And can we – especially we scientists – blame the politicians for taking this view with such a short time-horizon? For what has science done to give the political decision-maker rational instruments that would, for example, enable him to explain to the general public why, for the long-term benefit of society, certain decisions have to be taken now, although they might bring with them seemingly unnecessary disadvantages and costs in the short term? Under these circumstances, can science help here? But let us not be in doubt that the citizens of our Western nations expect such help, since we are the first civilization on earth that materially is based entirely on science and technology. I firmly believe that scientists have to make a strong attempt to provide such help to political decision-makers. Analysis and, when all goes well, understanding of the problems is not enough, although they are a necessary prerequisite and mostly quite difficult to attain. This is particularly because even understanding alone requires the linkage, the close co-operation of many – including widely diverging – scientific disciplines. For we are faced with the dilemma that such major problems – even major problems concerning technology – do not arise in the domain of the 'hard' sciences. As Karl Kaiser said on the concept of security in his commentary on Chapter 19, in most instances it is not technology itself that poses the problem, but the assessment of its impact. This holds not only for the military field, but just as well for the civilian application of modern technology. For example, in the concern about the generation of energy in nuclear reactors, and in particular about the disposal of some of the 'sensitive' waste components in the fuel cycle, we find that it is the different assessments on the one hand by various governments of the Alliance of the danger of proliferation of nuclear weapons and, on the other hand, by government and industry contrasting with those of the various national and trans-national pressure groups, of the environmental dangers that have led to a virtual halt of the — at first rapid — expansion of nuclear power in a number of our countries; although practically all truly informed experts are convinced that the technical problems of concern have either been overcome or will certainly be mastered.

In this situation it is clear to me that strong linkage and co-operation is urgently required between scientists of different disciplines covering the whole range of the 'hard' and 'soft' sciences, a co-operation where none has to take the back seat permanently, all of them being equally important in order to gain a comprehensive holistic view of the problem areas to be investigated. And as experience has shown that co-operation between scientists from different disciplines is best developed when they work together on specific projects, the NATO Science Committee should not hesitate in initiating projects concerning long-term problems of an international nature of particular concern to the Alliance. Personal experience has also shown me that the systems approach is necessary to facilitate a coherent purposeful co-operation between different scientific disciplines. And here I believe, in contrast to one of the authors here, that systems science in the conception of the system structure as well as in the construction of models should not retreat before ill-structured problems. If one wants to aid the political decision-maker, one must remember that it is just the essence of political life that decisions have to be made in the face of uncertainty. Hence, systems science must not shrink from uncertainty and from the lack of clearly definable structure that is bound to beset all important long-term political problems.

Is there a better way to cope with the problems of uncertainty and poorly defined structure than by studying possible alternative structures and parameters through corresponding system structures? If the system structures are 'concretized' in computer models, then we can get a feeling of what is important (for example identifying where we must initiate special research to remove uncertainty as far as possible) and what is unimportant (for example where we do not have to worry owing to an insensitivity to parameter change). Therefore the Science Committee should — possibly in a joint venture with other international bodies comprising the countries of our Alliance — initiate a project with the aim of probing the feasibility of fashioning a policy-planning tool based on the systems approach suitable to investigate alternatives for long-term planning; the sort of multidisplinary planning about which we heard from Eckaus (Chapter 17) and where Peccei and Mesarovic gave us a brief impression of its capabilities (Chapter 13).

Now when we talk about long-term planning, we cannot just think of developing planning-aiding instruments, we must also think about the formulation of the goals to be reached through such planning processes. Listening to the public debate in the past years and even of today, one cannot help but see that practically all governments set as one of their major goals the achievement of faster economic growth, measured by the rate of growth of GNP. And this not only because of their and the public's belief in the equation that increasing affluence means increasing happiness for the majority of people, but even more because they seem to believe that only this will enable them to solve their short-term as well as longer term problems. Since the publication of *Limits to growth* more and more people have become aware of the idea that material growth, hence also growth of GNP, cannot continue forever, the debate being not so much about whether this is inevitable, but when the time will have actually arrived for a deliberate slow-down. Now, I personally believe that man needs human development, as Peccei and Mesarovic pointed out, that man wants to grow, and that he wants to measure his growth. At present we seem to use the simplistic indicator of GNP-increase as the measure of successful growth, and hence, not surprisingly, we focus on the goal of further economic growth. I believe that it is high time to define additional social indicators and combine them – possibly in different manners to suit different ideas of what contributes most to human happiness or quality of life – to arrive at new yardsticks for the measure of our present state and of the further development of our well-being. Then we can also define far more comprehensive goals than just the material ones towards which we want to grow, and we will then be in a position to measure our success in making headway in this complex of goals. And while we grow towards such goals, we shall want to redefine them after having reached new vantage points in our progress. So this will not be a static process but a highly dynamic one. In the age of computers I think we do not have to, indeed we should not have our eyes fixed just on one single indicator, but on a composite of weighted social indicators that reflect the various ideas of what constitutes the quality of life. I believe that here also the NATO Science Committee could initiate – possibly through commissioning a competent working group, such as for example the one we had the privilege to have in the field of scarce metals led by Professor Ashby, or by way of a Science Committee Conference – such work by bringing together from the various nations of the Alliance social and natural scientists and engineers. This would most probably contribute positively to the often sterile debate on growth and to the setting of new and clearly defined dynamic goals for growth. In this way, the Science Committee would respond to the repeated request to shift the emphasis of its efforts, to build bridges for co-operation between different scientific areas besides – as it did so effectively in the past – building bridges for the co-operation of scientists from different countries of the Alliance. I might mention that the Science Committee has actually acquired quite some experience in such work ever since it set up its various Special Programme panels at the beginning of this decade, and saw to it that close co-operation came about between the panels on environmental sciences, human factors, and systems science.

Also, a number of suggestions made, for example, by Maddox (Chapter 1) and Hare (Chapter 3) concerning the question of criteria for the distinction between spurious and important environmental problems, could directly be taken up and dealt with in the panels just mentioned. The same could be said *vis-à-vis* the issue of CO_2 concerning all three levels of the problem as stated by Brooks and Skolnikoff (Chapter 18). In addition, the request by Duncan Davies in his commentary for a conference involving the less-developed countries and the multinational corporations on appropriate-technology transfer and co-operation, should and could, in my opinion, be easily accommodated within the already established framework of sponsoring instruments of the NATO Science Committee.

Last, a word about the energy issue. The papers written in the past five years on this subject could probably fill a very large room; and still, by the stretch of imagination, one can always find new aspects of the problem which could fruitfully be studied more closely. So far the Science Committee has dealt only with the aspect of energy conservation – in a delightful, concrete, down-to-earth manner however – even though it was only a comparatively small effort. At this conference we learned from Häfele and Sassin (Chapter 12) how big the problem is, when in the course of the next 50 years the present energy structure, based mainly on oil, has to be replaced by a new structure owing to the coming exhaustion of oil, and later of natural gas, for energy purposes. Even though there was sharp disagreement by Colombo in his commentary on their contribution concerning the eventual size of the problem, and the systemic investigation by Peccei and Mesarovic of the Häfele and Sassin scenaro reduced to some extent its immensity, one can safely say that it remains big enough. And, taking also the aspects of social resistance, difficulties in reaching international agreements, and environmental and organizational problems, etc. into account, one can safely say that in any case the energy issue is a truly formidable one, beset with tremendous uncertainties in the face of which it will be most difficult to implement such strategic plans as developed by Häfele and Sassin in time before the real pinch is here. But still we have to be prepared.

Now, preparation requires at least the development of suitable prototypes of different alternatives, mainly nuclear and solar, for energy generation. The corresponding environmental-impact studies have to be carried out, risks evaluated, siting possibilities investigated. Capital is required, and the necessary international agreements, for example concerning international nuclear energy plants and disposal sites, as well as the governmental access to the necessary resources, etc., has to be obtained. To strengthen the Alliance through such preparation, co-operation is necessary, and not inimical competition.

Co-operation requires also a division of labour in R and D, that is a departure from the way in which, for example, the fast-breeder development is presently handled, developing practically very similar types of sodium-cooled breeders in 4 or 5 NATO countries at the same time.

Now, I think the way in which we prepare for the necessary change-over in our energy structure in the not too distant future could be greatly facilitated if the NATO Science Committee could commission a study of the scope, the costs, the sharing of R and D work, etc. of such preparation. The study ought to be comprehensive, encompassing also the economic consequences, the requirements for social change, the legal aspects, the environmental impact, the resource problems, the risks, etc. Such a study would enable the countries of the Alliance – now in the possession of clear goals – to go ahead with the preparation, as described above, which is absolutely essential for the implementation of a new energy structure when the need actually arises, and cannot be put off any longer. Without such preparation I fear we shall stumble and fumble to such a degree as to disrupt our economies and consequently endanger our societies.

The foregoing have been a number of suggestions that I have found worthwhile to be considered by the Science Committee on the basis of the Twentieth Anniversary Commemoration Conference. Closer study than I was able to carry out in the few hours at my disposal, as well as the future deliberations in the Science Committee will certainly bring up new ideas, modify suggestions made by me in this brief chapter, and throw some out altogether. In the extreme, it might also be possible that most of the contributions and commentaries set out in those two volumes will turn out to be of value for us only as a sort of general education. But even then the information absorbed will help us in the Science Committee to make our further decisions with a possibly more comprehensive background and more complete insight into the problems of our time than before. The members of the Science Committee are therefore grateful to all the contributors for the great effort they made.

Résumés des contributions

Chapitre 11: Introduction: un contexte nouveau pour la science
Marcus C.B. Hotz

En guise d'introduction, l'auteur expose les faits qui sont à l'origine de la Conférence du Vingtième anniversaire du Comité scientifique de l'OTAN. Il analyse les principaux thèmes qui lient les divers rapports sur les réalisations et les perspectives scientifiques à ceux qui traitent des aspects sociaux, économiques et politiques de la science et de la technologie, en indiquant les orientations générales qui s'en dégagent pour l'avenir.

Chapitre 12: Ressources et disponibilités: aperçu sur les systèmes énergétiques de l'avenir
Wolf Häfele et Wolfgang Sassin

L'énergie est l'un des biens les plus importants de la civilisation moderne. Les difficultés croissantes que suscitent les fournitures de quantités suffisantes d'énergie, sont amplifiées par la rareté des ressources naturelles traditionnelles ainsi que par les effets secondaires liés au déploiement à grand échelle de techniques énergétiques.

Le rapport aborde tout d'abord le problème de la demande globale d'énergie au cours des quinze à cinquante prochaines années. Il examine la situation en matière de réserves fossiles et le potentiel de ressources énergétiques de remplacement. La structure des énergies nucléaire et solaire diffère de celle des ressources énergétiques actuelles; les premières représentent un potentiel infini d'énergie mais ont l'inconvénient de faire appel dans une large mesure aux autres ressources: matériaux, environnement et capital.

Ensuite, le rapport présente des réflexions plus approfondies sur la demande d'énergie. Il examine certains mécanismes qui tendent aussi bien à accroître qu'à réduire le coefficient énergétiques de l'économie. Le concept de neguentropie est introduit à partir du principe de base de conservation de l'énergie. La neguentropie (ou information) est un concept qui rend compte de l'usage de l'énergie et de sa consommation. L'analyse mène à la notion de source permanente d'énergie de durée infinie. L'analyse présente des éléments d'un scénario d'approvisionnement maximum et de demande minimum qui se caractérise par une consommation en l'an 2030 d'énergie de base en «équivalent pétrole» de 35 TW, c'est-à-dire 4,6 fois le niveau actuel. Un tel scénario offre aux programmes d'énergie nucléaire et solaire une perspective nouvelle.

Ce scénario montre que le passage du pétrole et du gaz à d'autres sources d'énergie modifiera l'interdépendance existant actuellement entre les régions du monde, sans cependant le faire disparaître. La résolution du problème global de l'énergie dépend de l'évolution de la coopération internationale. Le progrès technologique ne remplace pas cet effort; il en est cependant l'un des instruments.

Commentaire (I): Umberto Colombo

Le commentaire porte sur le problème des ressources énergétiques à long terme; il se réfère en particulier au document de MM. Häfele et Sassin, qui fournit un cadre quantitatif extrêmement utile pour l'évaluation des choix en matière d'approvisionnements énergétiques à long terme et vise à trouver une solution garantissant une dotation en énergie suivant le principe de la neguentropie.

Le présent document constitue un examen critique des hypothèses implicitement contenues dans l'ouvrage de MM. Häfele et Sassin, à savoir:

un schéma de développement essentiellement semblable à celui des dernières décennies, qui entraine un rythme de croissance quantitative difficile à soutenir; des systèmes énergétiques rigides et fortement centralisés; la difficulté de réaliser de vastes progrès énergétiques dans les PED en raison du manque de capitaux; et le peu d'intérêt accordé aux innovations dans le domaine social, ainsi qu'au rôle de l'évolution technologique dans la recherche d'une forme de développement plus modérée et conforme aux besoins de la société.

Le document de MM. Häfele et Sassin a incité l'auteur à projeter un scénario moins ambitieux en matière d'énergie de remplacement (15 TW pour l'année), qui implique la recherche de sources d'énergie «douces» et la mise au point de techniques appropriées, tout en maintenant un taux modéré de croissance économique.

Ce scénario est plus compatible avec les tendances socio-économiques souhaitables et pourrait apporter une réponse plus satisfaisante aux problèmes actuels et futurs, compte tenu également des aspects politiques et stratégiques.

Commentaire (II): Robert J. Uffen

Les matières premières traditionnelles sont peu à peu remplacées par de nouvelles ressources et de nouvelles techniques. Certaines matières premières d'importance stratégique resteront rares, mais de nouvelles sources, comme l'extractions des dépôts sous-marins, ou la substitution de matériaux anciens par des matériaux artificiels tels que les polymères ou les céramiques, sont capables de satisfaire la demande mondiale des cents prochaines années. Cela implique cependant une modification de nos concepts de propriété et de gestion en même temps que la prise en compte de facteurs non-économiques et non-techniques. On peut s'étonner de l'importance excessive accordée aux énergies de l'avenir, alors que peu d'attention est donnée au ressources renouvelables comme les forêts, la pêche, l'agriculture et les produits alimentaires. L'ensemble de ces matières premières peuvent devenir sources de conflit. Le contrôle des climats, la désertification et le traitement des déchets réclament plus d'attention.

La conservation des ressources naturelles deviendra un objectif essentiel à mesure que la population s'accroît entraînant une demande accrue de biens.

Les prévisions et la compréhension des conséquences des nouvelles techniques feront un appel plus large aux acquis de la science. Subsistera un conflit permanent entre les principes traditionnels de gestion et les besoins de nouvelles méthodes pour gérer les problèmes globaux.

Chapitre 13: Dynamique de la science, de la technique, et de la société: analyse et processus de prise de décision
Aurelio Peccei et Mihajlo Mesarovic

La pression de la science et de la technologie sur la société a été si constante, particulièrement au cours de ces dernières décennies, que cette société n'a pas été en mesure de s'adapter aux changements profonds ainsi engendrés. C'est pourquoi des milliards de personnes – et non pas seulement celles qui sont les moins instruites ou qui vivent dans les pays en voie de développement – sont plongées dans le désarroi et l'ignorance et ne savent quel comportement adopter: elles sont incapables de s'adapter á leur nouvel état et aux conditions du milieu. Il s'est produit une dangereuse rupture entre l'aptitude de l'homme à se rattacher à son milieu et les réalités de la vie artificielle instaurée par l'homme sur notre planète. C'est le fondement de ce que l'on a pu appeler «le dilemme de l'humanité» au stade actuel de son évolution sociale. Poursuivre la marche dans des directions qui aggraveraient encore cette «rupture» serait très imprudent, voire même insensé.

Deux conséquences importantes découlent de ces prémisses. Pour arriver à une évaluation exacte de la situation actuelle et des perspectives d'avenir, il nous faut examiner les avantages et inconvénients du mouvement technico-scientifique dans son ensemble sans nous limiter à certain aspects qui peuvent se traduire par des progrès dans certains domaines avec des effets très négatifs dans d'autres. D'une manière plus générale, le développement non harmonieux de la science et de la technologie risque de déséquilibrer la société – évidence qui n'est que trop démontrée par la situation actuelle.

Nous manquons d'une vision globale de l'état actuel de la planète, vision indispensable cependant pour pouvoir entrer dans l'avenir en prenant des décisions sensées. Ce défaut provient de notre tendance invétérée à analyser, par exemple, l'économie, indépendamment des problèmes de population ou de sécurité, et à examiner les situations régionales hors de leur «contexte» mondial. Il est donc indispensable d'interpréter de manière holistique l'essence et la dynamique du présent – quelles que puissent être l'étendue et l'imperfection de cette interprétation.

Nous pouvons tirer de ces considérations une conclusion importante: notre préoccupation majeure, en la circonstance, doit être la société dans son ensemble, et plus particulièrement sa capacité à sortir de la situation actuelle de crises à répétition pour pouvoir tirer un parti intégral et légitime de son énorme potential scientifique et technique.

Si nous envisageons l'humanité comme un tout interdépendant et en considérant tous les éléments de sa condition on peut penser que les quelques années à venir seront beaucoup plus difficiles que celles que nous venons de connaître et qu'elles seront probablement considérées comme une période décisive appelée à exercer une influence prolongée sur l'avenir. Nous n'estimons pas les vues exposées dans ce document comme pessimistes mais la preuve du contraire serait évidemment bienvenue. En effet, il est probable que seules les démocraties libérales sont à même de concevoir – et, dans certaines conditions, de proposer avec un certain crédit – une voie nouvelle, plus sensée, pour l'humanité. Si c'est bien le cas, elles se trouvent investies de la responsabilité historique de concevoir un tel dessein.

Le document propose deux voies principales pour, sinon résoudre, au moins améliorer, l'état de choses actuel si complexe. Il conviendrait tout d'abord rationaliser les processus actuels de prise de décision qui se sont révélés tout à fait inadaptés devant la nouvelle combinaison de changements, complexités, incertitudes et interdépendances qui caractérisent la problématique mondiale. Il existe dans ce domaine de nouveaux moyens qui sont déjà utilisés dans diverses parties du monde; il est probable que leur utilisation s'étendra rapidement au cours des années à venir. Le document décrit leur utilité et la diversité qu'ils présentent.

La seconde proposition est fondie sur la simple constatation que le monde ne peut être meilleur que les gens qui le peuplent. Il n'est virtuellement pas de lieux où l'homme ait pu s'adapter à l'évolution du monde matériel, et c'est cette inadaptation qui est génératrice de souffrances personnelles et qui met en danger la structure sociale. Reste donc à savoir pallier cette rupture de l'homme et de son milieu en faisant évoluer son aptitude à comprendre et à réagir de façon qu'il puisse toujours dominer le changement et sa propre condition. Il s'agit probablement de la plus fondamentale des questions que l'humanité doive se poser en cet âge de changement et de mutation généralisés et elle se posera à nous pendant longtemps. Le document lui donne une réponse positive car les êtres humains en général – et non seulement les élites – possèdent des capacités latentes, actuellement négligées et inutilisées, mais qui peuvent être développées. Une étude en cours, au Club de Rome, a précisément pour objet de déterminer si l'homme peut apprendre à devenir réellement «moderne», et par quelles voies. La discussion approfondie de cette question cruciale sera possible lorsque cette étude sera terminée, probablement à la fin de cette année.

Commentaire (I): Edmond Malinvaud

Éliminer la confusion qui s'est instaurée à propos de l'adaptation au changement est certainement l'un des problèmes majeurs de notre génération. S'il est incontestable que les processus de décision ne sont pas à la mesure de la complexité des décisions à prendre, il est néanmoins douteux qu'une logique du type décrit par Peccei et Mesarovic puisse jouer plus qu'un rôle secondaire dans les situations extrêmement diverses auxquelles les pays se trouvent confrontés, logique qui procède largement d'opinions totalement subjectives. De plus, il ne faut pas sous-estimer les forces que représentent, au sein de la population, les mouvements spontanés en faveur du changement.

Commentaire (II): Denos Gazis

Tout en reconnaissant que le progrès, stimulé par la science et la technologie, a dépassé la capacité de l'homme à s'adapter à son milieu et qu'il faut qu'il mesure mieux les effets de ses actes, que soit améliorée l'organisation de la société et qu'une plus réelle participation soit offerte à tous dans la prise des décisions, l'auteur considère que l'environnement est un capital exploitable, à la condition que les changements opérés par l'homme ne soient pas irréversibles ou qu'ils puissent être corrigés et compensés par les avantages qui en résultent. La faiblesse du système APT en tant qu'instrument de décision réside dans le fait qu'il sert trop souvent à établir des modèles de situations mal structurées. Les ordinateurs ne devraient être utilisés que pour les tâches qu'il exécutent avec succès, telles que la présentation de données condensées à un décideur chargé d'établir les modèles, l'interaction restant cependant utile dans ce domaine.

Chapitre 14: L'impact sociétal des technologies de l'information
André Danzin

Un faisceau convergent de technologies nouvelles nées dans l'après guerre et dont l'explosion des applications s'est manifestée au cours des 20 dernières années conduit à une évolution profonde de notre civilisation.

Naissance d'une industrie nouvelle, surtout faite de «matière grise», ayant un énorme poids économique et social.

Apparition d'oligopoles nouveaux dont la sur-puissance pose des problèmes d'éthique à l'économie libérale et remet en cause la distribution traditionnelle des forces en matière de télécommunications.

Redistribution des emplois par suite des progrès de la productivité des industries et des services et d'une nouvelle distribution internationale du travail.

Modification des concepts en sciences humaines, en médecine, en droit, en planification, en prévision économique.

Primauté du pouvoir que donne le contrôle de l'information sur les autres formes de pouvoirs, menace sur les libertés et le fonctionnement de la démocratie. Mais, simultanément, apparition d'outils nouveaux au profit de la concertation et de la participation aux décisions, de l'enrichissement des tâches de travail, de l'éducation et de la diffusion de la culture.

La technologie progresse suivant un processus quasi darwinien. Elle répond aux règles d'évolution du monde vivant: poussée constante de la complexité, montée du psychisme, méthode des essais et de la sélection par le marché. Tout se passe comme si l'évolution de ses techniques poussait l'homme vers une «civilisation de la connaissance» où ses efforts seraient essentiellement consacrés au traitement de l'information, débouchant sur une croissance non limitée par la rareté des ressources primaires.

Sur cette voie, les déséquilibres qui apparaissent entre les différentes régions du monde et, notamment au sein de l'OTAN, du fait du retard de l'Europe, deviennent préoccupants. Ils appellent, pour éviter leur retentissement sur le plan politique, un effort considérable des pays de la Communauté Européenne afin qu'ils deviennent capables de participer aux initiatives et d'apporter le concours de leurs laboratoires et de leurs industries à la création de cette nouvelle civilisation en voie d'élaboration.

Commentaire: Lewis M. Branscomb

Le Professeur Danzin et moi-même partageons le point de vue que les technologies de l'information représentent un élargissement des capacités humaines d'une nature suffisamment profonde pour avoir dans l'avenir une influence radicale sur le développement d'une société donnée. Je pense que les technologies de l'information sont un prolongement naturel de l'intelligence humaine et qu'elles affecteront le développement de l'humanité d'une manière qui n'est pas sans rappeler la révolution industrielle.

Je crois, cependant, que les technologies de l'information peuvent conférer une dimension encore plus importante aux capacités humaines que les machines ne l'ont fait à la force physique de l'homme, en permettant un partage suffisamment général de l'accès à la connaissance pour que celle-ci devienne réellement un «héritage commun de l'humanité».

Si l'on veut être sceptique sur les avantages à retirer des technologies de l'information, on peut y trouver maint sujet d'inquiétude. Il me semble, néanmoins que bon nombre des tensions et des anxiétés du monde d'aujourd'hui ne sont que les manifestations d'un processus d'adaptation à un monde dans lequel les aspirations à une vie meilleure sont partagées, sur une base réellement mondiale, par tout le monde dans toutes les formes de sociétés. Les aspirations précèdent nécessairement leur satisfaction. Les frustrations des non-privilégiés sont un témoignage plus éloquent des bienfaits potentiels de la technologie que des expressions d'inquiétude au sujet d'une interruption du progrès social dû à la technologie. Les technologies de l'information peuvent être, alors, une vigoureuse force de libération de l'ignorance, de l'exploitation sociale et de l'inégalité.

L'exposé du Professeur Danzin passe assez rapidement de ces observations générales à la considération particulière de la technologie des ordinateurs et, en fait à une vision assez étroite des questions économiques et politiques liées au développement de l'ordinateur et des industries connexes dans différents pays. En se limitant à ce point de vue, on risque, me semble-t-il, de sous-estimer gravement l'importance de l'aspect communications de la technologie de l'information.

L'une des conséquences profondes de la combinaison des technologies des télécommunications et de l'ordinateur est la tendance à accroître l'interdépendance des sociétés, à l'échelon mondial, à uniformiser l'aspiration humaine des différentes sociétés, étant donné la possibilité évidente pour toutes les sociétés de profiter de l'héritage du savoir humain.

Alors que, d'une part, les nouvelles technologies font progresser les concepts des droits humains tant en ce qui concerne le secret que l'accès à la connaissance, de nouvelles menaces apparaissent aussi, qui visent la libre circulation des renseignements au travers des frontières nationales. La balkanisation du savoir, pour des raisons politiques, pourrait nous priver des bienfaits universels d'une large diffusion de l'information et de l'avènement d'une coopération et d'une inter-dépendance à l'echelle mondiale. En dernière analyse, c'est la qualité, l'utilité et la disponibilité de l'information elle-même qui détermineront le caractère de l'apport que les technologies de l'information feront au progrès social. Les boîtes, les fils et les antennes ne sont, après tout, qu'un moyen en vue d'une fin et ne déter-minent pas en eux-mêmes la nature de cette fin.

Chapitre 15: Science, technologie, et l'homme: Population et éducation
Lord Vaizey of Greenwich

L'auteur part du principe que le taux de croissance de la population a été une des lignes de force dominantes de l'histoire de l'humanité. Au cours des 12 derniers millénaires, la population du globe est passée de 2 ou 20 millions d'êtres humains environ à presque 4 milliards; selon toute probabilité, elle sera de l'ordre de 6 milliards au début du vingt et unième siècle. L'attitude de l'homme envers la naissance, la vie et la mort est fondamentale pour la civilisation et ses valeurs. Discuter de démographie revient donc fréquemment à discuter des valeurs humaines fondamentales qui déterminent la nature de la société.

Les conséquences économiques des changements démographiques ont fait l'objet de nombreuses études; de façon générale, il semble que les grands changements dans le monde développé aient désormais tendance à disparaître (diminution de la nata-lité et stabilité relative d'une population âgée), équilibrant par la même occasion les rapports entre la population active et les personnes à sa charge. L'économie dépend plus de l'importance de cette population active – et en particulier de la partici-pation de groupes jusqu'ici dépendants, tels que les femmes – que des phénomènes démographiques.

Le document traite des questions essentielles relatives à la migration, particu-lièrement de la main-d'oeuvre hautement qualifiée, et aux déplacements des zones rurales vers la ville qui influencent la structure de l'économie et l'intensité des pressions sociales. Il prouve aussi en quelque sorte que l'évolution du taux de nata-lité dépend des classes sociales et ceci peut avoir d'importantes conséquences à long terme sur le pool génétique dont proviennent les différents groupes. L'équilibre entre pays développés et pays en voie de développement et entre pays de l'OTAN et pays communistes sera affecté dans une grande mesure par des mouvements démographiques qui tendront à influencer l'équilibre militaire, politique et écono-mique du globe.

La question cruciale posée par l'évolution démographique est celle de la place que doit occuper l'éducation. Au fur et à mesure que baisse la proportion d'enfants dans la population totale, l'éducation revêt une autre signification. Dans les pays développés, à la génération précédente, des changements radicaux sont intervenus dans la structure de l'éducation, en partie à cause de la brusque expansion démo-graphique qui a suivi la deuxième guerre mondiale mais essentiellement à la suite de la transformation du rôle de l'éducation dans la société et de l'évolution de l'opinion sur la pédagogie et sur la place du changement social proprement dit. Quant à l'avenir de l'éducation on s'est d'abord nettement prononcé en faveur d'une éducation centrée sur l'enfant puis on a repensé les critères plus classiques de succès et d'échec de l'éducation. Il ne peut toutefois pas y avoir de retour au système qui existait avant que l'on ait modifié radicalement le système d'éducation. Il s'agit en même temps de relever les normes et de modifier notre façon de percevoir l'éducation, surtout en ce qui concerne l'éducation permanente dans tous les domaines.

Commentaire: Sylvia Ostry
L'ampleur de la communication de Lord Vaizey est telle qu'elle rend impossible une critique à la fois sérieuse et brève. A défaut, certains thèmes et certains questions sont choisis pour être commentés et pour stimuler la discussion, par exemple: l'importante question de la modification des valeurs et des attitudes, en particulier l'idée d'une conception de plus en plus hédonistique de la vie et ses incidences sur les sociétés et les économies contemporaines, les conséquences économiques d'un changement du taux de dépendance, la notion de la «persistance des structures sociales» et son conflit possible avec le thème du changement des valeurs et du comportement, la critique du concept de «déscolarité» et l'importance que peut revêtir l'amplification et l'exploration des incidences économiques de la «scolarité permanente», enfin le rapport entre la recherche sociale et l'élaboration d'une politique appropriée.

Chapitre 16: Rapports de la technologie, du niveau de vie, de l'emploi, et de la main-d'oeuvre
Willem Albeda

On observe d'intéressants courants alternatifs d'optimisme et de pessimisme au sujet des effets de la technologie sur l'emploi. En période de dépression, la technologie est considérée comme une source spécifique de chômage. L'onde continue du développement économique s'explique par des poussées d'innovation technologique. Cette onde longue a des répercussions sur l'emploi, classées ici sous la rubrique structuralisme de la demande.

D'autres causes de chômage structurel sont des problèmes accrus de friction, classés sous la rubrique demande-chômage, et l'influence du coût élevé de la main-d'oeuvre sur la demande de main-d'oeuvre, appelée ici structuralisme du coût. Ces coûts élevés de main-d'oeuvre peuvent être considérés comme une conséquence à retardement du haut niveau de vie rendu possible par le progrès technique.

En dehors des causes structurelles, il existe, bien entendu, de nos jours une importante composante cyclique du chômage.

Les nombreuses causes du chômage font apparaître plusieurs remèdes possibles. L'organisation de la demande est une possibilité, mais limitée dans son application par les répercussions sur les coûts des dépenses administratives. Un renouveau de l'innovation technique serait un moyen de «casser» le structuralisme de la demande. Une modération dans l'évolution des coûts de main-d'oeuvre reduirait le structuralisme des coûts et encouragerait le processus de renouvellement technique. Pour une telle politique l'appui des partenaires sociaux est nécessaire. Cependant, les rapports avec la main-d'oeuvre deviennent de plus en plus difficiles, employeurs et employés essayant de renforcer leurs positions de défense. L'auteur de ce traité plaide en faveur d'une polique de négociations entre le capital et le travail, allant dans le sens d'une coopération qui rendrait possible un nouveau décollage à long terme.

Commentaire: Duncan Davies

Les questions posées par le Professeur Albeda s'adressent directement à notre volonté de survie politique et à notre capacité d'améliorations à long terme. Tout en suivant l'analyse du Professeur Albeda sur les politiques à court terme et à long terme une attention particulière doit être portée aux questions fondamentales que sa communication soulève:

(a) L'investissement technologique d'une façon générale influence-t-il la création d'emplois présents et la structure de l'emploi de l'avenir?
(b) Si l'investissement technologique agit sur l'emploi, ne faut-il pas substituer des emplois orientés vers la fourniture de services qui semblent apporter une plus grande satisfaction sociale et humaine, afin d'éliminer le chômage dont la persistance érode la confiance en soi et l'abilité professionnelle?

Il ne manque pas de preuves au niveau micro-économique que, dans le court terme au moins, l'investissement technologique réduit l'emploi traditionnel dans la production, et qu'il existe une tendance à une diminution de la quantité de travail par unité produite. On peut raisonnablement prévoir que la croissance globale de la production particulièrement dans les secteurs à forte intensité d'énergie et de capital continuera de décroître.

Les questions essentielles du Professeur Albeda, quel que soit le sens des réponses, ouvrent un débat important qui, à son tour, débouche sur des questions de nature philosophique. En vérité, il est difficile d'envisager que les problèmes posés trouvent une solution autre que celle proposée par le Professeur Albeda, c'est-à-dire des solutions dégagées peu à peu au cours d'un processus continu de consultation et de participation entre les travailleurs, les syndicats, les employeurs et la puissance publique.

Chapitre 17: Science, technologie, et société: problèmes à long terme des pays développés et des pays en développement
Richard Eckaus

Un grand nombre de problèmes économiques à long terme ne reçoivent pas actuellement l'attention qu'ils méritent. Ce sont des problèmes qui, s'ils ont des répercussions dans le présent, ne se prêtent pas à des solutions immédiates, sauf si l'on veut faire plus, financièrement, que des ajustements progressifs.

Cette note énumère quelques exemples spécifiques de problèmes économiques nationaux et internationaux à long terme. L'évolution radicale des coûts de l'énergie impose une restructuration majeure des économies des pays non producteurs de pé⁺role. Il se pose aussi, à long terme, des problèmes urbains, tels que la détérioration du parc de logements, la surutilisation sans cesse croissante de la capacité des infrastructures urbaines anciennes et la dégradation de la qualité des services urbains. Les tentatives qui sont faites pour établir un système permettant, en coopération, d'éliminer les difficultés de balance des paiements de tel pays ou groupe de pays constituent un exemple de problème à long terme de portée internationale. Le passage à un système de taux de change flottants n'a pas permis l'élimination automatique de telles difficultés, mais a plutôt fourni un nouvel instrument politique, utilisé dans un esprit de concurrence. Le recours aux institutions financières internationales s'est accru avec une extrême rapidité au cours de ces dernières années, mais l'absence de mesures de régulation appropriées à eu des conséquences virtuellement désastreuses. Les griefs des pays en développement, qui trouvent leur expression dans l'appel lancé en faveur d'un nouvel ordre économique international, traduisent la déception de ces pays devant l'incapacité du système d'échanges internationaux à assurer une répartition équitable des revenus de ces échanges et à leur apporter une aide adéquate dans leurs efforts de développement.

Les problèmes économiques à long terme ont des caractères trop spécifiques pour pouvoir être résolus par le recours au secteur privé et aux méchanismes du marché. C'est donc au secteur public qu'il appartient de mener des politiques appropriées. Si les hommes politiques n'envisagent généralement que l'avenir à très court terme, il conviendrait qu'ils modifient leur optique afin de résoudre les problèmes à long terme.

Commentaire : Edmond Malinvaud

Traitant de la croissance à long terme nous devons considérer les moyens et les fins de l'activité économique. Bien que le professeur Eckaus se soit limité à la discussion de quelques problèmes choisis à titre d'exemples, je trouve surprenant qu'il ne mentionne pas comme problème potentiel les ressources alimentaires futures ; mais je dois reconnaître que les économistes ont peu à dire à ce sujet. A propos de l'énergie je trouve Eckaus un peu trop optimiste : il considère seulement l'adaptation de nos économies à des prix plus élevés de l'énergie, alors que je préférerais insister sur le difficile choix de l'utilisation la meilleure des ressources actuelles et pourquoi celle-ci dépend de façon cruciale de prévisions technologiques à long terme. Enfin j'estime que notre conférence ne devrait négliger ni les doutes actuels sur ce que la croissance économique apporte réellement aux hommes, ni l'absence d'un projet communément accepté sur les finalités à long terme de la croissance.

Ceci dit, je m'associe chaleureusement au message principal de Eckaus, selon lequel «les problèmes économiques à long terme ont des particularités telles qu'ils ne peuvent pas être bien résolus par des actions privées opérant par l'intermédiaire des marchés» et qu'en conséquence des politiques publiques sont requises. Effectivement ce message constitue un des résultats principaux des progrès réalisés par la théorie économique au cours des dernières décennies. Dans chaque cas la forme la plus appropriée de l'action publique, laquelle peut opérer soit directement soit par l'intermédiaire des prix, a aussi été étudiée sans dogmatisme. De plus, la théorie de la planification décentralisée a profité des recherches actives dont certaines ne visaient ni les pays en développement ni ceux de l'Europe Occidentale.

Quant à la question pertinente de savoir pourquoi les problèmes et politiques à long terme se trouvent plutôt négligés dans nos pays, j'accepte les réponses provisoires données par Eckaus. Il faut aussi mentionner que des politiques délibérées ont le plus souvent des conséquences distributives qui rendent leur adoption difficile, surtout quand elles remettent en cause des normes anciennes. Enfin une question se pose: si les citoyens s'accomodent d'une certaine négligence collective vis-à-vis des problèmes du long terme, est-ce par manque d'information ou manquent-ils plutôt de la volonté que susciterait un projet clair sur l'avenir de l'humanité?

Je trouve triste mais réaliste que Eckaus considère l'établissement d'un nouveau système à long terme, alors que le besoin est immédiat. J'accepte la conclusion selon laquelle un contrôle monétaire central est requis au niveau international, comme il l'est dans chaque pays. J'accepte aussi les analyses et recommandations concernant un nouvel ordre économique mondial dans les relations commerciales. Si nous ne pouvons pas compter sur des progrès rapides, c'est sans doute que les deux parties, le Nord et le Sud, surestiment les avantages qu'elles tireront à l'avenir du commerce international.

Chapitre 18: Science, technologie, et relations internationales
Harvey Brooks et Eugene B. Skolnikoff

Le développement scientifique et technique est d'abord une responsabilité publique nationale, bien que certains progrès réalisés en dehors de tout contrôle public aient affecté profondément la nature et le contenu des relations internationales. De nombreux problèmes actuels ou à venir ont acquis une dimension internationale et même globale; leurs répercussions et les solutions à apporter diffèrent cependant selon les pays.

Plusieurs grands problèmes de nature globale mais intéressant le présent et l'avenir des pays de l'OTAN sont passés en revue pour à la fois comprendre leur importance et, chercher des moyens de solutions institutionnelles ou politiques; il s'agit notamment des transferts de technologie, de la recherche et développement appliquée aux problèmes transnationaux, des ressources naturelles et du caractère global de l'environnement.

Les rapports entre institutions et stratégies sont analysés, notamment la capacité des institutions internationales à aborder les problèmes posés par la prévision à long-terme dans un environnement technologique en évolution, et les relations avec les groupes de pression agissant en dehors du système représentatif. Le rapport conclu en soulignant le rôle des pays industrialisés de l'occident dans la solution des problèmes essentiels du présent et de l'avenir.

Commentaire: Alexander King

Le chapitre des Professeurs Brooks et Skolnikoff est clair et balancé, et son propos et ses analyses sont proches de mes propres préoccupations. Mes commentaires visent plutôt à élargir le propos dans plusieurs directions, ou à préciser quelques points, par exemple sur la nécessité dans le processus de transferts de technologies de préserver et de renforcer la richesse du patrimoine culturel des pays moins développés.

Le classement traditionel des pays en trois mondes me semble peu satisfaisant, et je propose une autre façon de considérer les choses, fondée sur les critères de performance économique et de contrôle des ressources. Cela offre l'avantage d'indiquer clairement quels seront les problèmes à venir dans les différents pays, en particulier si l'on tient compte des facteurs démographiques. Plus précisément, l'extrême diversité des situations des pays sous-développés conduit à penser qu'il existe un besoin de concevoir des approches différentes du développement. On souligne également que l'existence de pays développés riches ou pauvres en ressources naturelles agira sur l'évolution des alliances et des groupes de pays, ainsi que sur leurs priorités en matière de recherche scientifique et d'éducation.

Les problemes d'échelle, de complexité et d'incertitude présenteront de sérieux défis aux gouvernements et aux institutions mais stimuleront l'innovation. Au plan international, les problèmes globaux appellent le renforcement des mécanismes de coopération au sein desquels, au nom de l'efficacité, les considérations à long terme devront prendre le pas sur l'intérêt national immédiat. Il ne faut pas tarder à reconnaître que le tempo de la science est largement différent du tempo de la politique.

Chapitre 19: Les plans de défence et la sécurité de l'Ouest
Johan Jørgen Holst

Le but de l'essai est d'esquisser certains paramètres-clé de l'environnement extérieur impliqués dans les plans de défense à long terme assurant la sécurité de l'Europe.

L'essai débute par un exposé de base des principaux traits du système de sécurité en Europe. L'hégémonie potentielle soviétique constitue le principal défi structurel. La participation active de l'Amérique dans le processus de la sécurité de l'Europe est une condition *sine qua non* du maintien d'un équilibre de force. Cette participation est nécessaire également pour éviter une prolifération explosive des armements nucléaires. Le rôle que joue le contrôle des armements dans le processus de la sécurité de l'Europe est souligné en tant que variable importante.

Le rôle des armements nucléaires dans l'équilibre Est–Ouest est examiné en conjonction avec l'évolution de l'équilibre central et aussi de l'équilibre nucléaire sur le théâtre des opérations en Europe. Cette évolution est examinée sous deux aspects de la stabilité: la stabilité en temps de crise et la stabilité de la course aux armements. La contrôle des armements est aussi une variable-clé pour l'évolution future. Le lien qui existe entre l'équilibre des forces centrales et l'équilibré des forces sur le théâtre des opérations est souligné du point de vue fonctionnel et aussi en fonction des modalités de négociation.

L'essai esquisse dix catégories différentes de circonstances qui peuvent avoir des répercussions sur la sécurité en Europe. L'interaction de la souplesse des options et de la fluidité politique est soulignée. Le front central est toujours la clé de la sécurité en Europe, mais des circonstances impliquant des défis ambigus et différenciés peuvent confronter l'OTAN avec la nécessité de préparer des ripostes crédibles et adéquates.

Enfin, l'essai examine le rôle de la technologie dans l'évolution de l'équilibre

Est–Ouest. Les avantages inhérents de la culture industrielle de l'Ouest sont soulignés. L'introduction de munitions guidées avec précision peut permettre à l'Ouest d'exploiter des avantages comparatifs dans sa compétition avec l'Est.

Commentaire: Karl Kaiser
Parmi les principales tendances et les problèmes majeurs de la sécurité et de la défense, la question du champ d'application de la sécurité a pris une importance croissante au cours des vingt dernières années. Depuis que le «Rapport Harmel» a établi que la défense contre toute agression et la réduction des sources de conflit constituaient pour l'Alliance deux objectifs primordiaux, l'OTAN a joué le rôle d'une organisation de sécurité plutôt que celui d'une simple organisation de défense; ainsi s'est trouvée posée la question de savoir si les limites des institutions créées par le passé ne sont pas trop étroites pour que puissnet être traités les problèmes «non classiques» évoqués par Holst.

Pour ce qui est des progrès de la recherche et de la technologie dans les domaines spécifiques tels que les missiles guidés avec précision, la prolifération nucléaire et les nouvelles perspectives d'exploitation des mers et des océans résultant d'une révision du droit de la mer, c'est l'évaluation de l'incidence des changements et non pas la technologie elle-même, qui constitue le problème le plus important pour les études scientifiques.

Une politique scientifique axée sur la sécurité doit porter, non plus sur des domaines de recherche déterminés, mais sur les liens qui existent entre ces domaines, se concentrer sur les problèmes d'application, en faisant appel à des scientifiques extérieurs à la communauté de défense, et inclure une participation des sociologues à la recherche intéressant la sécurité, en mobilisant des ressources pour les travaux concernant des problèmes qui dépassent les limites des organisations traditionelles de recherche et de développement.

Chapitre 20: La troisième phase de l'OTAN
Harlan Cleveland

L'auteur considère que l'Alliance a traversé deux phases: la défense et la dissuasion, puis la détente, ou, en fait, la poursuite de la tension sous d'autres formes. Dans la troisième phase, qui s'ouvre maintenant, les problèmes de sécurité tiennent à la manière dont nous nous gouvernons. En réalité, alors que notre éthique passe d'une croissance aveugle à une croissance plus sélective, ou plus «humaine», le problème critique paraît porter sur les limites des institutions gouvernementales, plutôt que sur les limites de la croissance.

Huit éléments de cette éthique sont évoqués: le scepticisme à l'égard de la science et de la technologie, l'importance des causes et effets écologiques, l'information en tant que ressource fondamentale, la volonté d'adopter des approches globales et à plus long terme, le passage de critères de quantité à des critères de qualité dans l'optique du bien-être des individus, la recherche d'une formation permanente, la nécessité de pouvoir offrir des emplois intéressants, et de nouveaux modes de gouvernement permettant une plus large participation aux prises de décisions.

Le principal obstacle à la conversion en orientations politiques de ses valeurs nouvelles, inhérentes à la période de transition que traversent les démocraties industrielles, ne provient pas d'une limitation des ressources matérielles ou humaines; il tient à l'inadaptation des gouvernements, dont les institutions restent faites pour l'ère de croissance aveugle qui est maintenant derrière nous.

Chapitre 21: Problèmes socio-politiques et culturels d'importance majeure pour les pays occidentaux posés par les progrès scientifiques et technologiques
François de Rose

Le phenomène fondamental est l'explosion démographique qui conduira dans 40 ou 50 ans à une population de près de 10 milliards d'hommes. Cela pose deux problèmes: les nourrir et les faire travailler.

Une autre conséquence de l'explosion démographique sera de rendre de moins en moins tolérable la différence des niveaux de vie entre les Américains et les Européens d'une part, le reste du monde de l'autre.

Dans le domaine des relations internationales, la guerre entre les grands pays est devenue sinon impossible du moins irrationnelle. C'est la plus grande révolution jamais intervenue dans les rapports internationaux. Mais la guerre a longtemps servi d'exutoire à des situations économiques ou sociales devenues intolérables. Nous sommes donc tenus de résoudre nos contradictions internes sous peine de faire de la scepticémie économique ou sociale.

La pollution est un autre problème mondial. La sensibilisation de l'opinion envers la pollution est une bonne chose. Mais un excès n'en corrige pas un autre. Les solutions contre l'instinct de progrès sont aussi des solutions contre nature. La mission de l'écologie doit être la recherche d'un équilibre non entre la nature et l'homme-individu mais entre la nature et la société humaine avec ses caractères irréversibles et ses lois de developpement et de progrès.

Les nations occidentales ont à résoudre plusieurs problèmes. Elles sont vulnerables. Leur manque de ressources, notamment énergétiques met leur économie et leur niveau d'emploi à la merci des variations de prix. Elles sont menacées par concurrence des pays du Tiers Monde.

Les nations occidentales devraient se spécialiser dans les productions incorporant le plus de matière grise, mais cela pose le problème des reconversions et de l'amortissement des investissements. Nous sommes tenus à rester en tête sur les plans scientifiques et technologiques ou à ne peser que du poids de nos nombres et de nos ressources.

L'Unité de l'Occident qui est la question la plus grave ne maintiendra-t-elle face aux développements scientifiques et technologiques des nouveaux pays? L'osmose se fait mieux aux États-Unis entre les milieux universitaires et scientifiques et les milieux gouvernementaux et d'affaires. La science n'est pas admise au même titre dans tous nos pays comme une composante de l'humanisme ou mieux de la culture dont elle est le plus puissant agent de transformation. Cela tient pour partie à la superiorité des moyens mis à la disposition de chercheurs et l'education.

Pour certains, nous entrons dans une civilisation où la matière première de base ne sera pas l'énergie mais la connaissance. Or dans le domaine de l'informatique la supériorité des États-Unis est très marquée. Nous serions tous victimes d'un affaissement du poids relatif de l'Europe et d'un déclage intellectuel entre l'Europe et l'Amérique, et un redressement par un effort conjugué des Européens est nécessaire.

Un problème politique se pose également: comment gérer la communauté internationale? En présence de problèmes d'une complexité croissante, la loi du laisser faire, de l'économie de marché n'est plus efficace. Par ailleurs, le dirigisme est un échec sur le plan de ses résultats et, sous sa forme marxiste le socialisme engendre la dictature.

En conclusion, on retiendra que c'est l'Occident qui a donné au monde la science moderne et il est responsable et doit être à l'avant-garde de la recherche de solution; la connaissance de la nature est la plus exaltante des aventures; toute solution devra concilier notre instinct de conservation avec la poursuite de notre marche en avant.

Chapitre 22: L'avenir du programme scientifique
Eduard Pestel

Les données et l'analyse scientifiques ne sont pas en eux-mêmes des instruments suffisants pour aider les responsables politiques à prendre des décisions dont les avantages apparaitront dans le long terme seulement. Pour surmonter cette difficulté, il faut établir des relations solides entre l'ensemble des sciences naturelles et sociales, liens qui se créent par une approche en commun des problèmes et par une vision d'ensemble dans la formulation des objectifs et la mise en place des moyens de gestion.

Sont passés en revue les domaines pour lesquels la comité scientifique de l'OTAN peut encourager la collaboration internationale, notamment l'énergie, l'échange de technologies avec les pays en voie de développement, la croissance économique, et les problèmes majeurs de protection de l'environnement.

Summaries of contributions

Chapter 11: Introduction: an emerging context for science
Marcus C.B. Hotz

By way of introduction, the rationale for the NATO Science Committee's Twentieth Anniversary Conference is described. The major themes that link the papers on scientific achievements and prospects with those on the social, economic, and political aspects of science and technology are identified and explored, indicating broad directions for the future.

Chapter 12: Resources and endowments: an outline of future energy systems
Wolf Häfele and Wolfgang Sassin

Energy is one of the most important commodities supporting modern civilization. The growing difficulties in supplying adequate amounts of energy are manifest by the scarcity of traditional resources and the side-effects of the large-scale deployment of energy technologies.

The chapter first addresses the problem of global energy demand over the next fifteen to fifty years, reviewing the fossil reserves situation and the potential of alternative energy resources. Nuclear and solar power are seen to follow a pattern different from that of present energy sources in providing potentially infinite amounts of energy. In turn they heavily draw upon other resources, such as materials, the environment, and capital.

The discussion of energy demand then touches on some mechanisms which tend to increase as well as decrease the energy coefficient of an economy. They are analysed within the concept that energy observes a law of conservation and cannot be consumed, in contrast to negentropy (or information) that goes along with the use of energy and can be consumed. The analysis leads to the notion of endowment.

To provide some background, a minimum demand, maximum supply scenario is described that is characterized by an 'oil equivalent' primary energy consumption of 35 TW in the year 2030, which is 4.6 times the consumption level of today. It puts both nuclear and solar power into quite an unusual perspective.

The scenario demonstrates that switching from oil and gas to alternative energy sources will modify rather than discontinue the present interdependencies between world regions. It leads one to conclude that a solution to the global energy problem will depend on the further development of international co-operation, for which technological progress is not a substitute but a vehicle.

Commentary (I): Umberto Colombo

The comment is centred on the long-term energy resource problem, with particular reference to the chapter by Häfele and Sassin, which provides a most useful quantitative framework for the assessment of long-term energy supply options, and seeks a solution guaranteeing energy endowment based on the concept of negentropy.

The present paper offers a critical examination of the implicit hypotheses contained in the work of Häfele and Sassin:

essentially unchanged pattern of development with respect to that prevailing during the past decades, leading to a quantitative growth that is difficult to sustain;

highly centralized and rigid energy systems;

difficulty of achieving a massive energy development in the less developed countries because of capital shortage;

little attention paid to social innovation, and to the role of technological change in the interest of a more sober form of development and in keeping with the needs of society.

The Häfele–Sassin paper has stimulated the author to suggest a less ambitious alternative energy scenario (15 TW by 2030), which calls for the development of soft sources of energy and of appropriate technologies, while maintaining a moderate rate of economic growth.

This scenario is more consistent with desirable socio-economic trends, and could provide a more adequate response to present and future problems, in the light of political and strategic issues.

Commentary (II): Robert J. Uffen

Emphasis is shifting from traditional sources of natural resources to new sources and new technologies. Some strategic materials will remain scarce, but development of new sources of supply, such as ocean mining and substitution of new scientifically designed materials, for example, polymers and ceramics, should meet world demand for the next 100 years. However, traditional concepts of ownership and management will be severely challenged, and non-economic, non-technological considerations may prevail. While great attention is being given to alternative energy resources for the future, there is a dearth of contributions in this volume dealing with renewable resources, such as forest products, fishing rights, agriculture, and food production. These may well become the source of future conflicts. It is recognized that climate control, desertification, and waste management will command greater attention in the future. Conservation of resources will become a recognized goal as population growth continues to drive the demand for greater world consumption. Science will be used increasingly to predict and understand the possible consequences of alternative technologies. There will be a continuing conflict between traditional national methods of solving resource problems and a need to consider new approaches at a global level.

Chapter 13: Dynamics of science, technology, and society: analysis and decision-making
Aurelio Peccei and Mihajlo Mesarovic

The dynamic interaction of science and technology with society has been so un-relenting, particularly in the last few decades, that society has found itself unable to adapt to the profound changes thus engendered in the human condition. As a consequence, thousands of millions of people – and not just the less educated, or those living in the developing countries – are in a state of disarray, utterly con-fused by what is happening and about what they should do; they are incapable of adjusting to their new condition and environment. A dangerous gap has appeared between man's ability to relate to his surroundings and the man-made realities of' life on this planet. This is the root of what has been termed 'the predicament of mankind' at the present stage of its social evolution. To pursue directions that widen this 'human gap' any further would be highly imprudent: perhaps foolish is a more appropriate description.

Two important consequences flow from this basic premise. In order to make correct assessments of current situations and future prospects, we must consider the overall benefits and disbenefits of the entire techno-scientific enterprise, and not focus on certain aspects which may result in progress in some areas, but only at great cost to others. In more general terms, non-harmonious developments of science and technology may throw society off balance – a situation that is demon-stated all too well by present circumstances.

We lack a comprehensive view of the present state of the planet, though this is indispensable if we are to move into the future making intelligent decisions. This drawback is due to our inveterate tendency to analyse, for example, the economy independently from population or security problems, and to consider regional situations outside their global context. A holistic interpretation of the essence and dynamics of the present is therefore indispensable – however broad and imperfect this interpretation may be.

The important conclusion that follows from these considerations is that our major concern at this juncture must be society at large, and more specifically its capacity to emerge from its present crisis-ridden situation so as to be in a position to make good and full use of its tremendous scientific and technological potential.

Looking at mankind as an interdependent whole and considering all the ele-ments of its condition, we believe that the next few years are going to be much more difficult than the present, and that they will probably be seen as a decisive period that will influence the future over an extended period of time. We do not think that views expounded in the paper are pessimistic, but, of course, any proof to the contrary will be welcome. Indeed, probably only the liberal democracies have the capacity of devising – and, under certain conditions, credibly proposing – a new, saner course for mankind. If this is so, they are vested with the historic responsibility of conceiving such a design.

The chapter suggests two main ways in which this complicated state of affairs might at least be improved, if not resolved. One is to rationalize the current decision-making processes, which have shown themselves to be totally inadequate to cope with the new combination of changes, complexities, uncertainties, and interdependencies that characterize the global problematique. New tools are available for this purpose. They are already being used in various parts of the world, and their use will probably spread quickly in the years to come. A description of their usefulness and versatility is given.

The second way stems from the simple reflection that the world cannot be better than its inhabitants. Virtually nowhere have people been able to keep abreast of the changing real world; their maladjustment to the new realities causes personal suffering, and endangers social organisation. Thus the question becomes whether the human gap can be eliminated by upgrading their ability to understand and react, so that they can always be in control of change and of their own condition. This is probably the most fundamental question that mankind must pose to itself in this period of generalized change and mutation, and it will be with us for a long time. The answer given in the paper is a positive one, because human beings in general – not only the elite – are endowed with latent capacities which are at present neglected and unused, but which can be developed. The Club of Rome has a project under way focused precisely on the question of whether and how man can learn to be a truly 'modern'. A more thorough discussion of this crucial issue will be possible when this project is completed, probably at the end of this year.

Commentary (I): Edmond Malinvaud
The confusion resulting from adaptation to change is indeed one of the major problems of our generation. Decision processes are certainly inadequate for the complexity of decisions that have to be made, but it is questionable whether logic of the type described can play more than a secondary role in the wide variety of situations with which countries are confronted, many of which involve unrealistic subjective opinions. Furthermore, the role of spontaneous movements for change should not be underestimated as forces within the population.

Commentary (II): Denos Gazis
While agreeing that progress stimulated by science and technology has outrun man's ability to relate to his surroundings, and that better understanding of the effects of his actions, better social organization, and more effective public participation in decision-making are needed, the environment is viewed as a capital resource that can be exploited, provided that man-made changes are reversible or can be corrected and compensated for by the accrued benefit. The weakness of the APT system as a decision tool lies in the degree to which it attempts to model ill-structured situations. Computers should be confined to tasks that they do well, such as presenting digested data to a decision-maker for him to model, although interactive modelling is useful.

Chapter 14: The impact on society of information technology
André Danzin

During the last twenty years, the practical application of a whole series of new, post-war technologies has brought about radical changes in our society. We have seen the rise of a new industry, composed mainly of 'grey matter', with tremendous economic and social influence. New and over-powerful oligopolies have appeared, posing a moral challenge to the liberal economy and upsetting the traditional balance of forces in the field of communications. There has been a redivision of jobs as a result of productivity gains in industry and the services sector and of a new international division of labour. Conceptual modifications have come about in the humanities, medicine, law, planning, and economic forecasting. The power which control of the media confers surpasses other forms of power, threatens civil liberties and the democratic process. But at the same time new possibilities for concentration and participation in decision-making, for improving job satisfaction, for education and the promotion of culture have appeared.

The development of technology is almost Darwinian; it evolves like the living world with its growing complexity, the increasing importance of psyche and the process of market selection and tests. It looks very much as though the development of these techniques is propelling mankind towards a 'civilization of knowledge' in which his efforts will be concentrated chiefly on the processing of information leading to growth which is untrammelled by the scarcity of primary resources.

In this connection, the disparities noticeable between the different parts of the world, and particularly within NATO because Europe is lagging behind, give cause for concern. To prevent these disparities having political repercussions, the countries of the European Community will have to make a determined effort to catch up and so become able to take part, through their research and industry, in the shaping of this new emergent civilization.

Commentary: Lewis M. Branscomb

Professor Danzin and I share the view that information technologies represent an extension of human capability of a sufficiently profound character so that it will, in time, radically influence the development of a given society. I believe that information technologies are a natural extension of human intelligence, and will affect the development of humanity in ways not unlike the industrial revolution. However, I believe that information technologies can bring an even more important dimension to human capability than machines brought to man's physical strength, permitting a sufficiently wide sharing of access to knowledge, and that knowledge will indeed become 'common heritage of mankind'.

If one wishes to be sceptical about the benefits of information technologies, one can find much to be concerned about. Nevertheless, I believe that many of the tensions and anxieties of today's world are only evidence of the readjustment process toward a world in which aspirations for a better life are shared on a truly global basis, by people in all forms of societies. Aspirations necessarily precede their fulfilment. The frustrations of the underpriviliged are more eloquent testimony to the potential benefits of technology than expressions of concern about the disruption of technology-driven social change. Information technologies can, then, be a strong liberating force from ignorance, drudgery, and inequality.

Professor Danzin moves rather quickly from such general observations to a concern with computer technology specifically, and indeed a rather narrow focus on the economic and political issues associated with the development of the computer and related industries in different countries. This view, it seems to me, tempts one to underestimate seriously the importance of the communications side of information technology.

One profound consequence of the combination of telecommunications and computer technologies is their tendency to increase the interdependence of global societies and to equalize the human aspirations of different societies, given the evident possibility of all societies benefiting from the heritage of human knowledge.

While on one hand the new technologies are bringing advances in concepts of human rights of privacy and access, there are also new threats to the unimpeded movement of information across national borders. The Balkanization of knowledge for political reasons could frustrate the global benefits of enhanced diffusion of information and the development of world-wide co-operation and interdependence.

In the final analysis, it is the quality, appropriateness, and availability of the information itself that will determine the character of the contribution that information technologies make to society. The boxes, wires, and antennae are, after all, only means to an end and do not of themselves determine the character of the end to be served.

Chapter 15: Science, technology, and man: population and education
Lord Vaizey

The idea is introduced that the rate of growth of population has been a dominant force in human history. In the last twelve millennia the population of the world has risen from an order of 2 million or 20 million human beings to nearly 4 billion, and the probability is that in the early twenty-first century it will be of the order of 6 billion. Attitudes to birth, life, and death are fundamental to civilization and its values. Therefore the discussion of demography is frequently implicitly a discussion of the fundamental human values which determine the nature of society.

The economic consequences of demographic change have been much studied; generally speaking the impression has been given that the major changes in the developed world have now tended to stabilize, with a falling birth rate and the relatively stable elderly population thus stabilizing the relationship between the working population and dependants. Economic output depends more upon the size of the working population, and in particular upon the participation of hitherto dependent groups like women, than upon changes in demography.

There are central questions about migration, particularly the migration of high level manpower, and of movement from rural areas to the towns which affect the structure of the economy and growing social pressures. There is also some evidence that changes in the birth rate have been differential socially, and this too may have important consequences. The balance between the developed and developing countries and between the NATO countries and the Communist countries will be considerably affected by demographic movements which will tend to influence the military, political, and economic balance of the world.

The crucial question which is raised by demographic change is the place of education. As the child population tends to fall as the proportion of the total, so the significance of education changes. There have been radical changes in the education structure of the developed countries in the past generation, partly as a result of the demographic boom which succeeded the Second World War, but mainly as a result of the changed role of education in society and the changed views about pedagogy and the place of social change itself. The arguments about the future of education have swung from a substantial commitment to child-centred education to a re-evaluation of more classical criteria of educational success and failure. There can however be no going back to the system which existed before the radical educational changes. It is a question both of raising standards and of altering the perceptions of education, particularly with respect to lifelong education of all kinds.

Commentary: Sylvia Ostry

The scope of Lord Vaizey's paper is so broad as to preclude a meaningful and brief critique. Instead, certain themes or issues are selected for comment and to stimulate discussion. These include the important question of changing values and attitudes, in particular the idea of an increasingly hedonistic view of life and its implications for contemporary societies and economies, the economic consequences of a shift in the dependency ratio, the notion of the 'persistence of social structures' and its possible conflict with the theme of changing values and behaviour, the criticism of the idea of 'deschooling' and the potential importance of amplifying and exploring the economic implications of 'permanent schooling', and the relationship between social policy research and policy-making.

Chapter 16: Technology, standard of living, employment, and labour relations
Willem Albeda

There is an interesting alternating flow of optimism and pessimism as to the employment effects of technology. During a depression, technology is seen as an independent source of unemployment. The long wave in economic development is explained by outbursts of technological innovation. This long wave has its employment implications, here classed as demand structuralism.

Other causes of structural unemployment are increased frictional problems, classed as search-unemployment, and the influence of high labour costs on the demand for labour, classed as cost structuralism. These high labour costs may be seen as a lagged influence of the high standard of living made possible by technical change.

Apart from structural causes, there is of course today an important cyclical component in unemployment.

Many different causes of unemployment point to more possible remedies. Demand management is one possibility, but limited in applicability by the cost implications of government expenditure. A renewal of technical innovation would be a means of breaking through demand structuralism. A moderation of labour cost developments would reduce cost structuralism and would reinforce the process of technical renewal. For such a policy the support of the social partners is needed. However, labour relations are becoming more and more strained, with employers and employees trying to dig themselves into defensive structures. The author pleads for an incomes and employment bargaining policy directed to co-operation in order to make a new long-term take-off possible.

Commentary: Duncan Davies
The key questions in Professor Albeda's analysis pertain to political survival and long-term improvements. Alongside Professor Albeda's study of current and near-future policy, special attention might be paid to his fundamental questions, re-stated as follows:

(a) Does technological investment by and large, now and for the next few decades, reduce the total number of industrial jobs?
(b) If it does, then can we not substitute service jobs that carry general esteem and self-esteem for 'unemployment' that carries no general esteem and erodes self-esteem and job skill?

There is no shortage of *micro*-evidence that, in the short run at least, technological investment reduces operating employment per unit of output. It is reasonable to expect that overall manufacturing growth across the trade cycles, especially in energy-intensive or capital-intensive areas, may continue to decline.

Professor Albeda's key questions — no matter which answer he gives to them — have clearly raised wider questions in turn, some of which are rather philosophical. It is indeed difficult to see how the problems raised by these questions can be settled other than by something along the lines that Albeda proposes — that is a consultative process involving workers, unions, management, and governments.

Chapter 17: Science, technology, and society: Long-term economic problems of advanced and developing countries
Richard S. Eckaus

There are now many neglected long-term economic problems. These are problems which, while they have current consequences, do not have immediate solutions or, if there are such solutions, they have higher costs than slower adjustments.

A number of specific examples of domestic and international long-term economic problems are discussed in this chapter. The drastic changes in relative energy prices require major restructuring of the economies of the non-oil-producing countries. Long-term urban problems include the deterioration of the housing stock, the increasing strain on the capacity of old urban infrastructures, and the decline in the quality of urban services. The development of a system for co-operative adjustment of the balance of payments difficulties of any one or group of countries provides an example of an international long-term problem. The shift to a system of managed floating of exchange rates did not create the means for automatic adjustment of such difficulties but rather provided another policy instrument which has been used competitively. The use of international finance has grown extremely rapidly in recent years but its regulation has been neglected with consequences which are potentially disastrous. The complaints of the developing countries, embodied in their call for a new international economic order, reflects their view of the failures of the system of international commerce to provide an equitable distribution of benefits and satisfactory assistance in their development efforts.

Long-term economic problems typically have features which make it impossible for private, market-mediated actions to resolve them satisfactorily. Thus public policies are required. While the time horizon of policy-makers is typically quite short, this predilection should be changed to provide for the resolution of long-term problems.

Commentary: Edmond Malinvaud

Speaking of long-term growth we must pay some attention to the means and ends of economic activity. Although Professor Eckaus limited himself to the discussion of some problems taken as examples, I find it surprising that he did not mention the future food resources as a potential problem, but I must recognize that economists have little new to say about it. On energy I find him a little too optimistic because he considers only the adaptation of our economies to higher energy prices, whereas I would like to stress the difficult choice of the best utilization of present resources and why it crucially depends on long-term technological forecasts.

Finally, I think that we should not neglect consideration of the present doubts on what economic growth really does for people, and of the present lack of a commonly accepted long-term project in this respect.

This being said, I warmly associate myself with Professor Eckaus's main message, namely that 'long-term economic problems have features which make it impossible for private, market-mediated actions to resolve them satisfactorily' and that therefore public policies are required. Indeed, this message certainly follows from the progress of economic theory during the last few decades. The best policies for appropriate public intervention in each case, either by direct action or by indirect action through prices, have also been theoretically studied in a non-dogmatic way. Similarly, progress on the theory of decentralized economic planning has resulted from active research, some of which was not intended for developing or Eastern European countries.

As to the very relevant question why long-term problems and policies have been rather neglected in our countries, I accept the tentative answers given by Eckaus. I should also like to stress that deliberate policies have distributive consequences that make their adoption difficult when old patterns must be revised. Moreover, an important question is: if the citizens accept a relative neglect of the long-term, is it because they are poorly informed or because they lack the will to carry out the programmes that would result from a well-defined project that clearly describes their future?

I find it sad but realistic that Eckaus considers the establishment of a new international monetary and financial system as a long-term problem, whereas we badly need it in the short run. I agree that central monetary control is required, in the same way as it was required in each one of our countries. I also agree with his analysis and recommendations concerning the need for a new international economic order in trade relations. If we cannot expect quick progress in this area, it will probably result from the fact that both sides, the North and the South, overestimate the advantages that they will gain from international trade in the future.

Chapter 18: Science, technology, and international relations
Harvey Brooks and Eugene B. Skolnikoff

The support of science and technology is primarily a national enterprise, yet the effects of developments growing out of that support have profoundly altered the nature and substance of international relations. Many problems existing or in prospect now take on international or global dimensions, yet the problems and their possible solutions usually have quite varied implications for different nations.

Several international issue-areas of particular present and future concern for countries of the NATO Alliance are examined with a view to understanding the nature of the issues and the institutional and policy problems encountered in attempting to deal with them. Some of the issues examined are transfer of technology, national research and development versus transnational problems, natural resources, and global environmental issues.

Institutional/policy aspects are examined with respect to competence of international machinery, problems of planning for long-term impacts, policy-making in an uncertain technological environment, the problems and opportunities of disenfranchised electorates, transnational pressure groups, and professional networks. The conclusions underscore the role of Western industrialized countries in coming to grips with critical issues.

Commentary: Alexander King
The chapter by Professors Brooks and Skolnikoff is admirably clear and balanced, and its scope and analysis closely fits with my own views. The commentary seeks, therefore, to extend the arguments in a number of directions, and to add a few elements such as the desirability in the introduction of technology to the less-developed countries of maintaining and strengthening cultural diversity and integrity.

The traditional classification of countries into the three worlds seems to me unsatisfactory, and an alternative approach is proposed, based on the two variables of economic performance and resource possession. This provides a clear indication of future prospects and problems of different nations, especially when demographic considerations are taken into account. In particular, the vastly different conditions within the less developed group suggests the need for a more diversified approach to development. It is also pointed out that the problems of the resource-rich and resource-poor of the industrialized countries will favour the evolution of different types of groupings and alliances, as well as different priorities in their science and educational policies.

Problems of scale, complexity and uncertainty are likely to present serious challenges to government policies and structures, and to stimulate institutional innovations. On the international level, the global problems will necessitate a strengthening of co-operative machinery in which, if it is to be effective, politicization in terms of immediate national interest will have to give way to longer-term considerations. There is a particular need to recognize that the tempo of science is vastly different from the tempo of politics.

Chapter 19: Defence, planning, and Western security
Johan Jørgen Holst

The essay attempts to outline some of the key external parameters involved in long-range defence planning for European security.

It starts with a basic outline of the main features of the security system in Europe. Potential Soviet hegemony constitutes the main structural challenge. Active American involvement in the process of European security constitutes a *conditio sine qua non* for the maintenance of a balance of power. That involvement is also necessary for purposes of preventing an explosive proliferation of nuclear weapons. The role of arms control in the process of European security is stressed as an important variable.

The role of nuclear weapons within the East-West balance is discussed with reference to the evolution of the central balance as well as the theatre nuclear force balance in Europe. Developments are considered in relation to the two aspects of stability, crisis stability and arms race stability. Arms control is again a key variable with respect to future development. The linkage between the balance of central forces and theatre forces in the theatre of operations balances is stressed functionally as well as in terms of the modalities of negotiation.

The essay outlines 10 different classes of contingencies which may impact on the security order of Europe. The interrelationship between flexible options and political fluidity is stressed. The central front is still the key to European security, but contingencies involving ambiguous and differentiated challenges may confront NATO with the need to generate tailored and credible responses in the years ahead.

Finally, the essay discusses the role of technology in the evolution of the East-West balance. The inherent advantages of the industrial culture of the West are stressed. The introduction of precision guided munitions may enable the West to exploit comparative advantages in its competition with the East.

Commentary: Karl Kaiser

Of the central trends and issues of security and defence, that of the scope of security has expanded over the last twenty years. Since the Harmel Exercise decided that defence against aggression and attenuation of the roots of conflict were both major purposes of the Alliance, NATO has been a security organization, rather than a mere defence organization, a situation that has raised questions as to whether the institutions created in the past are too narrow to deal with the 'non-classical' issues raised by Holst.

With respect to research and technology in specific problem areas such as precision guided missiles, nuclear proliferation and changes in ocean regime brought about by new law of the sea, assessment of impact and not the technology itself poses the most important problem for scientific work.

A science policy that is security-oriented must shift emphasis from specific research areas to the links between them, concentrate on problems of application, involving scientists from outside the defence community, and involve social scientists in research on security, moving resources to support work on problems that cut across the divisions of typical research and development organizations.

Chapter 20: The third phase of NATO
Harlan Cleveland

The Alliance is seen as having passed through two phases, firstly defence and deterrence, followed by detente, or continuation of tension by other means. The third phase, now starting, is one in which the security problems are a product of the way in which we govern ourselves. The critical problem indeed seems to be one of limits to governance, rather than limits to growth as we move from an ethic of indiscriminate growth to more selective growth, or 'human growth'.

Eight elements of this ethic are cited — scepticism of science and technology, emphasis on ecological causes and effects, a concept of information as a basic resource, willingness to think in global perspectives and larger time-frames, a shift from quantitative to qualitative criteria as measures of personal well-being, interest in lifelong continuing education, insistence on the availability of interesting jobs, and new styles in governance involving greater participation in decisions.

In reviewing these new values, inherent in the transition in which the industrial democracies find themselves, the main obstacle to converting them into policies is not a limit to physical or human resources but limits to government, whose institutions are more appropriate for the indiscriminate growth era through which we have already passed.

Chapter 21: Major Socio-political and cultural problems created by scientific and technological progress facing the countries of the West
François de Rose

The basic problem is the population explosion which over the next 40 or 50 years will result in a world population of approximately 10 thousand million. This raises two problems: how to feed, and how to find work for this population. A further consequence of the population explosion will be to make the difference in living standards of Americans and Europeans on the one hand, and the rest of the world on the other, increasingly unacceptable.

In the field of international relations, war between the great powers has become, if not impossible, then at least illogical. This is the biggest revolution of all time in international relations. But war has for centuries been a safety-valve when economic or social situations have become intolerable. So we must solve our internal contradictions or risk economic or social decay.

Another global problem is pollution. The increased public awareness of pollution is a good thing but correcting one abuse will not cure another. Solutions which go against progress also go against nature. Ecology must seek a proper balance, not between nature and man as an individual, but between nature and human society, with its irreversible characteristics and laws of development and progress.

The Western world must resolve a number of problems. The Western countries are vulnerable; their lack of resources, particularly energy resources, places their economies and their job markets at the mercy of price variations. They are threatened by Third-World competition in areas which they themselves promote. Western countries should specialize in sectors requiring the maximum know-how, but this raises the problem of industrial conversion and the amortization of investments. We have to remain ahead scientifically and technologically: otherwise we will count only in terms of population and resources.

The most serious problem is that of Western unity — will it be maintained in the face of scientific and technological development in the emerging countries? There is better interface in the United States between academic and scientific circles and government and business, but science is not accepted in the same way in all our countries as an element of humanism, or indeed as the most powerful stimulant towards cultural development. This is partly because of differing facilities available to research workers and education.

Some believe that we are entering a civilization in which the basic raw material will not be energy but knowledge. And in the field of computer science, American superiority is very pronounced. We would all suffer from a drop in Europe's relative weight, and from an intellectual slippage between Europe and America, and a joint effort by Europeans will be necessary to redress the balance.

An important political question is also raised: how can the international community be administered? In the face of problems of growing complexity, *laissez-faire* and the market economy principle are no longer effective. But the command economy has not produced the goods, and the Marxist brand of socialism leads to dictatorship.

In brief, it must be remembered that: the West gave the world modern science and the West is responsible and must pioneer the quest for a solution; the exploration of nature is the most exciting of adventures; any solution must reconcile continued progress with our instinct for self-preservation.

Chapter 22: The implications for the future science programme
Eduard Pestel

Scientific analysis and understanding are by themselves not enough to help political decision-makers explain the rationale for decisions that are of long-term benefit but disadvantageous in the short term. To overcome these problems, strong linkages have to be forged across the whole range of natural and social sciences, which is only possible when scientists work together on specific projects, and is helped by taking an overall systems approach to the formulation of goals as well as the development of planning instruments.

Areas in which the NATO Science Committee might encourage international collaboration in addressing issues of this nature are reviewed, particular attention being paid to problems of energy, transfer of technology to developing countries, economic growth, and the identification of important environmental problems.

Index

Page numbers in italic refer to Volume 2.